Sustainable Intensification in Smallholder Agriculture

Sustainable intensification has recently been developed and adopted as a key concept and driver for research and policy in sustainable agriculture. It includes ecological, economic and social dimensions, where food and nutrition security, gender and equity are crucial components. This book describes different aspects of systems research in agriculture in its broadest sense, where the focus is moved from farming systems to livelihoods systems and institutional innovation.

Much of the work represents outputs of the three CGIAR Research Programs on Integrated Systems for the Humid Tropics, Aquatic Agricultural Systems and Dryland Systems. The chapters are based around four themes: the conceptual underpinnings of systems research; sustainable intensification in practice; integrating nutrition, gender and equity in research for improved livelihoods; and systems and institutional innovation.

While most of the case studies are from countries and agro-ecological zones in Africa, there are also some from Latin America, Southeast Asia and the Pacific.

Ingrid Öborn is Regional Coordinator for Southeast Asia at ICRAF, The World Agroforestry Centre, Bogor, Indonesia, and Professor of Agricultural Cropping Systems at the Department of Crop Production Ecology, Swedish University of Agricultural Sciences (SLU), Uppsala, Sweden.

Bernard Vanlauwe is R4D Director for the Central Africa hub and the Natural Resource Management Research Area of the International Institute of Tropical Agriculture (IITA), Nairobi, Kenya.

Michael Phillips is Director of the Sustainable Aquaculture Program at WorldFish, Penang, Malaysia.

Richard Thomas is Director of the CGIAR Research Program on Drylands Systems at the International Center for Agricultural Research in the Dry Areas (ICARDA), Amman, Jordan.

Willemien Brooijmans is a rural development sociologist and a consultant at Humidtropics, International Institute of Tropical Agriculture (IITA), Ibadan, Nigeria.

Kwesi Atta-Krah is Director of the CGIAR Research Program on Integrated Systems for the Humid Tropics (Humidtropics) at the International Institute of Tropical Agriculture (IITA), Ibadan, Nigeria.

Other books in the Earthscan Food and Agriculture Series

For further details please visit the series page on the Routledge website: www.routledge.com/books/series/ECEFA/

Sustainable Intensification in Smallholder Agriculture

An Integrated Systems Research Approach

Edited by Ingrid Öborn,
Bernard Vanlauwe, Michael Phillips,
Richard Thomas, Willemien Brooijmans
and Kwesi Atta-Krah

LONDON AND NEW YORK

First published 2017 by Routledge

2 Park Square, Milton Park, Abingdon, Oxfordshire OX14 4RN
52 Vanderbilt Avenue, New York, NY 10017

Routledge is an imprint of the Taylor & Francis Group, an informa business

First issued in paperback 2019

British Library Cataloguing-in-Publication Data
A catalogue record for this book is available from the British Library

Library of Congress Cataloging-in-Publication Data
Names: Öborn, Ingrid, editor.
Title: Sustainable intensification in smallholder agriculture : an
    integrated systems research approach / edited by Ingrid Öborn,
    Kwesi Atta-Krah, Michael Phillips, Richard Thomas, Bernard Vanlauwe
    and Willemien Brooijmans.
Description: London ; New York : Routledge, 2017. | Series: Earthscan
    food and agriculture series | Includes bibliographical references and
    index.
Identifiers: LCCN 2016037540 | ISBN 9781138668089 (hbk) |
    ISBN 9781315618791 (ebk)
Subjects: LCSH: Sustainable agriculture. | Farms, Small.
Classification: LCC S494.5.S86 S885 2017 | DDC 338.1—dc23
LC record available at https://lccn.loc.gov/2016037540

ISBN: 978-1-138-66808-9 (hbk)
ISBN: 978-0-367-22778-4 (pbk)

Typeset in Bembo
by Apex CoVantage, LLC

Cover images (clockwise from top left):

Bananas mixed in cocoa systems in Soubre, Cote d'Ivoire.
Photo credit: Ingrid Öborn, ICRAF & SLU

Aquatic agricultural system, Kampong Chhnang wetland, Cambodia.
Photo credit: Eric Baran

Smallholder agricultural landscape in the humid tropics, Uganda.
Photo credit: Neil Palmer, CIAT

Agro-pastoral and pastoral systems in semi-arid areas, West Pokot, Kenya.
Photo credit: Ingrid Öborn, ICRAF & SLU

Murambi, Rwanda, during a legume participatory variety evaluation event.
Photo credit: Bernard Vanlauwe, IITA

Farmer and her son against a backdrop of maize monoculture and degraded landscape
in Son La Province, Northwest Vietnam.
Photo credit: Lisa Hiwasaki, ICRAF

Contents

Figures

Tables

Boxes

Contributors

Editors

Ingrid Öborn, Regional Coordinator for Southeast Asia, at ICRAF The World Agroforestry Centre, Bogor, Indonesia, and Professor of Agricultural Cropping Systems, Department of Crop Production Ecology, Swedish University of Agricultural Sciences (SLU), Uppsala, Sweden, main editor, lead-author Chapter 1 and co-author Chapter 12, i.oborn@cgiar.org; Ingrid.oborn@slu.se

Bernard Vanlauwe, is R4D Director for the Central Africa hub and the Natural Resource Management Research Area of the International Institute for Tropical Agriculture (IITA), Nairobi, Kenya, editor, co-author Chapter 1 and lead-author Chapter 11, b.vanlauwe@cgiar.org

Michael Phillips, Director Sustainable Aquaculture Program, WorldFish, Penang, Malaysia, editor and co-author Chapters 1, 5, and 10, m.phillips@cgiar.org

Richard Thomas, Director of the CGIAR Research Program on Drylands Systems, International Center for Agricultural Research in the Dry Areas (ICARDA), Amman, Jordan, editor and co-author Chapter 1, r.thomas@cgiar.org

Willemien Brooijmans, Rural Development Sociologist, Consultant at Humidtropics, International Institute of Tropical Agriculture (IITA), Ibadan, Nigeria, editor, w.brooijmans@cgiar.org

Kwesi Atta-Krah, Director of the CGIAR Research Program on Integrated Systems for the Humid Tropics (Humidtropics), International Institute of Tropical Agriculture (IITA), Ibadan, Nigeria, editor and co-author Chapter 1 and 2, k.atta-krah@cgiar.org

Chapter authors

Victor Afari-Sefa, World Vegetable Center, Eastern and Southern Africa, Duluti, Arusha, Tanzania, co-author Chapter 14, victor.afari-sefa@worldveg.org

Conny Almekinders, Knowledge, Technology and Innovation Group, Wageningen University, the Netherlands, co-author Chapter 23, conny.almekinders@wur.nl

Stéphanie Alvarez, Farming Systems Researcher, Farming Systems Ecology Group, Wageningen University, Wageningen, co-author Chapters 13 and 18, stephanie.alvarez@wurl.nl

Simon J. Attwood, Agrocecology and Sustainable Intensification, Bioversity International, Rome, lead-author Chapter 5, s.attwood@cgiar.org

Edmundo Barrios, Soil Ecosystem Scientist, World Agroforestry Centre (ICRAF), Nairobi, Kenya, co-author Chapter 11, e.barrios@cgiar.org

Fenton Beed, World Vegetable Center, East and Southeast Asia, Kasetsart University, Nakhon Pathom, Thailand, co-author Chapter 14, fenton.beed@worldveg.org

Friederike Bellin-Sesay, Dr., University of Giessen, Institute of Nutritional Sciences, Giessen, Germany, co-author Chapter 16, frederike.bellin@ernaehrung.uni-giessen.de

Hervé D. Bisseleua, Systems Scientist, CGIAR Research Program on Integrated Systems for the Humid Tropics (Humidtropics), International Institute of Tropical Agriculture (IITA), Ibadan, Nigeria, and World Agroforestry Centre (ICRAF), Nairobi, Kenya, lead-author Chapter 7, h.bisseleua@cgiar.org

Jean-Marc Boffa, World Agroforestry Centre (ICRAF), Nairobi, Kenya, co-author Chapter 4, m.boffa@cgiar.org

Robin Bourgeois, Agricultural Economist, Cirad/UMR-ARTDev, and Senior Foresight and Development Policy Expert, GFAR, Rome, Italy, co-author Chapter 10, robin.bourgeois@fao.org

Inge Brouwer, Associate Professor, Division of Human Nutrition, Wageningen University, Wageningen, Netherlands, co-author Chapter 18, inge.brouwer@wur.nl

Moses Chamwada, Migori County Leader, Western Regional Agricultural Technology Evaluation (WeRATE), Kenya, co-author Chapter 15, moseskive@gmail.com

Linley Chiwona-Karltun, Researcher, Swedish University of Agricultural Sciences, Department of Urban & Rural Development, Uppsala, Sweden, lead-author Chapter 16, linley.chiwona.karltun@slu.se

Steven M. Cole, Gender Scientist, WorldFish, Lusaka, Zambia, co-author Chapter 18, s.cole@cgiar.org

Fabrice DeClerck, Landscape Management and Restoration Program, Bioversity International, Montpellier, France, co-author Chapter 18, f.declerck@cgiar.org

Ann Degrande, Socio-Economist Scientist, World Agroforestry Centre (ICRAF) West and Central Africa Regional Programme, Yaoundé, Cameroon, co-author Chapter 7, a.degrande@cgiar.org

Marie de Lattre-Gasquet, Foresight and Research Management, Coordinator of Agrimonde-Terra Foresight at Cirad (Centre de coopération internationale en recherche agronomique pour le développement) and Senior Foresight Scientist at the Consortium Office of the CGIAR, Montpellier, France, lead-author Chapter 9, m.delattregasquet@cgiar.org; marie.de_lattre-gasquet@cirad.fr

Katrien Descheemaeker, Assistant Professor, Plant Production Systems, Wageningen University, Wageningen, Netherlands, co-author Chapters 13 and 18, katrien.descheemaeker@wur.nl

John Dixon, Principal Advisor/Research Program Manager for the Cropping Systems and Economics Program, Australian Centre for International Agricultural Research (ACIAR), Canberra, Australia, co-author Chapter 4, john.dixon@aciar.gov.au

Natalia Estrada-Carmona, Postdoc Researcher, Bioversity International, Montpellier, France, co-author Chapter 18, nestradacarmona@gmail.com

Anne Floquet, Faculté des Sciences agronomiques, Université d'Abomey-Calavi, Cotonou, Bénin, co-author Chapter 23, anneb.floquet@gmail.com

Romain Frelat, PhD student at University of Hamburg, Hamburg, Germany [affiliated with International Livestock Research Institute (ILRI) and International Maize and Wheat Improvement Center (CIMMYT) at time of writing], co-author Chapter 6, romain.frelat@uni-hamburg.de

Dennis Garrity, World Agroforestry Centre (ICRAF), Nairobi, Kenya, lead-author Chapter 4, d.garrity@cgiar.org

Bruno Gérard, Agricultural Engineer, International Maize and Wheat Improvement Center (CIMMYT), Mexico, co-author of Chapter 11, b.gerard@cgiar.org

Jeroen C.J. Groot, Associate Professor, Farming Systems Ecology Group, Wageningen University, Wageningen, Netherlands, lead-author Chapter 18 and co-author Chapters 13 and 17, jeroen.groot@wur.nl

To Thi Thu Ha, World Vegetable Center, and Fruit and Vegetable Research Institute (FAVRI), Trau Quy, Gia Lam, Hanoi, Vietnam, lead-author Chapter 14, ha.to@worldveg.org

Helen Hambly Odame, Associate Professor, Capacity Development and Extension Program, University of Guelph, Canada, co-author Chapter 21, hhambly@uogelph.ca

Leif Hambraeus, Professor Emeritus, Karolinska Institute, Department of Biosciences & Nutrition, Stockholm, Sweden, co-author Chapter 16, leihamb@gmail.com

James Hammond, PhD student, World Agroforestry Centre (ICRAF), University of Bangor, UK, co-author Chapter 6, j.hammond@cgiar.org

Peter Hanson, World Vegetable Center, Shanhua, Taiwan, co-author Chapter 14, peter.hanson@worldveg.org

Lisa Hiwasaki, Humidtropics Central Mekong Action Area Coordinator, World Agroforestry Centre (ICRAF), Hanoi, Vietnam, lead-author Chapter 24, l.hiwasaki@cgiar.org

Emma Hollows, Environmental Scientist and Lawyer, Independent Consultant, Penang, Malaysia, co-author Chapter 10, contact@emmahollows.com

Lummina Horlings, Adjunct Professor Socio-spatial Planning, Rijks Universiteit Groningen, Netherlands, co-author Chapter 18, l.g.horlings@rug.nl

Bernard Hubert, Senior Scientist, Inra Ehess, and Agropolis International, lead-author Chapter 8 and co-author Chapter 9, bernard.hubert@avignon.inra.fr

Latifou Idrissou, Humidtropics West Africa Action Area Coordinator, International Institute of Tropical Agriculture (IITA), Ibadan, Nigeria, co-author Chapter 24, l.idrissou@cgiar.org

Ray Ison, Professor of Systems, ASTiP (Applied Systems Thinking in Practice Group), School of Engineering and Innovation, The Open University, UK, co-author Chapter 8, ray.ison@open.ac.uk

Celister Kaleha, M&E Specialist, Western Regional Agricultural Technology Evaluation (WeRATE), Kenya, co-author Chapter 15, kalehah@gmail.com

Geoffrey Kamau, Kenya Agriculture and Livestock Organization (KALRO) Headquarters, Nairobi, Kenya, co-author Chapter 23, gmkamau_1@yahoo.com

Gina Kennedy, Theme Leader Diet Diversity for Nutrition and Health, Bioversity International, Rome, Italy, lead-author Chapter 17 and co-author Chapter 18, g.kennedy@cgiar.org

Anne Kuria, Agroforestry Systems and Local Knowledge Scientist at the World Agroforestry Centre (ICRAF), Nairobi, Kenya, and PhD Agroforestry student at the School of Environment, Natural Resources and Geography (SENRGy), Bangor University, Wales, UK, co-author Chapter 12, a.kuria@cgiar.org

Genevieve Lamond, World Agroforestry Centre (ICRAF), Nairobi, Kenya, and Teaching Associate at the School of Environment, Natural Resources and Geography (SENRGy), Bangor University, Gwynedd, Wales, UK, corresponding author Chapter 12, g.lamond@bangor.ac.uk

Cees Leeuwis, Chair of Communication, Technology and Philosophy Department, Wageningen University, Wageningen, Netherlands, lead-author Chapter 22 and co-author Chapter 25, cees.leeuwis@wur.nl

Brigid Letty, Institute of Natural Resources, Scottsville, South Africa, co-author Chapter 23, bletty@inr.org.za

Nguyen Thi Tan Loc, Fruit and Vegetable Research Institute (FAVRI), Trau Quy, Gia Lam, Hanoi, Vietnam, co-author Chapter 14, nguyen.thi.tan.loc@ gmail.com

Jacqueline Loos, Agroecology, Department of Crop Sciences, Geirg-August University Göttingen, Germany, co-author Chapter 5, jloos@gwdg.de

Nester Mashingaidze, Postdoc Researcher, Farming Systems Ecology Group, Wageningen University, the Netherlands, and International Institute of Tropical Agriculture (IITA), Bujumbura, Burundi, co-author Chapter 18, n.mashingaidze@cgiar.org

Cynthia McDougall, Gender Leader, WorldFish, Penang, Malaysia, lead-author Chapter 19 and co-author Chapter 5, c.mcdougall@cgiar.org

Mirja Michalscheck, PhD Farming Systems Analysis, Farming Systems Ecology, Wageningen University, Wageningen, the Netherlands, co-author Chapter 13, mirja.michalscheck@wur.nl

David Mills, WorldFish and ARC Centre of Excellence on Coral Reef Studies, James Cook University, Australia, co-author Chapter 5, d.mills@cgiar.org

Jonathan Muriuki, Agroforestry Scientist, World Agroforestry Centre (ICRAF), Nairobi, Kenya, co-author Chapter 12, j.muriuki@cgiar.org

Mary Mutemi, Research Fellow at the World Agroforestry Centre (ICRAF) and MSc student at Jomo Kenyatta University of Agriculture and Technology, Nairobi, Kenya, co-author Chapter 12, m.mutemi@cgiar.org

Maureen Njenga, Research Assistant, World Agroforestry Centre (ICRAF), Nairobi, Kenya, co-author Chapter 12, m.njenga@cgiar.org

David Norman, Professor Emeritus, Agricultural Economics Department, Kansas State University, Manhattan, Kansas, USA, lead-author Chapter 2, dnorman@ksu.edu

Verena Nowak, Bioversity International, Rome, Italy, and Nutrition Consultant Diet Quality and Dietary Diversity, UN World Food Programme, Lusaka, Zambia, co-author Chapter 17, vevenow@gmail.com

Chris Okafor, East and Central Africa Action Area Coordinator Humidtropics, International Institute of Tropical Agriculture (IITA), Bukavu, DR Congo, co-author Chapter 24, c.okafor@cgiar.org

Bonface Omondi, Farm Liaison Officer, Western Regional Agricultural Technology Evaluation (WeRATE), Kenya, co-author Chapter 15, bonomondi@ gmail.com

Sarah Park, Discipline Director Natural Resource Management, WorldFish, Penang, Malaysia, co-author Chapters 5 and 10, spark_devsci@hotmail.com

Ranjitha Puskur, Senior Policy Advisor, WorldFish, Penang, Malaysia, lead-author Chapter 10, r_puskur@yahoo.com

Jessica Raneri, Nutrition Research Support Officer, Bioversity International, Rome, Italy, co-author Chapters 17 and 18, j.raneri@cgiar.org

Roseline Remans, Bio-engineer, Bioversity International and The Earth Institute of Columbia University, Addis Ababa, Ethiopia, co-author Chapters 17 and 18, remans.roseline@gmail.com

Trinidad del Río Mena, PhD student, Faculty of Geo-Information Science and Earth Observation (ITC), University of Twente, Enschede, the Netherlands, co-author Chapter 18, t.delrio@utwente.nl

Anne Rietveld, Bioversity International, Kampala, Uganda, lead-author Chapter 20, a.rietveld@cgiar.org

Randall S. Ritzema, Scientist – Systems Analysis, International Livestock Research Institute (ILRI), Hanoi, Vietnam, lead-author Chapter 6, r.ritzema@cgiar.org

Timothy Robinson, Livestock Systems Analyst, International Livestock Research Institute (ILRI), Nairobi, Kenya, co-author Chapter 11, t.robinson@cgiar.org

Silvia Sarapura Escobar, Research Associate, School of Environmental Design and Rural Development, University of Guelph, Canada, lead-author Chapter 21, ssarapur@alumni.uoguelph.ca

Murat Sartas, Innovation Systems Scientist, International Institute of Tropical Agriculture (IITA), Kigali, Rwanda; Wageningen University, Knowledge, Technology and Innovation Group, Wageningen, Netherlands; Swedish University of Agricultural Sciences, Uppsala, Sweden; lead-author Chapter 25, m.sartas@cgiar.org

Pepijn Schreinemachers, World Vegetable Center, Shanhua, Taiwan, co-author Chapter 14, pepijn.schreinmachers@worldveg.org

Marc Schut, Social Scientist, International Institute of Tropical Agriculture (IITA), Kigali, Rwanda, and Wageningen University, Knowledge, Technology and Innovation Group, Wageningen, co-author Chapters 1 and 25, m.schut@cgiar.org

Fergus L. Sinclair, Leader of Agroforestry Systems at the World Agroforestry Centre (ICRAF), Nairobi, Kenya, and Senior Lecturer at the School of Environment, Natural Resources and Geography (SENRGy), Bangor University, Gwynedd, Wales, UK, lead-author Chapter 3 and co-author Chapter 12, f.sinclair@cgiar.org

Ramasamy Srinivasan, World Vegetable Center, Shanhua, Taiwan, co-author Chapter 14, srini.ramasamy@worldveg.org

Minke Stadler, Researcher, Plant Production Systems group, Wageningen University, Wageningen, the Netherlands, co-author Chapter 18, minke.stadler@wur.nl

Sharon Suri, Project Coordinator, WorldFish, Penang, Malaysia, co-author Chapter 10, s.suri@cgiar.org

Amare Tegbaru, Gender Coordinator, Humidtropics, the International Institute of Tropical Agriculture (IITA), Dar Es Salaam, Tanzania, co-author Chapter 21, a.tegbaru@cgiar.org

Celine Termote, Agrobiodiversity and Dietary Diversity, Bioversity International, Nairobi, Kenya, co-author Chapter 17, c.termote@cgiar.org

Shakuntala H. Thilsted, Research Program Leader Value Chains and Nutrition, WorldFish, Phnom Penh, Cambodia, co-author Chapter 17, s.thilsted@cgiar.org

Le Thi Thuy, Fruit and Vegetable Research Institute (FAVRI), Trau Quy, Gia Lam, Hanoi, Vietnam, co-author Chapter 14, thuyfavri@gmail.com

Carl Timler, PhD student, Farming Systems Ecology Group, Wageningen University, Wageningen, the Netherlands, lead-author Chapter 13 and co-author Chapter 18, carl.timler@wur.nl

Bernard Triomphe, Cirad UMR Innovation, Montpellier, France, lead-author Chapter 23, bernard.triomphe@cirad.fr

Dang Thi Van, Fruit and Vegetable Research Institute (FAVRI), Trau Quy, Gia Lam, Hanoi, Vietnam, co-author Chapter 14, vandang2001@yahoo.com

Piet Van Asten, Systems Agronomist, International Institute of Tropical Agriculture (IITA), Kampala Uganda, co-author Chapter 11, p.vanasten@cgiar.org

Rein van der Hoek, Central America and Caribbean Action Area Coordinator Humidtropics, International Center for Tropical Agriculture (CIAT), Managua, Nicaragua, co-author Chapter 24, r.vanderhoek@cgiar.org

Mark T. van Wijk, Senior Scientist – Farming Systems Analysis, International Livestock Research Institute (ILRI), Turrialba, Costa Rica, co-author Chapter 6, m.vanwijk@cgiar.org

Joost Vervoort, Environmental Change Institute, Oxford University, Oxford, UK, co-author Chapter 9, joost.vervoort@eci.ox.ac.uk

Jaw-Fen Wang, World Vegetable Center, Shanhua, Taiwan, co-author Chapter 14, jawfenwang@gmail.com

Ann Waters-Bayer, Royal Tropical Institute (KIT), Amsterdam, the Netherlands, co-author Chapter 23, waters-bayer@web.de

Seerp Wigboldus, Knowledge, Technology and Innovation Group, Wageningen University, the Netherlands, co-author Chapter 22, seerp.wigboldus@wur.nl

Paul L. Woomer, N2Africa Kenya Country Coordinator, IITA-Kenya, lead-author Chapter 15, plwoomer@gmail.com

Shamie Zingore, Soil Fertility Specialist, International Plant Nutrition Institute (IPNI), sub-Saharan Africa Region, Nairobi, Kenya, co-author Chapter 11, szingore@ipni.net

Preface

The subject of systems research in agriculture is not new; however, the concept remains relevant and has evolved over time, building on the original thinking of "farming systems", and strengthening elements of agroecosystem sustainability, sustainable intensification, on-farm / off-farm interactions, value chain analysis, innovation systems and integrated agricultural systems for development. The key element in systems research is the need for holistic analysis and exploration of the totality of elements present within and impacting on the system. Equally important is the analysis of the interactions in terms of trade-offs and synergies between system components, leading to overall enhancement in the productivity and in the livelihoods of the farming community, and maintenance of environmental health through proper natural resources management.

This book is written to provide the state of the knowledge in integrated systems research for development, and how it contributes towards livelihoods enhancement for smallholder farmers and communities at large. It is written for researchers, students, development actors, decision and policy makers, donors and investors having an interest in the contribution of agricultural research to poverty reduction, food and nutrition security, and improvement of the natural resources base and ecosystem services. The book project started as a result of an international conference on "Integrated Systems Research for Sustainable Intensification in Smallholder Agriculture", held in Ibadan, Nigeria, on March 3–6, 2015, and organized by the CGIAR Research Program on Integrated Systems for the Humid Tropics (Humidtropics), in collaboration with the CGIAR Research Programs on Dryland Systems (Drylands) and Aquatic Agricultural Systems (AAS).

The overall questions that were addressed in the conference, and which are elaborated on in this book, are:

- How to move systems concepts into practice effectively and efficiently?
- How to move to scale and impact with systems research and development?
- What is the added value of that approach?
- Why is systems research needed to achieve development impact?

These and other relevant questions were elucidated through keynote lectures, plenary and parallel sessions, panel discussions, poster presentations and plenary discussions. A selection and synthesis of which forms the basis of this book, with references to existing literature, and practical applications and implications of systems research. The book is organized around four themes: (i) Conceptual underpinnings of systems research, (ii) Sustainable intensification in practice, (iii) Integrating nutrition, gender and equity in research for improved livelihoods, and (iv) Systems and institutional innovation.

The book has a unique composition, and is a comprehensive output of integrated agricultural systems research for development. It illustrates different aspects and dimensions of systems research in agriculture in the broad sense, where the focus is moved from farming systems to livelihood systems, where sustainable intensification includes ecological, economic, social and institutional dimensions, and where food and nutrition security, gender, equity and social inclusion are crucial components. It argues for strengthening and developing capacity in agricultural innovation systems to support stakeholder engagement, partnerships and collective action for achieving development outcomes across local and national levels. Furthermore, it claims that stronger innovation systems create an enabling environment for improved access to credit, input, services and markets, which enable farmers and other agricultural actors to explore new opportunities. The book reflects very recent, and on-going, integrated systems research cutting across traditional disciplines, and the majority of the material in the chapters are yet to be published in peer-reviewed journals.

Humidtropics, Drylands and AAS have contributed and collaborated in its development. The International Institute of Tropical Agriculture (IITA), who leads Humidtropics, provided hosting and technical direction for the conference and the development of the book. We believe the book will make a major contribution to the advancement of knowledge in integrated systems research for development, and to the enhancement of livelihoods for smallholder farmers in the tropics.

Kwesi Atta-Krah
Director, Humidtropics
Nteranya Sanginga
Director General, IITA

Acknowledgements

This book was developed under the CGIAR Research Program on Integrated Systems for the Humid Tropics (Humidtropics).

We would like to thank the CGIAR Research Programs on Integrated Systems for the Humid Tropics (Humidtropics), Aquatic Agricultural Systems (AAS) and Dryland Systems, the various other CGIAR Research Programs, bilateral research projects and other institutions who contributed to the research that led to the development of this book. We also wish to acknowledge the CGIAR Fund Donors for supporting this research. For a list of CGIAR Fund Donors please see: http://www.cgiar.org/about-us/our-funders/.

The contributions of all authors and reviewers are acknowledged and highly appreciated.

Further thanks go to Ingrid Öborn, who led the technical coordination of this book project, as well as to Willemien Brooijmans, who provided tireless support in coordinating the technical and structural details and providing administrative support throughout the project. Final thanks go to all the members of the Editorial Committee: Ingrid Öborn, Bernard Vanlauwe, Michael Phillips, Richard Thomas, Willemien Brooijmans and Kwesi Atta-Krah.

Finally, we would like to thank the Humidtropics Executive Office, Eric Koper, Valerie Poire and Oyewale Abioye for administrative support and their work with the conference that was the starting point for the book project.

Acronyms and abbreviations

AAS	CGIAR Research Program on Aquatic Agricultural Systems
ACIAR	Australian Centre for International Agricultural Research
AFLI	Agroforestry for Livelihoods of Smallholder Farmers in Northwest Vietnam
Africa RISING	Africa Research in Sustainable Intensification for the Next Generation
AFS	Agri-Food Systems
AHI	African Highlands Initiative
AI	Appreciative Inquiry
AIDS	Acquired Immune Deficiency Syndrome
AIS	Agricultural Innovation System
AKIS	Agricultural Knowledge and Innovation System
AKT	Agro-ecological Knowledge Toolkit
ANN	Artificial Neural Network
ANOVA	Analysis of Variance
AR&D	Agricultural Research and Development
AR4D	Agricultural Research for Development
AU-NEPAD	Africa Union New Partnership for African Development
AVRDC	World Vegetable Center
BMR	Basal Metabolic Rate
BNF	Biological Nitrogen Fixation
CA	Conservation Agriculture
CAADP	Comprehensive Africa Agriculture Development Programme
CAC	Central America and the Caribbean Action Area
CCA	Collaborative Case Assessment
CCAFS	Climate Change, Agriculture and Food Security
CEC	Cation Exchange Capacity
CFU	Colony-Forming Units
CGIAR	Consultative Group on International Agricultural Research
CIAT	International Center for Tropical Agriculture
CIMMYT	International Maize and Wheat Improvement Center

CIP	International Potato Center
Cirad	*Centre de Coopération Internationale en Recherche Agronomique pour le Développement*
CM	Central Mekong Action Area
COMESA	Common Market for East and Southern Africa
CRP	CGIAR Research Program
CSA	Climate-Smart Agriculture
DC Survey	Detailed Characterization Survey
DDS	Dietary Diversity Score
DHS	Demographic and Health Survey
DRC	Democratic Republic of the Congo
DRI	Dietary Reference Intake
ECA	East and Central Africa Action Area
EXTRAPOLATE	EX-ante Tool for RAnking POlicy ALTErnatives
EU	European Union
FAD	Fish Aggregating Devices
FA	Food Availability
FAI	Food Availability Index
FAO	Food and Agriculture Organization of the United Nations
FAR	Food Availability Ratio
FARA	Forum for Agricultural Research in Africa
FFS	Farmer Field School
FGD	Focus Group Discussion
FMFI	Farmer-Managed and Farmer-Implemented
FOVIDA	*Fondo de Vida* (an NGO in Peru)
FSR	Farming Systems Research
GENNOVATE	Innovation and Development through Transformation of Gender Norms in Agriculture
GHG	Greenhouse Gasses
GIS	Geographical Information Systems
GR	Green Revolution
GTA	Gender Transformative Approach
HFIAS	Hunger and Food Insecurity Access Scale
HICs	High-Income Countries
HIV	Human Immunodeficiency Virus
IAASTD	International Assessment of Agricultural Knowledge, Science and Technology for Development
IARCs	International Agricultural Research Centers
IAR4D	Integrated Agricultural Research for Development
ICARDA	International Center for Agriculture in the Dry Areas
icipe	International Centre of Insect Physiology and Ecology
ICRAF	World Agroforestry Centre
ICRISAT	International Crops Research Institute for the Semi-Arid Tropics

ICT	Information and Communications Technology
IDS	Institute of Development Studies
IEA	Independent Evaluation Arrangement
IFPRI	International Food Policy Research Institute
IITA	International Institute of Tropical Agriculture
IIED	International Institute for Environment and Development
ILRI	International Livestock Research Institute
INERA	*Institut National d'Etude et Recherche Agronomiques*
Inra	*Institut National de la Recherche Agronomique*
INRM	Integrated Natural Resources Management
IP	Innovation Platform
IPCC	Intergovernmental Panel on Climate Change
IPG	International Public Goods
IPM	Integrated Pest Management
IRRI	International Rice Research Institute
ISFM	Integrated Soil Fertility Management
ISR	Integrated Systems Research
IT	Information Technology
JOLISAA	JOint Learning in and about Innovation Systems in African Agriculture
KAP	Knowledge, Attitudes and Practices
LESARD	Learning System for Agricultural Research for Development
LICs	Low-Income Countries
LM1	humid Lower Midland agro-ecological zone
LSMS-ISA	Living Standards Measurement Study – Integrated Survey on Agriculture
M&E	Monitoring and Evaluation
MAAIF	Ministry of Agriculture Animal Industry and Fisheries in Uganda
MAE	Male Adult Equivalents
MAIZE	CGIAR Research Program on Maize
MDS	Minimum Dietary Diversity
MSP	Multiple Stakeholder Platform
MSPr	Multiple Stakeholder Process
NAFSIP	National Agriculture and Food Security Investment Plan
NAIDS	Nutritionally Acquired Immune Deficiency Syndrome
NARES	National Agricultural Research and Education Systems
NARS	National Agricultural Research Systems
NCD	Non-Communicable Disease
NFD	Nutritional Functional Diversity
NGO	Non-Governmental Organization
NRM	Natural Resources Management
NSL	Nutrition-Sensitive Landscapes
NSY	Nutritional System Yield

OECD	Organization for Economic Development and Cooperation
ParTriDes	Participatory Trial Design
PAL	Physical Activity Level
PASIC	Policy Action on Sustainable Intensification of Cropping Systems
PCA	Principal Component Analysis
PEM	Protein-Energy Malnutrition
PPA	Participatory Prospective Analysis
PPB	Participatory Plant Breeding
PRA	Participatory Rural Appraisal
PTPD	Percent of Total Production Diversity
PVS	Participatory Varietal Selection
R&D	Research and Development
R4D	Research for Development
RAAIS	Rapid Appraisal of Agricultural Innovation Systems
RBM	Results-Based Management
RCT	Randomized Control Trial
RHoMIS	Rural Household Multi-Indicator Survey
RinD	Research in Development
RMFI	Researcher-Managed and Farmer-Implemented
RMRI	Researcher-Managed and Researcher-Implemented
RRA	Rapid Rural Appraisal
RTB	CGIAR Research Program on Roots, Tubers and Bananas
SACCO	Savings and Credit Cooperative Organizations
SDG	Sustainable Development Goals
SI	Sustainable Intensification
SIMLEZA	Sustainable Intensification of Maize-Legume Systems in Eastern Province of Zambia
SOC	Soil Organic Carbon
SRT	Strategic Research Theme
SSA	Sub-Saharan Africa
SSA-CP	Sub-Saharan Africa Challenge Program
STEEP	Social, Technical, Economic, Environmental, Policy and Political
SUN	Scaling Up Nutrition
TLUs	Tropical Livestock Units
ToC	Theory of Change
TOSA	Tools for Systems Analysis
UM	Upper Midland agro-ecological zone
UN	United Nations
UNCCD	United Nations Convention to Combat Desertification
UNICEF	United Nations Children's Emergency Fund
USAID	United States Agency for International Development
VietGap	Vietnam Good Agricultural Practice

WA	West Africa Action Area
WASI	Western Highlands Agriculture and Forestry Science Institute
WeRATE	Western Region Agriculture Technology Evaluation
WLE	CGIAR Research Program on Water, Land and Ecosystems
WUR	Wageningen University and Research Centre
ZIP	Zonal Investment Plans

1 Integrated systems research for sustainable intensification of smallholder agriculture

Ingrid Öborn, Bernard Vanlauwe, Kwesi Atta-Krah, Richard Thomas, Michael Phillips and Marc Schut

Sustainable development goals and smallholder agriculture

The Sustainable Development Goals (SDGs) 2030, endorsed by the Heads of States in the United Nations (UN) 2015, and the national discussions and implementation plans that followed, have put light on how intertwined and interdependent the various aspects of sustainability and sustainable development are (UN, 2015; van Noordwijk et al., 2015). This book on sustainable intensification of smallholder agriculture is relevant for many of the SDGs and in particular for achieving the following goals: reducing poverty (#1), achieving food security, improved nutrition and sustainable agriculture (#2), gender equity and empowering women (#5), conserving and sustainably using aquatic resources (#6, #14), and promoting sustainable use of terrestrial ecosystems and reversing land degradation (#15). There are trade-offs and synergies between the SDGs and there is a need to balance the economic, social and environmental dimensions of sustainable development. The scale at which the SDGs are implemented also matters for agricultural development: global (e.g. climate agreement), regional (e.g. trade pacts), national (e.g. policies and incentives) or local level (e.g. innovation platforms and networks).

Building on experiences from research in sub-Saharan Africa, South and Southeast Asia and Latin America, this book elaborates on different aspects of sustainable intensification and diversification of smallholder agriculture and livelihood systems leading to *systems intensification*. The aim is to illustrate different aspects and dimensions of integrated agricultural systems research where the focus is moving from farming systems to livelihood systems and institutional innovation. Sustainable systems intensification includes ecological, economic and social dimensions where food, income and nutrition security and reduced natural resources degradation are the main focuses of agricultural interventions. Combined with supporting innovation systems and capacity to innovate that are vehicles for stakeholder engagement and partnerships, they are the basis to achieve development outcomes inclusive of gender and equity improvements. This chapter provides a summary and synthesis of the other chapters, putting the methods, approaches, analyses, experiences and research and development findings in different contexts (Figure 1.1).

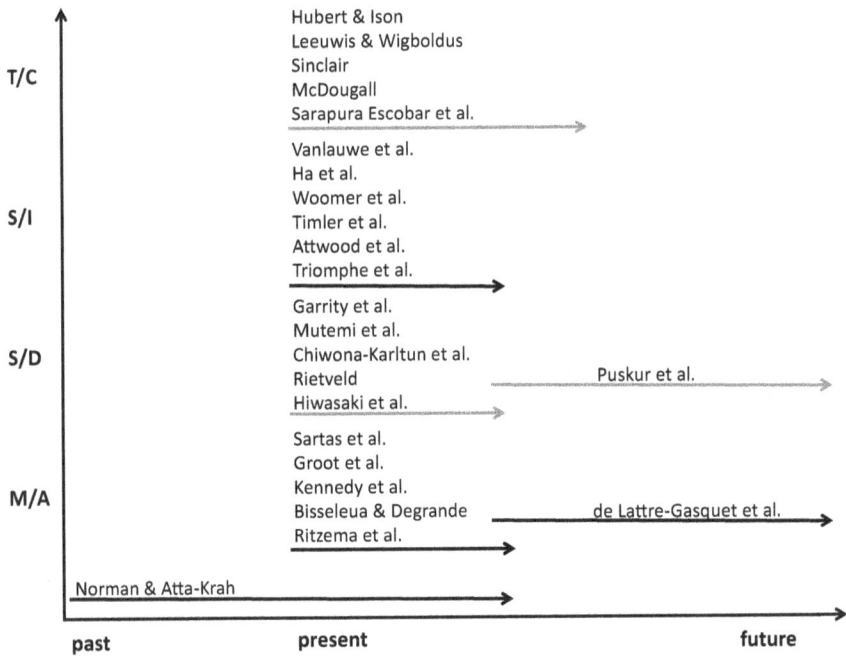

Figure 1.1 The chapters of the book are covering different aspects of integrated systems research (*y*-axis): methods and approaches (M/A), situation analysis and diagnosis (S/D), systems improvement (S/I), and transformation and change (T/C). They deal with past experiences and trends, recent research achievements and work directed towards understanding the future, i.e. foresight (*x*-axis). The author names are indicated.

What is integrated systems research?

Integrated systems research involves the management and improvement of the system based on the holistic analysis of its components within a defined agro-ecological space, their interactions, trade-offs and synergies aimed at livelihoods enhancement for farmers and communities and agro-ecological sustainability. This adds a level of complexity similar to what many farmers face daily. They manage farms that have multiple crops, livestock, soil and water management challenges and make frequent decisions to minimize trade-offs and optimize synergies. Farmers, for example, decide on utilization of their labour, income and savings considering on- and off-farm needs such as fertilizer; improved breeds, seed and education; and opportunities for short-term cash income and longer term food and income security. Agricultural research needs to support farmers managing their farming and livelihood system with all its complexity and complications and therefore adopts a systems approach aiming at systems improvement and rural transformation.

The focus on production of agricultural commodities that characterizes much of the agricultural research over the last 40–50 years paid insufficient attention to the need to co-develop the social, economic, environmental, cultural, technological, infrastructural and institutional contexts including impacts on gender and generations. In addition investment opportunities, externalities and trade-offs, non-linearity and tipping points in the relationships among the social, natural, economic and production environments were often insufficiently studied (van Ginkel et al., 2013). This has repeatedly led to poor adoption rates of innovations, particularly among the poor and vulnerable. It also challenges scaling-up and scaling-out and neglects other income-generating opportunities the system potentially provides. A comparison between the systems and more 'conventional' approaches is provided in Table 1.1.

A value proposition to guide systems research (Thomas, 2015) was presented at the conclusion of the international conference on 'Integrated Systems Research for Sustainable Intensification in Smallholder Agriculture' in Ibadan, Nigeria, 3–6 March 2015. The conference was organized by Humidtropics, a CGIAR Research Programme (CRP) on integrated systems for the humid tropics in collaboration with the Dryland Systems and Aquatic Agricultural Systems CRPs and brought together biological and social scientists to present and discuss their research that is reflected in this book. It provides a solid basis to advance the scientific base and skills in agricultural systems research and its contribution to the SDGs.

Table 1.1 Conventional and systems approaches to smallholder farming

Conventional approach	*Systems approach*
Focus on single commodity and single livelihood component	Focus on multiple commodities and livelihood components
Aimed at improving productivity and closing yield gaps, regardless of risk	Aimed at improving whole farm productivity with explicit consideration of trade-offs and social, economic and environmental sustainability. Targets multiple wins where possible; balances trade-offs where not
Focus on discrete value chains, overlooking externalities	Attention to interactions between value chains, explicitly considering externalities
Focus on innovations and investments responding to specific drivers of change within sectors at discrete scales	Focus on interactions between multiple drivers of change and innovation and investment within options across sectors and scales
Linear approaches	Dynamic iterative approaches
Discrete research disciplines	Blended biological with social research
Scientific knowledge transferred to stakeholders	Local and scientific knowledge combined, co-generated and embedded in the broader community
Gender equality and social justice as isolated outcomes of the research process	Disadvantaged groups involved and empowered throughout

Source: Adapted from Dryland Systems Task Force, 2015.

Value proposition to guide systems research

There is a need for integrated systems research that improves the understanding of place-based social, financial, technical and environmental contexts and provides a knowledge resource to enhance the targeting and relevance of potential systems interventions with an aim to scale these out to similar extrapolation domains (van Ginkel et al., 2013). Integrated systems research then develops and tests, with farming households and development partners, feasible combinations of technical, market, governance and policy options capable of improving livelihood systems. It helps to improve total farm productivity, including closing yield gaps of system components with greatest relevance to smallholder farmers. A fully integrated systems approach requires further development of monitoring and evaluation systems with indicators that can show whether systems approaches are working, for whom, where, to what extent and how, and fast enough to support adaptive management.

To reach scale, systems research has to be better embedded in development where it:

- Fosters partnerships that better target social, institutional and technical options;
- Creates hybrid knowledge that builds science onto local knowledge to reduce yield gaps of systems components and enhance multiple value chains;
- Improves capacities of households and institutions to innovate;
- Improves the effectiveness of development spending through enabling research embedded in development programmes;
- Realizes social, economic and environmental co-benefits;
- Creates platforms where outputs of other research programmes can be delivered at scale; and
- Identifies diversification opportunities in agriculture for investments.

A major challenge is to ensure that systems research strengthens the science–policy interface that has prevented governments and international bodies to contribute to transformational change on the ground to rural populations. Without meeting the pre-conditions for change that are often outside the control of the smallholder such as land access, capital, seeds, fertilizer and agro-chemicals (Sumberg, 2005; van Ginkel et al., 2013), the uptake of interventions usually stalls. Identification of diversification opportunities and new combinations of systems components such as cereals with legumes, livestock and trees can act as a magnet for the agricultural sector and draw in diverse parties and increase investments in rural areas.

The prospect for using new science, big data and information, ease of access to geographic information systems, better understanding and application of the 'options by context' concept and heterogeneity in both biophysical and socio-economic factors (Sinclair, *this volume*) will also allow systems researchers to deal with 'wicked' problems, productivity trade-offs and synergies, climate change, land degradation, gender inequities and youth unemployment at the expected scale of impact, that is with millions of farmers across millions of hectares. At the same time, systems research may directly improve the effectiveness of development spending at local scales, produce generalizable knowledge and forge new partnerships to improve livelihoods and human well-being.

Methods and approaches for integrated systems research

Agricultural systems research started in the late 1960s and 1970s with farming systems research that had a focus on looking into the farms to better understand limited resourced smallholder farmers. Norman and Atta-Krah (*this volume*) provide the historical perspective and experiences of systems research during 40–50 years characterized by farmer participatory approaches and inter-disciplinarity. Focus shifted towards on- and off-farm dimensions (livelihoods), interactions and trade-offs, and involved multi-stakeholders in the research–development continuum. Systems science at the scale of impact is discussed further by Sinclair (*this volume*), reconciling bottom up participation with the production of widely applicable research outputs. This requires moving away from a notion that there is a dichotomy between participatory bottom up approaches and comparisons of options across locations and different contexts being a prerequisite for large scale adoption and impact at scale. The FAO farming systems classification (Dixon et al., 2001) has been updated and taken forward for sub-Saharan Africa by Garrity and co-authors (*this volume*). The five major farming systems in sub-Saharan Africa provide the principal livelihoods of more than two-thirds of Africa's poor. Garrity et al. (*this volume*) conclude that bold initiatives are needed to drive sustainable intensification in African farming systems, underpinned by new ways of organizing and governing the innovation process.

Innovation and its role for transformation (change) is the topic of the chapter by Hubert and Ison (*this volume*) who have developed a theoretical framework for systems thinking and innovation in praxis and situations. Foresight, systems thinking and institutional change is the theme of the chapter by de Lattre-Gasquet and co-authors (*this volume*) where they examine three examples of foresight exercises and their contribution to institutional innovation and policy making, including (1) the direction of the cocoa and rubber sectors and related research (Cirad), (2) scenarios and challenges for feeding the world in 2050 (Agrimonde), and (3) agriculture in the face of climate change (CCAFS).

The Systems CRPs implemented from 2012 to 2016 are examples of integrated systems research in practice. Humidtropics is used as example by Bisseleua and Degrande (*this volume*) who describe approaches to operationalizing integrated systems research. Figure 1.2 illustrates the backbone of the programme with systems analysis and synthesis, integrated systems improvement, and institutional innovation and scaling as three pillars (strategic research themes) interlinked through the research cycle and cross-cutting activities on gender, nutrition and capacity development.

Although most chapters focus on terrestrial agricultural systems, some attention is given to aquatic systems. 'Does sustainable intensification offer a pathway

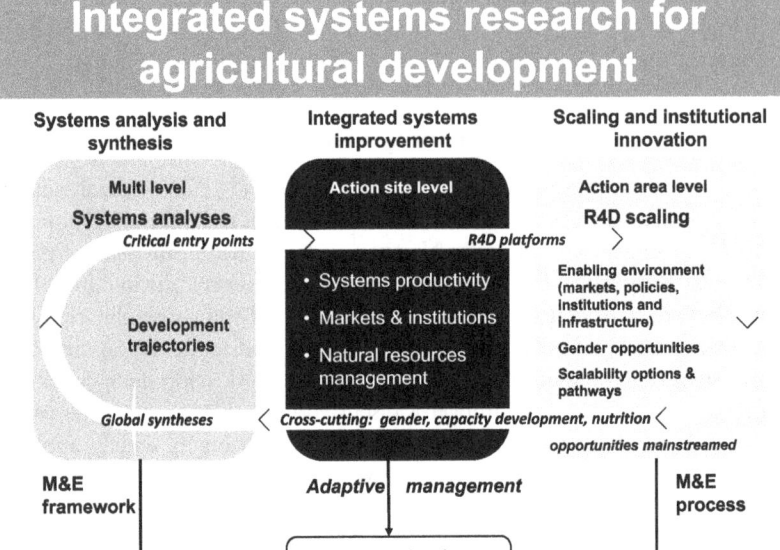

Figure 1.2 Conceptual framework for the agricultural systems research programme on integrated systems for the humid tropics (Humidtropics) funded through the CGIAR research programmes 2012–2016

Source: Humidtropics.

to improved food security for aquatic agricultural system-dependent communities?' – is the question Attwood and co-authors (*this volume*) pose. They share experiences from the aquatic systems research in sub-Saharan Africa and Asia, using examples from Bangladesh, Cambodia and Zambia, and define and discuss the difference between 'intensification' and 'sustainable intensification'. The futures of aquatic agricultural food systems in Southern Africa (Madagascar, Malawi, Mozambique and Zambia) are explored by Puskur et al. (*this volume*) and the drivers to future-smart research and policy options are discussed, including multi-stakeholder involvement and empowerment of local communities.

There is a need to experiment with different sets of entry points for different farmer typologies to enhance the targeting and relevance of potential systems interventions for sustainable intensification. This is exemplified by Ritzema et al. (*this volume*) who describe a quantitative approach for characterizing livelihoods and assessing potential wide-scale impact of interventions that complements detailed household modelling in informing intervention strategies.

Sustainable intensification in practice

Conceptualization and building frameworks are important to help understand the factors that need to be considered to implement integrated systems research, but are insufficient to create positive change at the level of fields, farms, farming communities and landscapes. Traditionally, after clearing natural fallows, nutrient mining is the first degradation process kick-starting a number of other degradation processes. Declining soil fertility generates declining crop yields and triggers a mutually reinforcing vicious cycle of resource degradation. Enhancing farm-level productivity while improving the natural resource base is essential for smallholder farmers' livelihoods. Sustainable intensification encompasses the need to enhance productivity whilst maintaining or improving ecosystem services and system resilience to shocks.

However, Vanlauwe et al. (*this volume*) recognize that increased system productivity and improvement in natural resource integrity do not necessarily go hand-in-hand. Pathways need to be identified to move from current smallholder farming systems with low productivity and degraded natural resources towards productive systems with the provision of soil-based ecosystem services preserved. Various intensification paradigms are evaluated through an analysis of the dynamics of crop yields (proxy for system productivity) and soil carbon contents (proxy for natural resource integrity) in long-term trials. External nutrient inputs are clearly needed to trigger farming systems productivity and break the downward spiral of soil degradation, especially when land is in short supply. Extra crop residues need to be recycled to maintain soil carbon stocks, thus gradually moving up the path towards sustainable intensification. Such paths are intersected by trade-offs between investments in space and time by smallholder farming families who most often lack the necessary resources to simultaneously obtain short-term crop productivity increases and maintain favourable production and ecosystem conditions for the longer term.

Mutemi et al. (*this volume*) used participatory approaches to understand fine-scale variation and entry points for sustainable intensification in Western Kenya. Knowledge about agro-ecological interactions in mixed farming systems was elicited from smallholders across four villages in two counties using a knowledge-based systems approach. The study revealed common challenges across the four villages related to land scarcity, decreased soil fertility and pests and diseases in staple crops and fruit trees. However, each village had its own natural resources management constraints and dynamics requiring customized approaches to intensification of mixed crop-tree-livestock systems. Farmers had detailed knowledge of the challenges faced in crop production, but had significant knowledge gaps in terms of pest and disease identification and control. The study demonstrates the importance of integrating local knowledge with scientific knowledge to better understand fine-scale variation in farming contexts and the need of farmers to identify locally relevant entry points for sustainable intensification.

The research by Timler et al. (*this volume*) in Eastern Zambia further explores the entry points for sustainable intensification. The entry points in their study focussed on legume interventions, identified *ex ante* for various farmer typologies, as defined by the presence of legumes and livestock, labour use and off-farm income, through the use of the bio-economic model FarmDE-SIGN. Taking into account farmers' structural constraints and their objectives to (1) maximize operating profit and organic matter added and (2) to mini-mize labour requirements, this model revealed different trade-offs and syner-gies between objectives and confirmed that specific interventions were more suitable for different farm types. For instance, soybeans were found beneficial to low resource-endowed households that spent most labour on land preparation, to medium-to-high resource-endowed households with substantial off-farm income, and to high resource-endowed households with high crop and animal income. Only medium-to-high resource-endowed households with substantial off-farm income could benefit from maize–cowpea intercropping.

In their chapter, Ha et al. (*this volume*) give examples of diversification of mono-cropping with vegetable production as a means of sustainable systems intensification. Because land use is dominated by monocultures of starchy staple crops in Northwest Vietnam, agricultural diversification could improve the live-lihoods of farmers in the region. Vegetable production in particular could offer the opportunity to increase household incomes and nutrition from relatively small plots of land and from integration with livestock and agroforestry systems. However, basic production constraints will need to be addressed, including the supply of quality seed of a diverse range of vegetable varieties and integrated pest management. Providing training while addressing production constraints could have a large impact on rural livelihoods in Northwest Vietnam.

While the authors of the chapters mentioned above used various argu-ments and approaches to define entry points towards sustainable intensification, Woomer et al. (*this volume*) used the legumes soybean and climbing bean as entry points in Western Kenya and evaluated them for their yield, biological nitrogen (N) fixation and potential benefits to farming livelihoods. They identi-fied the following elements to include in future on-farm trials: (1) comparing short- and long-rains performance, (2) evaluating additional new, rust-tolerant soybean varieties, (3) examining different rates of legume-specific fertilizer, and (4) comparing standard and experimental formulations of legume inoculants. Stakeholder-identified issues such as Striga weed elimination, crop diversifica-tion (away from maize) and animal enterprise intensification were also found to be key to understanding how improvements in legume enterprise interact with other components of the small-scale maize-based farming and livelihood systems in Western Kenya.

Although authors in general agree that sustainable intensification requires the right entry points, various approaches were used to determine these, vary-ing from open-ended, over classes of entry points (legumes, vegetables), to spe-cific crops (soybean, climbing beans). Recognition that not all entry points will suit all farming families in a specific environment was explicitly addressed in one chapter (Timler et al., *this volume*). The need for varying approaches to experiment with different sets of entry points for different farmer typologies is highlighted by the fact that only two chapters include applied research to validate entry points and just for specific crops. Entry points in all chapters focussed – rightfully, one could argue – on increasing and diversifying farm productivity to address food and nutrition insecurity, but natural resource integrity dimensions were absent from most of the chapters, or implicitly present at best. Pathways towards sustainable intensification will also require investments in natural resources management. Long-term trials are one of the few options to validate the nature (productivity, natural resource status, resilience) of interventions claiming to deliver sustainably intensified smallholder farming systems.

Nutrition as element in food security

Food security is important for livelihoods improvement and includes nutritional security. Food must reflect the essential elements, nutrients and vitamins that are needed for healthy living. In the case of smallholders, such dietary balance can begin from the diversity in crops, trees and livestock that are managed as part of the integrated agricultural landscape. Farm production must not only be seen in terms of major staples like maize, cassava or rice, but also in the mix of nutritional diversity options it provides. Chiwona-Karltun et al. (*this volume*) addresses the importance of balancing agri-food systems for optimal global nutrition transition, including the importance of nutrition of women and children. The shift towards agri-food systems is in line with integrated agricultural systems thinking, and incorporates food value chain dimensions, with an increasing focus on health and nutrition through agriculture. Chiwona-Karltun et al. (*this volume*) look at the priority of nutritional needs and the specific roles that each group of nutrients have within this context. This is followed by an examination of the evidence around the issues of food security and food safety. The authors emphasize that malnutrition is no longer limited to undernutrition, but also includes over-nutrition in energy intake leading to obesity and concomitant non-communicable diseases.

At the heart of systems orientation is an emphasis on understanding relationships between changing factors. This necessity has led to the emergence of the nutrition-sensitive landscape approach, which addresses the relationship between nutrition, agriculture and the environment, and aims to identify, quantify and tackle unsustainable trade-offs while generating synergies. This subject is addressed by Kennedy et al. (*this volume*), relating to the conceptual underpinnings, and followed with an overview of approach and methods to assess food availability and diversification of diets. Further analysis of use of the nutrition-sensitive landscape approach for exploring trade-offs and synergies between food and nutrition security, agricultural production, market interactions and natural resources management across temporal (seasons) and spatial (farm to landscape) scales is provided by Groot et al. (*this volume*). It entails multi-disciplinary analyses of how choices of women and men regarding land

and farm management and their food acquisition and consumption patterns affect the food system, nutrition adequacy and ecosystem services.

The gender and equity dimensions in systems research

Growing acknowledgement among some scholars and practitioners that both agriculture and gender are embedded in how societies and their institutions function provide openings for advancing a more complex, systemic understanding of gender within agriculture. This includes understanding of the relationships and interface between gender and nutrition, and the dynamics embedded in their interactions across varied agro-ethno-ecological zones. This is particularly vital given the differential roles of women and men in relation to addressing nutrition issues in the household and farm systems. Improving household nutrition and health requires a multi-faceted approach dealing with nutrition, income and social aspects such as gender dynamics. Overall developments in economic circumstances and transformational changes in the form of improvements in social inclusion and changes in gender norms and agency require further analysis and study.

There are two levels at which gender analysis is reported in this book. The first level relates to understanding gender norms and agency and involvement of women and men in agricultural research and development, as influenced by traditions, cultures and social regulations. The second level involves the mainstreaming and incorporation of gender dimensions in systems research and technology development, such as in nutrition-based research, through gender-linked treatments or the collection of sex-disaggregated data. Such incorporation and analysis is aimed at better targeting technology development and ensuring that women's and men's roles and benefits are built into the conduct of the research. Aspects and examples of both levels of gender research and analysis are addressed in this book.

A case study that analyzes two cases in gender norms and agency in Uganda out of a large-scale global study is reported in Rietveld (*this volume*). The analysis addresses the question "how do gender norms shape poor men's and women's abilities to adopt and benefit from agricultural and natural resources management (NRM) innovations?" A second analysis of gender research is provided by McDougall (*this volume*), who makes the point that understanding the role of gender in agricultural development research offers much needed new insights for making significant increases in productivity, food security and livelihoods. The chapter by McDougall explores the significance of gender in agricultural systems research for achieving global sustainable development outcomes, including poverty reduction, and increased food and nutrition security. It also focuses on a 'more novel role' for gender as a leverage point for innovations in systems research.

Further analysis and examples linking gender to agricultural systems research are analyzed in other chapters. For example, Sarapura Escobar and co-authors (*this volume*) present the Papa Andina Initiative in Peru as illustrating the role of gender transformative approaches in agricultural innovation. The case suggests that gender transformative outcomes occur when a gender neutral programme design is abandoned in favour of gender responsive processes achieved through participatory and applied methodologies that foster collective work, communication and individual and group learning among diverse groups of stakeholders. All of these processes influence changes in gender norms, perceptions and relations entrenched within social systems, in this case the Central Andes of Peru.

Systems and institutional innovation

Systems research and development approaches to sustainable agricultural intensification require integrated technological and institutional innovations (Schut van Asten et al., 2016). Examples of technological innovations are new or improved crop varieties, animal breeds, appropriate mechanization, information and communication technology (ICT) and new (farm) management practices. The effective development and uptake of such technological innovations require institutional innovations, for example new forms of stakeholder collaboration and novel policy, business or development strategies. Furthermore, lifting technological and institutional barriers will reveal new limitations (e.g. capacity of the market to absorb increased produce) and unintended consequences (e.g. herbicide resistance of crops), which require further technological and institutional progress. This shows the importance of iterative innovation processes of continuous reflection, learning and adaptation, which is exactly how systems have evolved historically (Geels, 2002). It is therefore no surprise that the performance of an agricultural innovation system – one of the more integrated and holistic systems approaches – is often expressed in the capacity of the system to continuously identify and overcome challenges and proactively explore new opportunities (Foran et al., 2014).

In agriculture, the concept of 'systems' and 'systems research' has different meaning for different people. Leeuwis and Wigboldus (*this volume*) have positioned this query at the centre of their chapter. They argue that the type of systems thinking (e.g. hard versus soft systems; functionalist versus political systems) to a large extent determines what types of research questions, intervention or change strategies, monitoring and evaluation frameworks, and scaling pathways are deemed credible, legitimate and effective. Rather than favouring one systems research approach over another, the authors propose enhancing the leverage and 'actionability' of 'systems research' through the participatory experimentation and systemic evaluation of combined technological and institutional options. This is in line with scientific theories on how change in complex configurations happens, and leaves sufficient space for different conceptualizations of systems research and systems boundaries.

The broadening of systems boundaries and the challenges that creates for measuring the effectiveness of system innovation and multi-stakeholder innovation processes form the starting point of the contribution by Sartas et al. (*this volume*). The authors problematize the lack of generalizable evidence on the effectiveness of multi-stakeholder processes to systems research and development, and present results of the development and testing of a "Learning System for Agricultural Research for Development (LESARD)". LESARD uses online open access tools and data repositories to document and analyze – amongst others – divergence or convergence of stakeholder perspectives; representation of different stakeholder, gender and age groups; and the 'actionability' in the multi-stakeholder process. It provides short-term feedback to facilitators and researchers on whether the multi-stakeholder innovation process is contributing to achieving a diverse range of development outcomes and impacts that allows for critical reflection and increased effectiveness.

Multi-stakeholder platforms are a popular vehicle for supporting multi-stakeholder innovation processes. Especially, so-called 'innovation platforms' are increasingly used in the implementation of agricultural research for development programmes (Dror et al., 2016; Schut, Klerkx et al., 2016). Hiwasaki et al. (*this volume*) reflect on the constraints and opportunities of using interlinked local and subnational multi-stakeholder platforms by presenting experiences from the East and Central Africa, West Africa, Central America and Central Mekong regions. They identify common objectives, appropriate representation of stakeholders and stakeholder engagement building on existing partnerships, secured resources and capable facilitation as success factors. Challenges include: demonstrating 'quick wins' for stakeholders in subnational platforms, strong focus on research over hands-on business and development approaches, limited ability of platforms to steer research for development agendas and resource allocation, limited practical support to implement the platforms, ensuring local ownership and sustainability beyond donor funded projects, and the tailoring of key platform concepts to specific social, political and institutional contexts.

The chapter by Triomphe and colleagues (*this volume*) illustrates the dichotomy between local 'organically-evolving innovation processes' and 'externally-induced innovation interventions' by building on the Joint Learning in and about Innovation Systems in African Agriculture (JOLISAA) programme experience. The chapter concludes that 'externally-induced innovation interventions' through public agricultural research and development organizations dominate the African innovation landscape. However, many farmers actively innovate individually and collectively in 'the social wild', without support through the agricultural research and development systems and often under the radar of researchers, policy or development actors. Rather than institutionalizing specific types of local innovations, the authors propose to strengthen agricultural innovation systems and – in doing so – the capacity to innovate for stakeholders across different systems levels. In that sense, Triomphe and colleagues (*this volume*) provide empirical support to the claim made by Leeuwis and Wigboldus (*this volume*) that understanding the nature and features of 'real' innovation processes can provide an important basis for strengthening innovation systems.

Lessons learnt and ways forward

Integrated agricultural systems research has gone through progressive steps from its early days as farming systems research to recent studies and analysis of sustainable intensification and diversification of livelihood systems taking into account the variability in context such as farmer typology (resources), bio-physical (soil, climate, etc.), socio-economic and institutional aspects, as well as the nested scales where interventions and change will have to take place (field, farm, community, district/landscape, national, etc.). Farmers are natural experimenters and farmer empowerment emerges across the many examples in this book as a guiding principle. This suggests the need for alignment with national and regional development agendas and policy frameworks. Systems research for impact requires multi-stakeholder engagement and capacity development with the systems researchers well embedded within development processes towards co-learning. The examples also suggest the need for appropriate methods and approaches to identify entry points for interventions aimed at sustainable systems intensification and diversification. Special attention is required on identifying leverage points for change for social equity and gender.

Integrated systems research has an important role in agricultural research for achieving development outcomes and impact based particularly around the goals of food and nutrition security, sustainable intensification and diversification of livelihood systems, gender and social inclusion and enhancing the natural resource base (e.g. soil health, water quality and availability, biodiversity) forming parts of the SDGs towards 2030 and the strategy and results framework of the CGIAR 2016–2030 (CGIAR, 2015). Whilst traditionally the focus has been on productivity enhancement or natural resources management approaches, a wider integrated systems perspective is required. This includes the need for trade-off analysis, working across scales from field-farm-household to socio-economic (institutions, markets, policy), human nutrition and biophysical landscapes (ecosystem services, soil and water management).

The following chapters provide a rich set of approaches to systems research operating at various scales from household to landscape to global scales. These approaches aim at delivering research and development outcomes and impacts. The challenge is now to move forward with the wider application of such approaches. One crucial aspect to successfully implement integrated systems research is further development of monitoring and evaluation systems with indicators that can show what is working, for whom, where, to what extent and how, and with feedback that is fast enough to support co-learning, adaptive management and development of options for specific contexts.

References

Attwood, S., Park, S., Loos, J., Phillips, M., Mills, D. and McDougall, C. *Does sustainable intensification offer a pathway to improved food security for aquatic agricultural system-dependent communities?* chapter 5 (this volume).

Bisseleua, H.D. and Degrande, A. *Approaches to operationalizing integrated systems research*, chapter 7 (this volume).

CGIAR (2015) *CGIAR strategy and results framework 2016–2030. Redefining how CGIAR does business until 2030.* http://www.cgiar.org/our-strategy/ (accessed 2 July 2016).

Chiwona-Karltun, L., Hambraeus, L. and Bellin-Sesay, F. *Balancing agrifood systems for optimal global nutrition transition*, chapter 16 (this volume).

CRP Dryland Systems Task Force (2015) *Drylands and mission critical research areas for the CGIAR*. http://drylandsystems.cgiar.org/sites/default/files/DrylandSystems_Mission-CriticalResearchAreas_TaskForce.pdf.

De Lattre-Gasquet, M., Hubert, B. and Vervoort, J. *Foresight for institutional innovation and change in agricultural systems: three examples*, chapter 9 (this volume).

Dixon, J., Gulliver, A. and Gibbon, D. (2001) *Farming Systems and Poverty: Improving Farmer's Livelihoods in a Changing World*, FAO & World Bank, Rome, Italy & Washington, DC.

Dror, I., Cadilhon, J.-J., Schut, M., Misiko, M. and Maheshwari, S. (2016) *Innovation Platforms for Agricultural Development: Evaluating the Mature Innovation Platforms Landscape*, Routledge, Oxon.

Foran, T., Butler, J.R.A., Williams, L.J., Wanjura, W.J., Hall, A., Carter, L. and Carberry, P.S. (2014). 'Taking complexity in food systems seriously: an interdisciplinary analysis', *World Development* 61: 85–101.

Geels, F.W. (2002) 'Technological transitions as evolutionary reconfiguration processes: a multi-level perspective and a case-study', *Research Policy* 31(8–9): 1257–1274.

Ha, T.T.T., Schreinemachers, P., Beed, F., Wang, J.F., Loc, N.T.T., Thuy, L.T., Van, D.T., Srinivasan, R., Hanson, P. and Afari-Sefa, V. *Sustainable intensification of smallholder agriculture in Northwest Vietnam: exploring the potential of integrating vegetables*, chapter 14 (this volume).

Hubert, B. and Ison, R. *Systems thinking: towards transformation in praxis and situations*, chapter 8 (this volume).

Garrity, D., Dixon, J. and Boffa, J-M *Understanding African farming systems as a basis for sustainable intensification*, chapter 4 (this volume).

Groot, J.C.J., Kennedy, G., Remans, R., Carmona, N.E., Raneri, J., DeClerck, F., Alvarez, S., Mashingaidze, N., Timler, C., Stadler, M., del Rio Mena, T., Horlings, L., Brouwer, I., Cole, S.M. and Descheemaeker, K. *Integrated systems research in nutrition-sensitive landscapes: a theoretical methodological framework*, chapter 18 (this volume).

Hiwasaki, L., Idrissou, L., Okafor, C. and van der Hoek, R. *Constraints and opportunities in using multi-stakeholder processes to implement integrated agricultural systems research: the Humidtropics case*, chapter 24 (this volume).

Kennedy, G., Raneri, J., Termote, C., Nowak, V., Remans, R., Groot, J.C.J. and Thilsted, S.H. *Nutrition-sensitive landscapes: approach and methods to assess food availability and diversification of diets*, chapter 17 (this volume).

Leeuwis, C. and Wigboldus, S. *What kinds of 'systems' are we dealing with? Implications for systems research and scaling*, chapter 22 (this volume).

McDougall, C. *Gender and systems research: leveraging change*, chapter 19 (this volume).

Mutemi, M., Njenga, M., Lamond, G., Kuria, A., Öborn, I., Muriuki, J. and Sinclair, F.L. *Using local knowledge to understand challenges and opportunities for enhancing agricultural productivity in Western Kenya*, chapter 12 (this volume).

Norman, D. and Atta-Krah, K. *Systems research for agricultural development: past, present and future*, chapter 2 (this volume).

Puskur, R., Park, S., Bourgeois, R., Hollows, E., Suri, S. and Phillips, M. *Exploring futures of aquatic agricultural food system in Southern Africa: from drivers to future-smart research and policy options*, chapter 10 (this volume).

Rietveld, A. *Gender norms and agricultural innovation: insights from Uganda*, chapter 20 (this volume).

Ritzema, R.S., Frelat, R., Hammond, J. and van Wijk, M.T. *What works where for which farm household: rapid approaches to food availability analysis*, chapter 6 (this volume).

Sarapura Escobar, S., Hambly Odame, H. and Tegbaru, A. *Gender transformative approaches in agricultural innovation: the case of the Papa Andina initiative in Peru*, chapter 21 (this volume).

Sartas, M., Schut, M. and Leeuwis, C. *Learning system for agricultural research for development (LESARD): documenting, reporting, and analysis of performance factors in multi-stakeholder processes*, chapter 25 (this volume).

Schut, M., Klerkx, L., Sartas, M., Lamers, D., Mc Campbell, M., Ogbonna, I., Kaushik, P., Atta-Krah, K. and Leeuwis, C. (2016) 'Innovation platforms: experiences with their institutional embedding in agricultural research for development', *Experimental Agriculture*. http://dx.doi.org/10.1017/S001447971500023X.

Schut, M., van Asten, P., Okafor, C., Hicintuka, C., Mapatano, S., Nabahungu, N.L., Kagabo, D., Muchunguzi, P., Njukwe, E., Dontsop-Nguezet, P.M., Sartas, M. and Vanlauwe, B. (2016) 'Sustainable intensification of agricultural systems in the central African highlands: the need for institutional innovation', *Agricultural Systems* 145: 165–176.

Sinclair, F.L. *Systems science at the scale of impact: reconciling bottom up participation with the production of widely applicable research outputs*, chapter 3 (this volume).

Sumberg, J. (2005) 'Constraints to the adoption of agricultural innovations', *Outlook on Agriculture* 34: 7–10.

Thomas, R. (2015) *Response to task force on mission critical research areas for drylands.* http://cgiarweb.s3.amazonaws.com/wp-content/uploads/2015/04/Responses-to-Task-Force-on-Mission-Critical-Research-Areas-for-Drylands-17.4.15.pdf.

Timler, C., Michalscheck, M., Alvarez, S., Descheemaeker, K. and Groot, J.C.J. *Exploring options for sustainable intensification through legume integration in different farm types in Eastern Zambia*, chapter 13 (this volume).

Triomphe, B., Floquet, A., Letty, B., Kamau, G., Almekinders, C. and Waters-Bayer, A. *How can external interventions build on local innovations? Lessons from an assessment of innovation experiences in African smallholder agriculture*, chapter 23 (this volume).

United Nations (UN) (2015) *Sustainable development goals. 17 goals to transform our world.* http://www.un.org/sustainabledevelopment/sustainable-development-goals/ (accessed 2 July 2016).

van Ginkel, M., Sayer, J., Sinclair, F., Aw-Hassan, A., Bossio, D., Craufurd, P., . . . and Ortiz, R. (2013) 'An integrated agro-ecosystem and livelihood systems approach for the poor and vulnerable in dry areas', *Food Security* 5: 751–767.

Vanlauwe, B., Barrios, E., Robinson, T., van Asten, P., Zingore, S. and Gérard, B. *System productivity and natural resource integrity in smallholder farming: friends or foes?* chapter 11 (this volume).

van Noordwijk, M., Mbow, C. and Minang, P. (2015). Trees as nexus for sustainable development goals (SDGs): agroforestry for integrated options. *ASB Policy Brief* 50, Nairobi: ASB Partnership for the Tropical Forest Margins.

Woomer, P.L., Omondi, B., Kaleha, C. and Chamwada, M. *On-farm testing of grain legumes and their management in West Kenya*, chapter 15 (this volume).

Part I

Conceptual underpinnings of systems research

2 Systems research for agricultural development

Past, present and future

David Norman and Kwesi Atta-Krah

Introduction

In the mid-1960s the Green Revolution (GR), based on fertiliser-responsive, high-yielding varieties of rice, wheat and maize, began having a major impact in favourable (including both biophysical (technical) and socioeconomic (human) elements) production environments (Norman et al., 1982), particularly in Asia, and parts of Latin America and North Africa. Since the GR technologies were scale-neutral, 'revolutionary' and robust, their adoption benefited all types of farmers even if they did not exactly follow the prescribed recommendations.[1] The key components of the GR technologies were (i) good seed (high-yielding varieties), (ii) availability of needed fertiliser, (iii) well-developed rural infrastructure, (iv) enabling policies and (v) availability of good quality land and water. However, the heavily 'supply-driven' and single commodity reductionist research approach was less successful in addressing the needs of farmers in less favourable and more heterogeneous production environments so common in Sub-Saharan Africa (SSA) and non-GR parts of Asia and Latin America.[2] Smallholder farms in Africa are extremely heterogeneous and diverse in both crop and livestock components, and are less compatible with the classic GR model, which explains why the GR had little impact in these areas.

Farmers in high-income countries (HICs) have historically been effective in articulating their needs – not only technological but also institutional – via more responsive research and extension systems, commodity-based groups, lobbying platforms and so forth.[3] However, because farmers in low-income countries (LICs) – the major focus of this chapter – have generally benefitted much less in terms of education and linkages with research, extension and policy institutions, their views have historically not been adequately taken into account, nor have they been involved in shaping policy on issues that concern agriculture and their livelihoods. Generally they have had no 'voice' in the kind of research that is supposedly done on their behalf. Mostly such research had been based only on researcher perspectives, and usually focused on individual commodities or specific components of the system, rather than embracing the diverse and integrated nature of the farming system and its linkage to livelihood systems that support smallholder farmers. However, since the 1970s the 'top-down' agricultural research/developmental paradigm has been criticised for excluding the farmers. It was this need that led to the emergence of farming systems research (FSR).

This chapter initially examines the origins of FSR and the path and reasons for its evolution and transformation into what is now called integrated systems research (ISR). In doing so we stress the need for more integrated system approaches, aimed at enhancing the livelihoods of smallholder farmers. The specific experiences of the international agricultural research centers (IARCs) under the umbrella of the Consultative Group of International Agricultural Research (CGIAR) (http://www.cgiar.org) with ISR are used to illustrate this evolution from FSR.

Building on the specific experiences of the CGIAR, and also based on FSR transitioning to a broader integrated livelihood systems focus, the chapter identifies a number of elements and pre-requisites that are considered essential for the conduct of ISR with a livelihoods orientation. The chapter concludes with an affirmation of the continued need for systems research in agriculture and for increased integrated and holistic analysis, taking into account the research-development continuum involving multiple stakeholders. In this regard, mechanisms for enabling multi-stakeholder processes in agricultural research, together with the need for farmer empowerment to ensure their full participation, are emphasised as essential.

Evolution of farming systems research and farmer participatory approaches[4]

Until the 1960s, there was little research collaboration between technical agricultural scientists (usually located on experiment stations), economists (mainly in planning units) and anthropologists/sociologists (generally in academia). However, in the early 1960s and later, many village-level studies were undertaken, initially by anthropologists/sociologists (i.e. especially in francophone SSA) and later by agricultural economists (i.e. mainly in Anglophone countries), that involved elements of whole farm dynamics and their relationship to the farming community. The major conclusion (Collinson, 1972; Spencer et al., 1979; Norman et al., 1982; Walker and Ryan, 1990) was that resource-limited farmers and their households had a very good understanding of the variable and risky production environments in which they operated and adopted farming systems (i.e. combining crop, livestock and off-farm enterprises) that were fundamentally sound (i.e. in terms of their goal(s) and resources (inputs) available) and that, historically at least, were sustainable.

Such positive conclusions about the rationality of farmers and the farming systems they operated (e.g. mixed cropping (Norman, 1974)) began to throw light on why many recommended research technologies were not adopted by farmers. Until the 1960s, most recommendations were derived from station-based trials, using technical evaluation criteria (e.g. yield increases) with practically no involvement of farmers. Farmer-implemented evaluation exercises indicated that many existing recommendations were inappropriate

(Norman et al., 1982). In fact, many of the recommendations compatible with farmers' biophysical environments were found to be incompatible with their socioeconomic environments. Consequently using technical evaluation and standard conventional economic criteria alone were inadequate for evaluating the suitability of technologies for resource-limited farmers, often operating in unfavourable and heterogeneous production environments. Farmers and their households usually had goals that go beyond profit maximisation, such as being risk averse in operational situations where markets for capital, land and labour worked very imperfectly. In spite of this, farmers were not inherently conservative since they were natural informal experimenters (Biggs and Clay, 1981). Another important conclusion was that it would be desirable to introduce some flexibility in formal recommendations enabling farmers to better respond to location-specific differences rather than relying on a few fixed technological packages (i.e. one size fits all syndrome) (Norman et al., 1982).

Two other factors fuelling a change in the conventional 'top-down' (i.e. supply driven) research paradigm were that:

- Those of us working the field increasingly recognised that the neoclassical economic paradigm training approach that most of us agricultural economists received had limitations in addressing all issues faced by resource-limited farmers, including that they operated in dynamic and very uncertain production environments.
- We also realised that there would be synergistic benefits in directly interacting with farmers in the technology design and development process itself (i.e. *ex ante* involvement), rather than treating them only as persons from whom data is extracted and whose only involvement occurred *ex post* after the recommendations had been formulated.

Thus a radical change in the research approach took place, with greater emphasis on the importance of taking a whole farm analytical approach, zooming out from plot level to farm/household level, and seeking to better understand socio-technical interaction issues such as labour/land and input/output dynamics between different crops/commodities at the farm level. This new thinking also required involving farmers in the research design and implementation, leading to the emergence of farmer participatory approaches as a critical component of agricultural research. This required an interdisciplinary[5] strategy and involving farmers throughout the technology design/evaluation process. Consequently in the mid-1970s the FSR approach emerged in response to the need for a more integrated 'bottom-up' and 'demand-driven' paradigm for agricultural research, focusing on the farmer in his/her environment/context. The farming systems perspective required new types of relationships between farmers and researchers (technical and social scientists). Over time, this expanded into farmers interacting with agricultural development stakeholders, and the incorporation of farmer participatory methodologies into national and international agricultural research programs.

The farming systems research approach process began with an understanding, from the perspective of the farmers and farm households, of their problems and opportunities, and using those as an input in the design/evaluation of solutions compatible with their objectives and production environments. In addition to farmers' involvement being critically important in the technology development/evaluation process (Matlon et al., 1984; Chambers et al., 1989), other significant characteristics of farming systems research were: its holistic perspective, the iterative nature of the process and the involvement of both technical and social scientists. One of the early schematic frameworks for implementing FSR was developed at a Ford Foundation-sponsored workshop at the *Institut d'Économie Rurale* in Bamako, Mali, in 1976 (Figure 2.1) (Norman et al., 1982).

Operationalising FSR first required classifying farming households into different farm types in which those within one type had analogous resources, basically experienced similar problems and opportunities, and hence would likely benefit from adopting the same potential solutions (i.e. recommendations). These different types/groups of farming households in essence constituted what later became known as recommendation domains. Through selecting farming households, representative of the different farm types, it was then possible to target activities while the recommendation domains provided a basis for scaling-up or introducing any recommendations to other farming households of a similar farm type.

Farming systems research has undergone a major evolution over time. Generally, four phases can be differentiated (Figure 2.2) based on the ability to deal with progressively higher ratios of variables to parameters. Over time, thanks to methodological developments, evolution through the four phases occurred, making it possible to handle increasingly complex situations in the later phases.

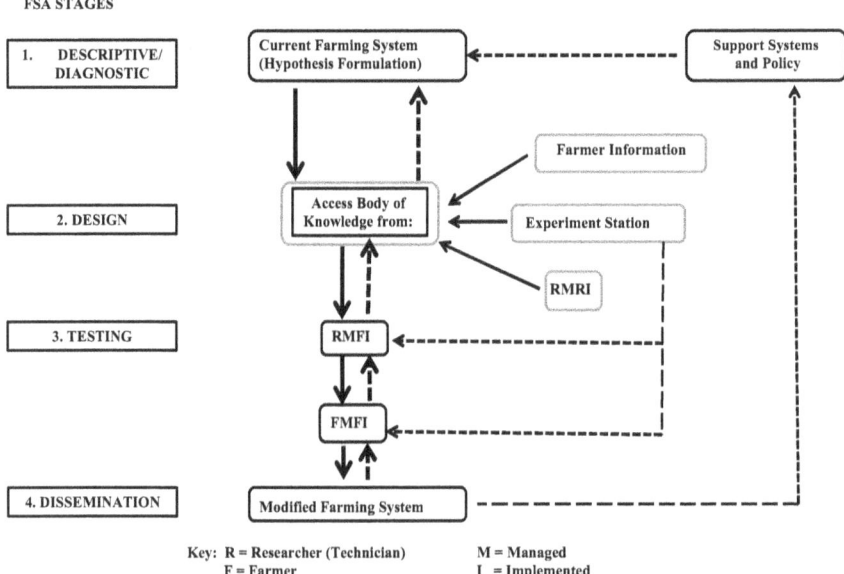

Figure 2.1 The farming systems approach in technology generation

Source: Norman et al., 1982.

A brief overview of the different phases of FSR since its beginning in 1970s is presented below, and illustrated in Figure 2.2.

1. ***Farming systems with a predetermined focus.*** Initially FSR considered only one specific commodity, with focus on identifying improvements within that commodity that were compatible with the whole farming system. For example, given their specific crop mandates, the International Maize and Wheat Improvement Center (CIMMYT) and the International Rice Research Institute (IRRI) focused on maize-, wheat- and rice-based systems, thereby introducing a systems perspective to commodity-based research programs – which by themselves often had a strong reductionist orientation. This approach was relevant to farming systems dominated by one crop, since improving the productivity of that enterprise would have the greatest impact on the productivity of the overall farming system. Because of their networks and training programs in Africa and Asia, these two IARCs were very influential in disseminating FSR principles in the early years, though the principles were coloured to a large extent by the commodities that these institutions focused on.

2. ***Farming systems with a whole farm focus.*** Although the IARCs, as peer research institutions, were important in popularising FSR, national

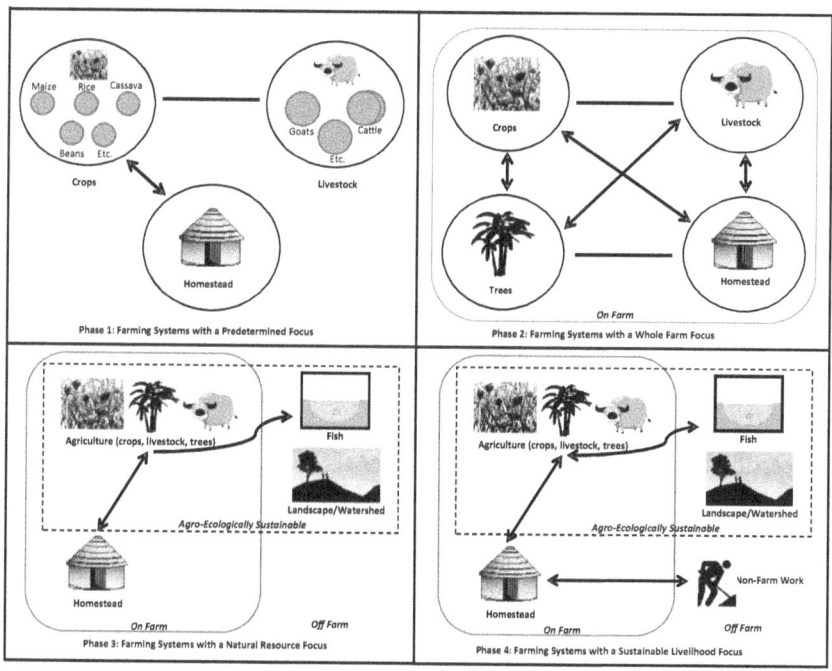

Figure 2.2 Progression in the farming systems approach

Source: Modified from Norman (2015).

agricultural research systems (NARS), with their multi-commodity mandates, were primarily responsible for the approach becoming more holistic. This required that focus on any commodity and its linkages with other commodities had to be based on the needs articulated by farmers and their households. Substantial donor funding supported the promotion of this FSR approach with early examples occurring in Guatemala, Thailand, Senegal and Nigeria. Its growth within the NARS resulted in distinct area-based farming system teams working directly with farmers. This phase encouraged greater farmer participation, as they found research targeting their needs.

Early experiences with these two phases of farming systems research encouraged technical scientists to design more flexible technological components and packages suitable for different types of farmers. Increased respect for the roles of, and tools used by, different disciplines resulted in improved cooperation between technical and social scientists. Finally, there was increased appreciation of the critical importance of appropriate policy/support systems (i.e. input distribution systems and product markets) in determining the relevance of new technologies.[6] These developments did not occur overnight, but were evolutionary based on growing confidence, discovery and learning within FSR.

3. *Farming systems with a natural resource focus.* In the late 1980s, ecological sustainability issues became more important. It was realised that a number of practices within traditional systems of agriculture (e.g. shifting cultivation, bush fallowing and ring cultivation) ensured ecological stability, but that these systems were under pressure and no longer feasible because of exogenous influences such as rapidly increasing populations (human and/or animal) and climatic changes. Short-term survival concerns were increasingly forcing farmers to adopt strategies ensuring short-term food supplies, such as continuous cropping (without fertiliser inputs), which had detrimental long-term environmental consequences. In SSA, increasing human population densities meant that lands unsuitable for cultivation were being cropped, raising issues of land degradation. Generally, cropping systems were being intensified, but without the means to supply plant nutrients to maintain soil fertility (such as through recycling crop residues or adding organic/inorganic fertilisers). In GR areas (e.g. the intensive rice–wheat systems of South Asia), farmers using high levels of inorganic fertiliser and other external inputs were also experiencing decreasing productivity. Although scientific analysis can foresee the challenge of ecological degradation, farmers, mainly driven by short-term survival, might not see such an issue as a major concern, unless it threatens immediate survival (Fujiska, 1989). Consequently, it is unlikely that such issues would be addressed in phases 1 and 2 of FSR. It therefore became necessary for external stakeholders (e.g. researchers and development practitioners) to study possible conflicts between strategies designed to improve short-run productivity and those ensuring long-run ecological sustainability. These concerns led to the introduction of an ecological sustainability dimension into FSR, mainly driven by researchers.

Two ways in which this FSR phase has been operationalised are as follows:

- Introduction/adoption of new methodologies that involved working with farmers to evaluate bio-resource (nutrient) trends and flows on the farm. These enabled the identification of vulnerable parts of the farming system and helped researchers, in collaboration with farmers, determine how current practices could be modified to promote ecological sustainability (Lightfoot et al., 1991; Defoer and Budelman, 2000). This approach helped farmers in transforming ecological problems from being a *foreseen* to a *felt* problem (Norman, Umar et al., 1995). Since solutions were usually farm-specific, with implementation being the primary responsibility of farmers themselves, it was important for them to assume ownership for the necessary changes identified during the participatory design exercises.

- Shift towards eco-regional research, undertaken by some IARCs in association with national research and development institutions. The focus of this kind of research was on priority eco-regional problems, with elements of productivity enhancement combined with natural resources management and environmental sustainability. The operational mode required the collaboration of all the stakeholders in the agricultural research/development continuum. A good example of eco-regional research was the African Highlands Initiative (AHI) project, which was conducted in the highlands of East and Central Africa (ECA) in the period 1995 to 2011 (http://www.worldagroforestry.org/programmes/african-highlands/evolution.html) with the aim of developing methodologies for integrated natural resources management (INRM) and their institutionalisation in partner NARS in the humid highlands of ECA. The AHI worked with teams from NARS, extension and non-governmental organisations (NGOs), and hence maintained a network of partner organisations on research for development in natural resources management (NRM). Using both action and empirical research, AHI developed several methods, tools and approaches for INRM, and at the pilot project level demonstrated an INRM approach that works. The strategy was to demonstrate it on a larger scale via cross-scale interventions and adaptive management. This constitutes the essence of eco-regional research, tying NRM issues into FSR.[7] This study contributed to the development of a guideline on the implementation of INRM research for development (Campbell et al., 2006).[8]

Three important challenges in implementing this phase of FSR are worth noting. The first is that significant resources (i.e. time and financial) were required to address the complex processes and to show results especially related to the sustainability element. Secondly, assessment of progress in improving ecological sustainability required a long time-frame. Thirdly, because of the poverty level of many farming households, ecological sustainability initiatives were likely to be attractive only if they simultaneously improved short-run welfare,[9] or had some incentives built into them.

4. ***Farming systems with a sustainable livelihood focus.*** In this fourth phase of FSR, of which ISR is an example, the ratio of variables to parameters is the highest.[10] This phase explicitly involved linking change at the household level with complementary changes at the meso- and macro-levels. The objective is to strengthen the abilities of households and communities to use existing and new knowledge in analysing their circumstances, ascertain problems and opportunities, evaluate possible strategies and, consequently, plan and implement action. Ideally, emphasis is on designing interventions improving productivity and income (i.e. reduce poverty) while simultaneously protecting the environment.[11] Preferably they should also strengthen the coping and adaptive strategies of the most vulnerable groups in the community. New technologies that fulfil such conditions are likely to have the following properties: be flexible through increasing the ability of farmers to adapt their production/livelihood systems to stochastic shocks and to a constantly changing economic environment (Chambers, 1991); reduce risk, such as the new more resistant/tolerant crop varieties and agronomic practices that reduce the impact of biotic and abiotic stresses, that promote enterprise diversification, etc.; and complement the complex livelihood systems of poor households.

Since assets, entitlements and social relationships of households vary according to household and socioeconomic stratum, resulting in different livelihood strategies (Chambers, 1989; Frankenberger and Coyle, 1992), attention is usually focused on the most vulnerable (i.e. poorest) households facing chronic or temporary food insecurity. A combination of analytical methods, including conventional farming system (farmer participatory) approaches, political economics, anthropology and environmental science, are used with the involvement of interdisciplinary teams working in conjunction with local communities. This phase of FSR includes and combines elements of all three phases described above, and thus constitutes the most advanced manifestation of FSR. It has often been said that in spite of the holistic characteristic of this phase, its application has been a challenge due to the greater complexities involved. A principal question, however, is 'adoption of what?' It can be argued that the application of such integrated system approaches should not viewed simply as the wholesale adoption of processes and products. The particular research processes and the outcomes resulting will be situation-specific. Among other challenges relating to scaling-out and dissemination are the location specificity of solutions, and the skill sets required of the interdisciplinary teams involved in implementation. As mentioned earlier, the CGIAR developed guidelines for this (Campbell et al., 2006) that stressed the need to focus on the weaker aspects of a proposed 'learning wheel', thus avoiding becoming bogged down in too much detail and complexity.[12] Unfortunately experience in applying this approach has been insufficient to determine its usefulness.

Methodological developments

Methodologies for eliciting the attitudes and expertise of smallholder farmers have evolved greatly since the 1970s, thus creating more avenues for their involvement in identifying and implementing relevant research and development initiatives. Initial methodological developments occurred in response to a specific need/stage in implementing the FSR, namely: description of the situation; diagnosis of problems and/or opportunities; testing/evaluation of solutions/opportunities; and dissemination (Figure 2.1). Over time, many of the techniques were found to be useful in addressing or operationalising more than one stage of the approach.

Among the most important methodological developments have been the following:

- Rapid rural appraisal (RRA) and later participatory rural appraisal (PRA), which were methods developed for obtaining information from farmers (Program for International Development, 1994; Pretty et al., 1995). Such techniques provided a means of ascertaining how farmers interpreted their production environments and could help them articulate their constraints and needs to researchers. It thus enabled them to contribute more directly and effectively to the design and evaluation of new technologies. PRA techniques, in particular, improved the potential effectiveness of farmers' participation through greater systematisation of their knowledge and opinions.
- Farming systems research dispelled the simplistic notion of the farming household being monolithic, with one decision-maker pursuing one goal, making all farming decisions, and every household member benefitting equally from the results. Many researchers helped develop techniques for examining intra-household relationships, particularly those that were gender-related (Feldstein and Jiggins, 1994). Consequently there evolved greater sensitivity to incorporating gender and other intra-household related issues in the research and development process.
- Prior to FSR, technology evaluation was usually accomplished through experimental station trials, where design, management and implementation were all done by the researcher. This was tagged as researcher-managed and researcher-implemented (RMRI) with the researcher (R) being responsible for deciding the treatments (i.e. management (M)) and their implementation (I). However, FSR encouraged use of two other types of trials, specifically researcher-managed and farmer-implemented (RMFI) and farmer-managed and farmer-implemented (FMFI). These three trial types differ according to several characteristics (Table 2.1) (Norman, Worman et al., 1998). RMRI trials generally dominating in the technology design stage are later increasingly substituted with RMFI and FMFI trials. This farmer 'learning by doing approach' is important in improving farmers' assessment of, and potential commitment to, adopting technological components/packages.
- Prior to FSR, farmers' participation in the technology development process was only at the adaptive end of the research spectrum. However, Sperling

Table 2.1 Expectations of different types of trials

Item	Researcher-managed and researcher-implemented (RMRI)	Researcher-managed and farmer-implemented (RMFI)	Farmer-managed and farmer-implemented (FMFI)
Experimental:			
Stage:	Design	1st stage testing	2nd stage testing
Design:			
Complexity		Less	Least
Type	Standard	Simple standard	With and without
Replication	Within and between sites	Usually only between sites but can also be within	Between sites only
Levels of treatment	Most	Less	Least
Standardised level of non-experimental variables	Most	Less	Least
Plot size	Smallest	Larger	Usually largest
Who selects technology?	Researcher	Researcher/farmer	Farmer
Who shoulders risk?	Mainly researcher	Researcher/farmer	Mainly farmer
Main discipline of researcher	Mainly technical	Technical/social	Mainly social
Participation by:			
Farmer	Least	More	Most
Researcher	Most	Less	Least
Number of farmers	None	Some	Most
Farmer groups	Least	More	Most

	Researcher		Farmer
Potential:			
"Yield"	Most	Less	Least
Measurement errors	Least	More	Most
Degree of precision	Highest	Less	Least
Data:			
"Hard" (objective)	Most	Less	Least
"Soft" (subjective)	Least	More	Most
Determination of cause/ effect relationships	Easiest	Less easy	Least likely
Incorporation into farming system	Least	More	Most
Evaluation:			
Who by?	Mainly researcher	Researcher/farmer	Mainly farmer
Nature of test	Assess technical feasibility	Technical feasibility plus economic evaluation	Validity for farmers – practicality, acceptability
Appeal to: Researchers	Most	Less	Least
Extension staff	Usually least	More	Most
Farmers	Least	More	Most
Ease of acceptance of trial results	Researcher	Researcher/farmer/extension	Farmer

Source: Adapted from Norman, Worman et al., 1995.

and Berkowitz (1994), working with beans in Rwanda, demonstrated farmers could make uniquely valuable contributions in the evaluation through, for example, participatory varietal selection (PVS) of suitable bean germplasm. This concept was further developed with farmers' involvement in participatory plant breeding (PPB) of improved varieties (Joshi and Witcombe, 1996; Witcombe et al., 1996) in both IARCs and NARS.

- Two approaches developed for analysing results of on-farm research and making recommendations based on them were: adaptability (formerly modified stability) analysis, a statistical tool for analysing RMFI and FMFI on-farm trials (Hildebrand and Russell, 1996; Sall et al., 1998); and PRA techniques – in particular matrix ranking and scoring – enabling farmers' criteria to be systematized,, that is ranked, both in designing and evaluating on-farm trials. Another less common approach developed was a quasi-arbitrary ordinal weighting approach for determining criteria farmers use in deciding, for example, which rice crop varieties to adopt (Sall et al., 2000).

Such methodological developments, greatly improving farmers' effectiveness in FSR, have been accompanied by other very positive changes, namely: more collaborative and collegial relationships between farmers themselves and with researchers, extension and other developmental stakeholders; and initiatives to improve the efficiency and potential multiplier impact of FSR activities. Examples of initiatives accomplishing this have been the following:

- Farmer groups (both formal and informal) have been extensively used, enabling researchers and developmental stakeholders to interact efficiently with farmers (Heinrich, 1993; Norman, Worman et al., 1995). These are effective in influencing research/development agendas, in testing/evaluating and in disseminating relevant technologies/strategies. A less common but potentially even more powerful means of farmer empowerment has been for farmer groups having a say in the allocating of research funds, thus helping in tailoring the research agenda to their needs (e.g. Colombia (Ashby et al., 1995); Mali and Senegal (Collion and Rondot, 1998)).
- The farmer field school (FFS) approach was developed by the Food and Agricultural Organisation (FAO) and partners nearly 25 years ago in Southeast Asia as an alternative to the prevailing top-down extension method of the GR, which failed to work in situations where more complex and counter-intuitive problems existed, such as pesticide-induced pest outbreaks (http://www.fao.org/agriculture/ippm/programme/ffs-approach/en/). FFSs have increasingly been used for encouraging farmer interaction, direct involvement as trained 'farmer researchers', and for disseminating technologies via FFS trained farmers.
- Somewhat later in the FSR era, in recognition of the importance of interactive linkages between the research and developmental stakeholders, committees were sometimes established at the national, regional and district levels, consisting of representatives of the different stakeholder groups (including farmer representation), for the purpose of exchanging and

disseminating information, and improving the coordination and the design and implementation of collaborative initiatives. Sometimes, decentralisation of governance and 'local' approval of technological recommendations have facilitated this process. Currently major emphasis is being placed on encouraging interaction between all agricultural development stakeholders via 'innovation platforms' (IPs). These are discussed later in this chapter.

Evolution in agricultural innovation thinking

A discussion on the evolution and development of FSR would not be complete without reference to agricultural innovation systems approaches, which are rapidly gaining recognition in systems research, both conceptually and programmatically. A paper tracking the evolution of systems approaches to agricultural innovation (Klerkx, van Mierlo et al., 2012) provides a lens through which such evolution can be seen. The paper asserts that innovation is not simply about adopting new technologies. Instead, agricultural innovation is presented as a co-evolutionary process that combines technological, social, economic and institutional change resulting from multiple interactions between components of farming systems, supply chains and economic systems, policy environments and societal systems. For these 'agricultural innovation systems', a wide range of analytical approaches have emerged, such as the Agricultural Knowledge and Information Systems (AKIS) (Röling, 2009) and Agricultural Innovation Systems (AIS) (Hall et al., 2001). These agricultural innovation systems can be viewed as the most recent manifestation of systems approaches in agricultural research. In the context of the broader evolution of FSR described earlier in this chapter, it is associated with the fourth phase of FSR, namely farming systems with a sustainable livelihood focus.

The CGIAR experience and role in farming systems research

In the early years (mid-1970s and 1980s), FSR was a strong component of the research portfolio of the CGIAR Centers. The roles played by two Centers, CIMMYT and IRRI, in the early years of FSR have already been mentioned. FSR was also very important within agro-ecology or region-based research Centers of the CGIAR, such as the: International Institute for Tropical Agriculture (IITA), focusing on agro-ecologies and farming systems within SSA; International Center for Tropical Agriculture (CIAT), focusing on agricultural systems within tropical America; International Crops Research Institute for the Semi-Arid Tropics (ICRISAT), focusing on agricultural systems within the semi-arid tropics; International Center for Agriculture in the Dry Areas (ICARDA), focusing on agricultural systems in drylands; and the International Livestock Research Institute (ILRI), focusing on integrated livestock systems in humid, sub-humid and semi-arid areas.

For example, in IITA, one of four research programs during the 1970s was the 'Farming Systems Program', which later became the 'Resource and Crop

Management Division'. This division, conducting research in resource management, involving soil and water management interactions, incorporated various dimensions of systems research, both on-station and on-farm and involved an interdisciplinary team of scientists (such as agronomy, sociology, economics, anthropology, ecology, etc.). The other three divisions were commodity-defined and focused mainly on breeding and research for enhancing the productivity of major food security crops – cereals, grain legumes, and root and tuber crops. In an article reviewing 40 years of research functioning and governance of the CGIAR, McCalla (2014) indicated that the early successes in commodity breeding (i.e. semi-dwarf rice and wheat) skewed donor interest strongly towards commodity breeding/productivity improvement at the expense of farming systems research/productivity. McCalla went on to postulate that "promising systems programs at IITA (understanding and managing cleared tropical soils) and CIAT (understanding complex crop/livestock systems using systems modeling) were abandoned and the Institutes were quickly converted into commodity focused Centers" (2014, 16).

Consequently, interest in systems research within the CGIAR Centers waned from the mid-1990s into the new millennium (21st century). A number of Centers actually reformed their research programs during this period, placing greater emphasis on commodity-based programming, and de-emphasising or, sometimes, eliminating FSR programs *per se*, apparently on the understanding that systems research dimensions would be integrated in the commodity research programs. This integration, however, rarely functioned optimally and the emphasis of research in the commodity programs continued to be dominated by breeding and crop improvement interests, often with little involvement of farmers and communities for whom the technologies were supposedly developed. Consequently the needs of most smallholder farmers with their diversified farming systems and specific socioeconomic, environmental and productivity challenges were once again not adequately addressed.

A review of the CGIAR research structure between 2008 and 2010 concluded that greater research coordination, integration and collaboration was needed between the various Centers, to enhance the overall effectiveness and productivity of research. A consortium of CGIAR Centers was created under one governance mechanism, with one Chief Executive Officer and one Consortium Board. Research was to be developed in an integrated manner across the various Centers to tackle identified global development challenges. In 2011–2012 fifteen cross-Center CGIAR Research Programs (CRPs) were created which constituted a new research portfolio for the entire CGIAR system. The CRPs were to contribute directly to agricultural development through partnerships and collaboration across very diverse groups of research and development actors (Sumberg et al., 2013). In this new iteration of research within the CGIAR, systems research re-surfaced as part of the research portfolio, not in the framework of farming systems, but more as integrated systems with a livelihoods focus. Three of 15 CRPs created were systems CRPs, with agro-ecological mandates. These were the: Integrated Systems for the Humid Tropics (Humidtropics), focusing on the humid and sub-humid tropics region; Dryland Systems

(Drylands), focusing on the drylands ecosystem; and Aquatic Agricultural Systems (AAS), focusing on farming and fishing systems around natural freshwater and/or coastal ecosystems. Systems research in the CGIAR was therefore reborn, and systems research processes and activities were initiated in various locations with partners from different agricultural stakeholder organisations.

This revival of systems research was, however, short-lived. In yet another review of the CGIAR research structure undertaken in 2015, barely 3–4 years after initiation of the first phase of the CRPs, it was decided that the research portfolio needed to be reformed to make it better aligned to a new Strategy and Results Framework (SRF) of the CGIAR (CGIAR, 2015a). Consequently, the CRP portfolio was re-structured into two key domain groups: (i) Agri-Food Systems CRPs (AFS-CRPs), consisting of eight CRPs; and (ii) Global Integrative CRPs, consisting of four crosscutting CRPs (CGIAR, 2015b). A key consequence of this reformulation was that the three systems CRPs would not continue to exist as separate CRP entities beyond 2016. The understanding was that systems thinking and approaches would continue through direct integration into the eight defined commodity CRPs, now branded as AFS-CRPs.

Many have questioned whether this latest development signals yet another downturn for agricultural systems research within the CGIAR. The answer to this question will depend to a large degree on how the implication of systems integration is understood and implemented, and what real integration and systems reform takes place within the AFS-CRPs. However, it could be argued that this new development might be positive, as systems research now moves from being in the periphery of agricultural research, where it has been over the years, into mainstream research. The idea of 'systems research' seen as being in one camp, and 'commodity research' in the other camp, will be eliminated, and new efforts can now relate to implementing core research agendas involving major commodity crops within the systems research framework. Only time will tell if this is successful. The hope is that everything necessary will be done to ensure proper alignment and integration of systems thinking into the development of the AFS-CRPs.

Looking forward: Key-lessons and methodological implications

The experiences with systems research in the CGIAR, exemplified through the Humidtropics and other system CRPs, provide a good example of seeing how the science and practice of FSR has evolved over the years from its inception in the 1970s to its role today. The question often asked is 'What is different between FSR at its inception and as it is currently practiced?' Using Humidtropics as an example to illustrate this, four key differences are worth mentioning:

- The emphasis has shifted from 'farming systems' to 'livelihood systems'. This is not just a terminology issue. The original FSR was inward (farm) looking. It was designed to focus almost exclusively on the farm and the components within it, with little attention to the outside realities. The ISR approach sees the farm as one component of a larger system influencing the livelihoods of the farmer households and therefore integrates on-farm and off-farm developments and their implications.

- Emphasis in FSR was on farmer participatory approaches and multidisciplinary interactions within the farm setting. Now with ISR the emphasis has broadened beyond the farm with off-farm dimensions seen as key for influencing the livelihoods of the farmers. Multi-sector and multi-stakeholder involvement and analysis, and linking research to major transformation goals, are important. This also implies that constraints and opportunities for innovation above the farm level need to be taken into account (Giller et al., 2008; Schut et al., 2014).

- Early FSR sometimes focused on the harvested yield of commodities as the principal determinant of the productivity of the system. Currently the emphasis goes much beyond yields of the specific commodities to whole systems performance (e.g., http://mel.cgiar.org/xmlui/handle/20.500.11766/4505) and explorations on value addition through value chain analysis and processing, linking farm produce to off-farm interventions that add value to the commodity, and assistance to link it to markets and income.

- There is now a much stronger emphasis on innovation systems, involving creating a platform that enables farmers and other stakeholder groups to be involved in innovation systems. This also includes the desired objective of fostering the capacity to innovate among farmers and other agricultural stakeholders at different levels (Hall et al., 2003; Klerkx, Schut et al., 2012; Adekunle et al., 2013; Leeuwis et al., 2014; Schut et al., 2016).

Ingredients of a systems approach

Building on experiences and lessons spanning the evolution of FSR in its various forms, and combining this with the experience of implementation of ISR, Humidtropics has established a set of ingredients required for a systems approach in the conduct of ISR. These can be used as a guide in the establishment of new systems research undertakings, or for assessment of existing projects with respect to their systems research considerations and opportunities. It can also be used to identify areas for capacity development for strengthening the integrated systems elements in agricultural research. Essential elements in ensuring a systems approach are the following:

- **Research team**: Systems research is never a one-person or one-discipline undertaking. It requires team effort, ensuring not only multi- but preferably inter-disciplinarity in the conduct of research. This is necessary to ensure that problems and opportunities are analysed from multiple perspectives, incorporating both socioeconomic and biophysical considerations. Thus disciplines such as economics, sociology and anthropology need to be considered along with agronomy, soil science, animal science and ecology. The right combination of the team will of course depend on the research issues at hand. This is not to say that all these positions need to be available in every systems research activity, or in any one institution before systems research can be embarked upon. This can form the basis for partnerships among institutions and inter-sectorial collaboration.

- **The role of the farmer in systems research**: The centrality of the farmer and his/her community in the conduct of systems research cannot be over-emphasised. Farmers have been operating complex farming systems over many generations; they are indeed 'system researchers' in their own right, and need to be viewed as such. Since the inception of FSR, farmer participation has been considered important. However, in reality farmers have often not been adequately recognised in such research. They have often been seen more as objects of study, or as participant observers, or simply considered as ultimate beneficiaries, rather than as full partners in the conduct of research. The reason for this may be that most of these farmers are poor smallholder farmers, who are often uneducated and generally powerless. Their participation in research is often for them to do what the researchers want them to do, rather than contribute their knowledge and experience built up over generations (i.e. including an intimate knowledge of their production environments). In this context there is a crucial need for farmer empowerment, which enables farmers to see themselves as *bona fide* members of the research team. This involves giving more authority to farmers, such as in being able to identify markets and influence prices for their commodities (Norman, 2004).[13] Recently Lundy et al. (2012)[14] have developed important participatory methodological guides for linking limited resourced farmers to markets. This is a good example of farmer empowerment.

- **Multi-stakeholder processes**: Multi-stakeholder engagement is not easy and does not just happen on its own. It requires processes and instruments to bring it about. Systems research therefore needs to incorporate mechanisms that enable engagement of stakeholders in the process of developing the research agenda and in the implementation of research activities. Examples of such mechanisms are Research for Development (R4D) platforms and IPs, used in programs such as Humidtropics (Schut et al., 2016) and the Forum of Agricultural Research in Africa (FARA) (Tenywa et al., 2010). These two mechanisms are interrelated. The R4D platform brings together stakeholders from the broader dimension of systems research covering the key components and sectors within the system, and helps in the confirmation of entry points, intervention domains or work packages, upon which research can be undertaken. On the other hand, IPs are specific platforms developed to undertake analysis and action research on specific constraints, challenges or opportunities (entry points) identified through the R4D platform. It can therefore be said that IPs are often spawned from R4D platforms, and involve partners and stakeholders in specific innovation domains. Membership in R4D platforms is generally much more diverse than for IPs, which usually tends to be focused on a particular issue such as a value chain for a particular commodity. However, in both cases membership will include various combinations of researchers, farmers, developmental NGOs, extension departments, private sector, traders and policy makers at different levels.

- **System diagnosis and analysis**: Implementation of systems research in a location-based context must begin with an understanding of the key components within the system, and more importantly also of the interactions, synergies and trade-offs, as well as the constraints and opportunities faced by the smallholder farmers in the area. Various tools and methods are available for addressing this, ranging from formal (structured or semi-structured) questionnaire surveys in some instances to more informal participatory methods such as RRA and PRA, discussed earlier in this chapter. Recently even newer methods have evolved such as Rapid Appraisal of Agricultural Innovation Systems (RAAIS) for creative diagnosis, observation and analysis (Schut et al., 2015), which incorporate an innovation systems dimension. Within Humidtropics, this is done in the context of situation analysis, using a variety of tools, and leads to identification of baselines and typologies, and indicates priority interventions and entry points to be explored in research.
- **Systems improvement orientation**: A key expectation in systems research is the accruing of benefits to farmers within the system through both sustainable productivity increases and livelihoods enhancement. Systems diagnosis and analysis is therefore expected to lead to technology development research addressing biological, socioeconomic and policy constraints that result in improvements in the system, and in livelihood conditions of smallholder farmers. This research includes an assessment and analysis of best-bet technologies, and ultimately leads to identification of best-fit technologies for further testing and eventually dissemination via development initiatives. The technologies address productivity enhancement, natural resources management, market linkage development and institutional dynamics, all focused on improving the system as a whole. An essential ingredient of this research domain is the analysis of trade-offs and synergies among key components, and the effect and impacts on overall productivity and sustainability of the targeted system.
- **Institutional and technological innovation**: Systems are not static but dynamic and constantly evolving. For this reason it is important to ensure that ISR always has an element of innovation built into it. Here, innovation is seen as embodying both institutional innovation and socio-technical innovation, requiring creating mechanisms to be able to encourage and recognise innovation at different levels among system actors. The IP is an example of a mechanism for institutional innovation, and for triggering the capacity to innovate among all agricultural stakeholders within the system (Adekunle et al., 2013; Schut et al., 2016).
- **Gender**: A central dimension in ISR are the people themselves. It is therefore essential to understand the people within the system, typologies and roles they play, the desires and constraints they face, and how the system impacts on their lives. In this connection the importance of women and the gender dimension, in general, cannot be over-emphasised. A good systems research program must have a built-in element of gender analysis and mainstreaming with respect to all key components, as well as including research analysing gender norms and facilitating positive transformative

change in the targeted communities. This requires involvement of women and youth in the research process.

- **Capacity development**: Undertaking ISR often involves a mind-set change and building capacities of the different agricultural stakeholders involved in the development process. This is particularly important as most researchers who end up engaging in systems research and development come, for example, with a background in commodity research, heavily focused on reductionist approaches. They have rarely been specifically trained in systems thinking and methodologies, and in complex interaction analytical techniques. Of course a notable exception, as indicated earlier, are the smallholder farmers themselves, who have applied systems think-ing to their farming practices over the generations (i.e. traditional wis-dom). However, for other agricultural stakeholders capacity development is essential to avoid the situation where people 'talk the talk' of systems research, but continue to do research in a business-as-usual fashion, based on prior experience and familiarity with specific disciplines and a tradi-tional research orientation. This essential element must be built into the process of integrated systems research and development.

- **Scaling-out and dissemination towards impact at scale**: One of the continuing challenges of FSR is the transferability of the results of place/ location-based systems research to other geographical areas. It has been said that systems research is context-specific and that the contextual differences from one site to another make it difficult to have effective scaling-out and dissemination of the results of systems research. As indicated earlier this becomes particularly challenging with the more complex phases of FSR involving agro-ecological and livelihood components. This challenge has stimulated research targeted to determining and synthesising the conditions and in what configurations different models, approaches and strategies are likely to be effective in generating positive impact at scale. Special attention is given to assessing the comparative value of different configurations and relative added value of different multi-stakeholder approaches. Such analysis can potentially help uncover the mechanisms, processes and contextual fac-tors that influence the effectiveness of such approaches at different stages of the impact pathway. Considerations on scaling-out require that partnerships in systems research need to include both research and development/exten-sion partners. New approaches and methodologies being used to enhance targeting and dissemination of systems research experiences include tools such as suitability (or similarity) analysis, which produces maps indicat-ing varying degrees of suitability or similarity of particular areas for the technologies developed within the system (Pfeifer et al., 2014). However, further research and development work is needed in this area.

Conclusion

Farming systems research has evolved significantly since its inception in the early 1970s and now has a broader integrated systems dimension that recognises the importance of viewing the farm as part of a larger integrated whole, with interactions between on-farm and off-farm entities, and incorporation of elements such as value chain, innovation systems and institutional and policy analysis. The central focus remains on the limited resource farmer and farming household and his/her livelihoods. The need for a more holistic approach to agricultural research in addressing the realities faced by smallholder farmers, in farming with mixed, diverse and multi-component entities, often in difficult, heterogeneous environments, remains as relevant today as it was in the early days of FSR. Integrated systems research has built on the foundation initially established by FSR.

We have argued that addressing and fulfilling the productivity and sustainability requirements of this century will necessitate greater focus on ISR with a livelihoods orientation in order to effectively address the needs of smallholder farmers. This will need multi-stakeholder involvement and participation, through using mechanisms such as R4D platforms and IPs. Farmer participation and empowerment will be a critical component in unleashing the full potential of systems research, through a complete inversion of the agricultural development paradigm to one with a 'demand-driven' orientation.

Although we emphasise the need of an ISR approach and its demand-driven and multi-stakeholder participation as being critically important in successfully addressing the challenges facing smallholder farmers in LICs, we recognise that other agricultural research approaches still have major contributions to make, particularly more reductionist-oriented types of research.[15] Examples include breeding improved crop varieties or livestock breeds, soil fertility management, plant nutrition, integrated pest management, etc. One of the major issues in the implementation of systems research, and agricultural research for development in general, has been the apparent disconnect and 'tension' that often exists between the two dimensions of research, categorised as 'systems research' versus 'commodity research'.

Systems and situation analytical techniques can help in identifying opportunities, challenges, trade-offs and also potential entry points, on the basis of which more targeted research initiatives can be developed. Thus systems perspectives help prioritise the problems and relationships to be addressed. Nevertheless, addressing global and local agricultural challenges in the 21st century will require placing greater emphasis on integrated agricultural systems for development, requiring not only systems research but also component research.

We have shown in this chapter the evolution of partnerships and involvement in systems research evolving from an early emphasis on interdisciplinary and farmer participatory approaches to a broader engagement, based on multi-stakeholder, multi-sector approaches, using instruments such as R4D platforms and IPs, now being advocated in the integrated systems approaches. This, however, does not in any way dilute the centrality of smallholder farmers playing major participatory roles. In essence, farmers' minds provide critically important informal modelling simulation functions in identifying and evaluating relevant pathways to improving agricultural productivity and sustainability. Capacity development will be needed across all the partnership and stakeholder categories. Fostering the capacity to innovate, a key element in ISR, must focus not only on the farmer but also on other agricultural stakeholders such as the private sector, development partners, advisory services, policy makers, etc. The current popularity of IPs provides a promising avenue for addressing such problems and in helping to improve the efficiency and payoff from ISR.

Notes

1 However, larger-scale farmers did benefit more than more limited resource farmers.
2 The world's drylands occupy 40 percent of the farming area and house the majority of poor smallholder/limited resource farmers.
3 The United States University Land Grant System was set up to foster close links between education, research and extension. The Netherlands and Australia (e.g., Birchip Cropping Group (BCG) in New South Wales, Australia (http://www.bcg.au)) are examples of other countries where linkages have been strong.
4 See also Norman (2015) for material discussed in this and the next section.
5 Interdisciplinary in the sense of different disciplines collaborating on solving an identified problem rather than different disciplines working independently on an identified problem (i.e. a multidisciplinary approach).
6 Thus justifying closer linkages with policy and planning units (Upton and Dixon, 1994)
7 An example of some of the results arising out of the AHI project can be found in Pender et al. (2006).
8 This CGIAR guideline resulted from a series of workshops and represented a culmination of the CGIAR's work at that time. We acknowledge the significant contribution of the late Ann Stroud in developing this guideline.
9 Unfortunately, contradictory policy frameworks often arise from conservation/ecological sustainability policies being separated from those targeting short-run productivity.
10 In fact, another FSR phase could be the application of farming systems to targeting, planning and policy making, an example being the FAO/World Bank study (Dixon et al., 2001), currently being updated.
11 However, this is not always easy or even possible. Later in the chapter we indicate the need for evaluating the relative merits or trade-offs of different scenarios for improving the overall productivity and sustainability of the targeted system.
12 See also the CGIAR 2012 Stripe Review of Natural Resources Management Research (CGIAR, 2012).
13 Also see http://www.fao.org/ag/ags/agricultural-marketing-linkages/en.
14 See http://dapa.ciat.cgiar.org/methodologies-to-make-market-linkages-work. Other organisations focusing on linking limited resource farmers to markets include the International Institute for Environment and Development (IIED) (http://www.iied.org/group/sustainable-markets) and the Institute of Development Studies (IDS) (http://www.future-agricultures.org/research/agricultural-commercialisations).
15 However, systems research at the farm level can help in prioritising topics for reductionist type research.

References

Adekunle, A.A., Fatunbi, A.O., Buruchara, R. and Nyamwaro, S. (2013) '*Integrated Agricultural Research for Development . . . from Concept to Practice*', Forum for Agricultural Research in Africa (FARA), Accra, Ghana.

Ashby, J., Gracia, T., del Pilar Guerrero, M., Quiros, C.A., Roa, J.I. and Beltran, J.A. (1995) 'Institutionalising farmer participation in adaptive technology testing with the CIAL', *Agricultural Research and Extension Network, Network Paper 57*, Overseas Development Institute, London.

Biggs, S.D. and Clay, E.J. (1981) 'Sources of innovation in agricultural technology', *World Development*, vol. 94, pp. 321–326.

Campbell, B.M., Hagmann, J., Stroud, A., Thomas, R. and Wollenberg, E. (2006) *Navigating Amidst Complexity: Guide to Implementing Effective Research and Development to Improve Livelihoods and the Environment*, Center for International Forestry Research Bogor, Indonesia. http://www.cifor.org/publications/pdf_ files/Books/BCampbell0602.pdf.

CGIAR (2012) *A Stripe Review of Natural Resources Management Research in the CGIAR*, Independent Science and Partnership Council Secretariat, Rome. http://ispc.cgiar.org/sites/default/files/ISPC_StrategyTrends_NRM_StripeReview_0.pdf.

CGIAR (2015a) *Strategy and Results Framework (SRF)*. https://library.cgiar.org/bitstream/handle/10947/3865/CGIAR%20Strategy%20and%20Results%20Framework.pdf.

CGIAR (2015b) *CGIAR Research Program Portfolio (CRP2) 2017–2022: Final Guidance for Full Proposals*. https://library.cgiar.org/bitstream/handle/10947/4127/CGIAR-2ndCall-GuidanceFullProposals_19Dec2015.pdf?sequence=1.

Chambers, R. (1989) 'Editorial introduction: vulnerability, coping and policy', *IDS Bulletin*, vol. 2, no. 2, pp. 1–7.

Chambers, R. (1991) 'Complexity, diversity and competence; toward sustainable livelihood from farming systems in the 21st century', *Journal of the Asian Farming Systems Association*, vol. 1, no. 1, pp. 79–89.

Chambers, R., Pacey, A. and Thrupp, L.A. (eds) (1989) *Farmer First: Farmer Innovation and Agricultural Research*, Intermediate Technology Publications, London.

Collinson, M.P. (1972) *Farm Management in Peasant Agriculture: A Handbook for Rural Development Planning in Africa*, Praeger, New York.

Collion, M.H. and Rondot, P. (1998) 'Partnership between agricultural services institutions and producer organizations: myth or reality', in Enserink, H., Cisse, A., Kiriro, F. and Roeleveld, A. (eds) *Shaping Effective Collaboration among Stakeholders in Regional Agricultural Research and Development in Sub-Saharan Africa*, Royal Tropical Institute (KIT), Amsterdam, 9 pages.

Defoer, T. and Budelman, A. (eds) (2000) *A Resource Guide for Participatory Learning and Action: Managing Soil Fertility in the Tropics*, Royal Tropical Institute (KIT), Amsterdam.

Dixon, J., Gulliver, A. and Gibbon, D. (eds) (2001) *Farming Systems and Poverty: Improving Farmers' Livelihoods in a Changing World*, FAO and World Bank, Rome and Washington, DC.

Feldstein, H.S. and Jiggins, J. (eds) (1994) *Tools for the Field: Methodologies Handbook for Gender Analysis in Agriculture*, Kumarian Press, West Hartford.

Frankenberger, T. and Coyle, P. (1992) 'Integrating household food security into farming systems research-extension', Presented at the *12th Annual Farming Systems Symposium, 13–18 September 1992*, Michigan State University, East Lansing.

Fujiska, S. (1989) 'A method for farmer-participatory research and technology transfer: upland soil conservation in the Philippines', *Experimental Agriculture*, vol. 25, pp. 423–433.

Giller, K.E., Leeuwis, C., Anderson, J.A., Andriesse, W., Brouwer, A., Frost, P., Hebinck, P., Heitkönig, I., van Ittersum, M., Koning, N., Ruben, R., Slingerland, M., Udo, H., Veldkamp, T., van de Vijver, C., van Wijk, M. and Windmeijer, P. (2008) 'Competing claims on natural resources: what role for science?', *Ecology and Society*, vol. 13, no. 2, p. 34.

Hall, A., Bockett, G., Taylor, S., Sivamohan, M. and Clark, N. (2001) 'Why research partnerships really matter: innovation theory, institutional arrangements and implications for developing new technology for the poor', *World Development*, vol. 29, no. 5, pp. 783–797.

Hall, A., Sulaiman, V.R., Clark, N. and Yogonand, B. (2003) 'From measuring impact to learning institutional lessons: an innovation systems perspective on improving the management of international agricultural research', *Agricultural Systems*, vol. 78, pp. 213–241.

Heinrich, G.M. (1993) 'Strengthening farmer participation through farmer groups: experiences and lessons from Botswana', *OFCOR Discussion Paper No. 3*, International Service for National Agricultural Research, The Hague.

Hildebrand, P.E. and Russell, J.T. (1996) *Adaptability Analysis: A Method for the Design, Analysis and Interpretation of On-Farm Research*, Iowa State University Press, Ames.

Joshi, A. and Witcombe, J.R. (1996) 'Farmer participatory crop improvement. II: participatory varietal selection, a case study in India', *Experimental Agriculture*, vol. 32, pp. 461-477.

Klerkx, L., Schut, M., Leeuwis, C. and Kilelu, C. (2012) 'Advances in knowledge brokering in the agricultural sector: towards innovation system facilitation', *IDS Bulletin*, vol. 43, pp. 53–60.

Klerkx, L., van Mierlo, B. and Leeuwis, C. (2012) 'Evolution of systems approaches to agricultural innovation: Concepts, analysis and interventions', in Darnhofer, I., Gibbon, D. and Dedidieu, B. (eds) *Farming Systems Research into the 21st Century: The New Dynamic*, Springer, Dordrecht, pp. 457–483.

Leeuwis, C., Schut, M., Waters-Bayer, A., Mur, R., Atta-Krah, K. and Douthwaite, B. (2014) 'Capacity to innovate from a system CGIAR research program perspective', *Program Brief AAS-2014–29*, CGIAR Research Program on Aquatic Agricultural Systems, Penang, Malaysia.

Lightfoot, C., Axinn, N., Singh, P., Botrall, A. and Conway, G. (1991) *Training Resource Book for Agro-Ecosystem Mapping*, IRRI and the Ford Foundation, Manila and Delhi.

Lundy, M., Becx, G., Zamierowski, N., Amrein, A., Hurtado, J.J., Mosquera, E.E. and Rodríguez, F. (2012) *LINK Methodology: A Participatory Guide to Business Models that Link Smallholders to Markets*, CIAT, Cali, Colombia.

Matlon, P.J., Cantrell, R., King, D. and Benoit-Caitlin, M. (eds) (1984) *Coming Full Circle – Farmers' Participation in the Development of Technology*, IDRC, Ottawa.

McCalla, A.F. (2014) 'CGIAR reform – why so difficult? Review, reform, renewal, restructuring, reform again and then "The new CGIAR" – so much talk and so little basic structural change – Why?' *Agriculture and Resource Economics Working Paper No. 14–001*, University of California, Davis.

Norman, D. (1974) 'The rationalisation of a crop mixture strategy adopted by farmers under indigenous conditions: the example of Northern Nigeria', *Journal of Development Studies*, vol. 11, no. 1, pp. 3–21.

Norman, D. (2015) 'Transitioning from paternalism to empowerment of farmers in low-income countries: farming components to systems', *Journal of Integrative Agriculture (Formerly Agricultural Sciences in China)*, vol. 14, no. 8, pp. 1490–1499.

Norman, D. (ed) (2004) 'Helping small farmers think about better growing and marketing: A reference manual', *Pacific Farm Management and Marketing Series Part 3*, FAO, Apia, Samoa.

Norman, D., Simmons, E. and Hays, H.M. (1982) *Farming Systems in the Nigerian Savanna: Research and Strategies for Development*, Westview Press, Boulder, CO.

Norman, D., Umar, M., Tofiunga, M. and Bammann, H. (1995) *An Introduction to the Farming Systems Approach to Development (FSD) for the South Pacific*, Institute for Research, Extension and Training in Agriculture, University of the South Pacific and FAO, Apia, Samoa and Rome.

Norman, D., Worman, F.D., Siebert, J.D. and Modiakgotla, E. (1995) 'The farming systems approach to development and appropriate technology generation', *Farming Systems Management Series No. 10*, AGSP, FAO, Rome.

Pender, J., Place, F. and Ehui, S. (eds) (2006) *Strategies for Sustainable Land Management in the East African Highlands*, International Food Policy Research Institute, Washington, DC.

Pfeifer, C., Omolo, A., Kiplimo, J. and Robinson, T. (2014) *Similarity Analysis for the Humid-tropics Action Area*, Humidtropics Report, Nairobi, Kenya.

Pretty, J., Guijt, I., Thompson, J. and Scoones, I. (1995) *Participatory Learning and Action: A Trainer's Guide*, IIED, London.

Program for International Development (1994) *Participatory Rural Appraisal Handbook*, Clark University, Worcester.

Röling, N. (2009) 'Conceptual and methodological developments in innovation', in Sanginha, P.C. et al. (eds) *Innovation Africa, Enriching Farmers' Livelihoods*, Earthscan, London, pp. 9–34.

Sall, S., Norman, D. and Featherstone, A.M., (1998) 'Adaptability of improved rice varieties in Senegal', *Agricultural Systems*, vol. 57, no. 1, pp. 101–114.

Sall, S., Norman, D. and Featherstone, A.M. (2000) 'Quantitative assessments and the adoption of improved rice varieties in Casamance, Senegal: The farmers' perspective', *Agricultural Systems*, vol. 66, pp. 129–144.

Schut, M., Klerkx, L., Rodenburg, J., Kayeku, J., Hinnou, L.C., Raboanarielina, C.M., Adegbola, P.Y. and Bastiaans, L. (2015) 'RAAIS: Rapid Appraisal of Agricultural Innovation Systems (part I). A diagnostic tool for integrated analysis of complex problems and innovation capacity', *Agricultural Systems*, vol. 132, pp. 1–11.

Schut, M., Klerkx, L., Sartas, M., Lamers, D., McCampbell, M.M.C., Ogbonna, I., Kuashik, P., Atta-Krah, K. and Leeuwis, C. (2016) 'Innovation platforms: experiences with their institutional embedding in agricultural research for development', *Experimental Agriculture*, FirstView Articles.

Schut, M., van Paassen, A., Leeuwis, C. and Klerkx, L. (2014) 'Towards dynamic research configurations. A framework for reflection on the contribution of research to policy and innovation processes', *Science and Public Policy*, vol. 41, pp. 207–218.

Spencer, D., Byerlee, D. and Franzel, S. (1979) 'Annual costs, returns, and seasonal labor requirements for selected farm and nonfarm enterprises in rural Sierra Leone', *African Rural Employment Paper No. 27*, Department of Agricultural Economics, Michigan State University, East Lansing.

Sperling, L. and Berkowitz, P. (1994) *Partners in Selection: Bean Breeders and Women Bean Experts in Rwanda*, CGIAR Gender Program, CGIAR and World Bank, Washington DC.

Sumberg, J., Thompson, J. and Woodhouse, P. (2013) 'Why agronomy in the developing world has become contentious', *Agriculture and Human Values*, vol. 30, no. 1, pp. 71–83.

Tenywa, M.M., Rao, K.P.C., Tukahirwa, J.B., Buruchara, R., Adekunle, A.A., Mugabe, J., Wanjiku, C., Mutabazi, S., Fungo, B., Kashaija, N.I., M., Pali, P., Mapatano, S., Ngaboyisonga, C., Farrow, A., Njuki, J. and Abenakyo, A. (2010) 'Agricultural innovation platform as a tool for development oriented research: Lessons and challenges in their formation and operationalization', *Learning Publics Journal of Agriculture and Environmental Studies*, vol. 2, no. 1, pp. 117–146.

Upton, M. and Dixon, J. (eds) (1994) 'Methods of policy analysis for agricultural programmes and policies: a guideline for policy analysis', *Farming Systems Management Series Number 9*, AGSP, FAO, Rome.

Walker, T. and Ryan, J. (1990) *Village and Household Economies in India's Semi-Arid Tropics*, Johns Hopkins University Press, Baltimore and London.

Witcombe, J., Joshi, A., Joshi, K.D. and Sthapit, B.R. (1996) 'Farmer participatory crop improvement. I: varietal selection and breeding methods and their impact on biodiversity', *Experimental Agriculture*, vol. 32, pp. 445-460.

3 Systems science at the scale of impact

Reconciling bottom up participation with the production of widely applicable research outputs

Fergus L. Sinclair

Introduction

This chapter is about conducting agricultural systems research at the scale needed to address the aspirations of the first two Sustainable Development Goals (SDGs) to end world poverty and hunger (United Nations, 2015). There are two distinct concepts of scale important in international agricultural research. The first is the operational scale at which research is conducted (the field, farm, landscape, nation, region or planet) and how connections are made across these operational scales (Coe et al., 2014). The second use of the term relates to the scale at which innovations emerging from research at any of these operational scales are adopted by farmers or other stakeholders along the food chain, generally referred to as scaling up and scaling out. These terms are variously used in the literature but here, scaling up refers to where more people adopt an innovation within a particular geography, or scaling domain, with a boundary that may be specified in both biophysical and socio-economic terms. For example, a scaling domain may include migrant people in the Afromontane zone of North Kivu in Eastern DRC (Smith Dumont et al., 2016). Large scaling domains comprise millions of people and millions of hectares and are the scale at which the donors who fund agricultural research expect it to make impact. Scaling out refers to where innovations generated in one scaling domain are promoted and adopted in another scaling domain. So, innovations developed in North Kivu may have relevance for bordering territories in Rwanda or Uganda. The chapter is divided into two main sections. The first part looks at issues associated with operational scale, while the second part explores how to ensure that innovations generated from research make large scale impact in terms of reducing hunger and poverty.

Scale of operation

There has been a long tradition of agricultural systems research aimed at supporting development in terms of improving food security and reducing rural poverty (Spedding, 1996). This has its foundations in understanding local agriculture (Allen, 1965; Ruthenburg, 1971; Grigg, 1974) and more recently mapping the extent of different broad types of farming practice at global and continental scales with a view to informing policy making about agricultural development (Dixon et al., 2001; HarvestChoice, 2015). These broad agricultural assessments have been complemented by grassroots farming systems research, aimed at improving yields and profitability in particular locations (Collinson, 2000). Formal systems methods involve applying the principles of general system theory (von Bertalanffy, 1968) and have been progressively replaced in grassroots research with farmers over the past quarter of a century by a thrust towards participatory research, where it is assumed that participating farmers bring an implicit understanding of their system to the research process (van Ginkel et al., 2013). Associated with the rising prominence of participatory approaches has been a parallel recognition that farming is often only part of what comprises rural people's livelihoods, and agriculture cannot, therefore, be usefully studied in isolation of other livelihood components (Carney, 2002).

Despite the effectiveness of systems research in agricultural development, particularly relating to smallholders, which is discussed in more detail below, the vast majority of agricultural research for development does not explicitly follow a systems approach, although some techniques, such as systems modelling, are widespread. Rather, the predominant mode of thinking about agriculture focuses on the crop field as a system (Figure 3.1). If we consider a system as converting inputs into outputs, then for monocultures in an industrialized agriculture this 'cropping system' model is appropriate and it served agricultural research for development well in the green revolution (Conway, 2012). In this model the objective of research for development is to increase crop yield, usually grain in the case of food staples. Key inputs like seeds, fertilizer, pesticides, irrigation and agronomy (that encompasses labour used in cultivation, planting, weeding and harvesting) are converted to grain yield. The efficiency of the system in terms of yield per unit land area has been the main, easily understood

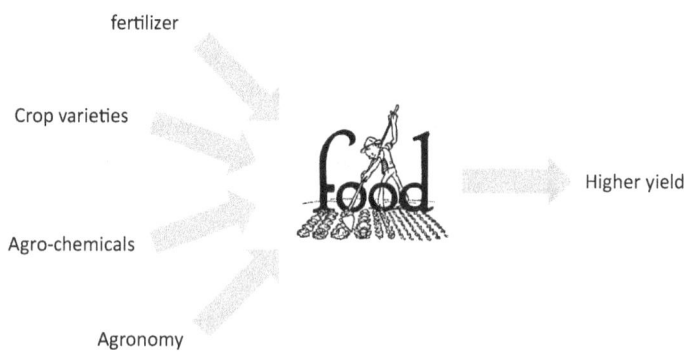

Figure 3.1 The cropping system concept

indicator of performance. Agricultural improvements initially involved increasing the harvest index (the proportion of the plant that is harvested product) coupled with packages of measures to increase productivity by using inputs to make the environment as suitable as possible for the crop. Markets and institutions associated with input supply and the sale of products were also considered.

Application of this 'cropping system' concept has led to spectacular yield increases for many crops in many places, as well as some major issues of environmental degradation and biodiversity loss (MA, 2005). It is also clear that the green revolution only worked in some places and for some farmers (notably more effective in Asia and Latin America than in Africa) and that yield increases from pursuing this approach ultimately level off (Keating et al., 2010), although widespread yield gaps (where yields are below what is calculated as an achievable maximum) remain for staple crops in many parts of the world (Mueller et al., 2012). For many smallholder farmers, the 'cropping system' concept is not a useful starting point for agricultural improvement because the crop field is not a system in its own right but, in fact, one component of a more complex livelihood system (Figure 3.2). The crop field itself may be a polyculture, involving more than one crop; there may be trees in fields or along boundaries providing fodder, fruit and fuelwood; and the crop field may interact strongly with other farm components, for example providing fodder to livestock and receiving manure from them. The livelihood system may involve the processing and marketing of products as well as their production, and household members may

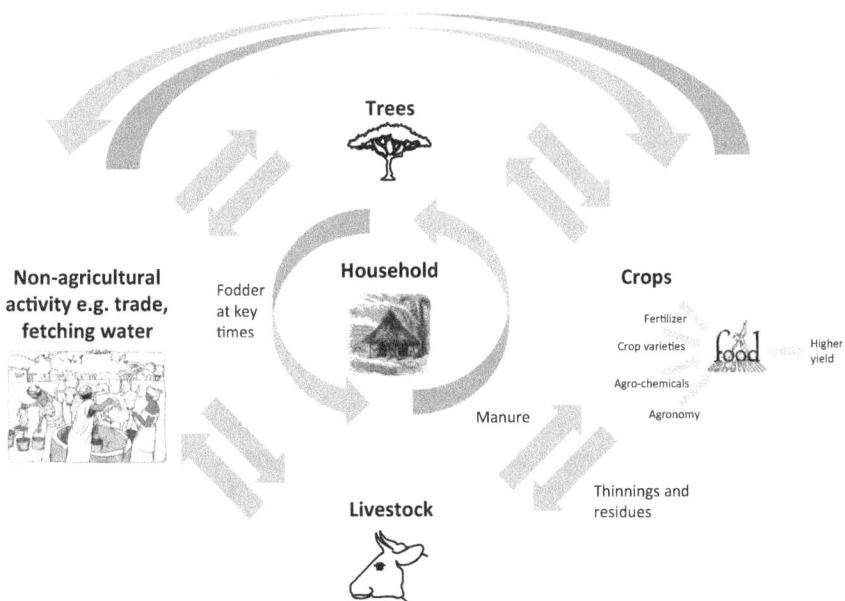

Figure 3.2 Crops within a livelihood system reality

engage in off-farm employment and remittances may be received from outside the immediate household. This means that decisions about how much cash or labour to invest in the crop field and how to cultivate the crop will involve complex trade-offs amongst the various livelihood components. The farmer or farm household members are not necessarily trying to maximize grain yield of a particular crop, but to manage the whole suite of livelihood components to achieve the standard of living to which they aspire, with grain yield of one crop being an important part of that but not the whole (Tiwari et al., 2004). In this respect, total factor productivity of the livelihood system is a more appropriate performance measure than grain yield of a particular crop (Antle and Capalbo, 1988).

While livelihood systems themselves typically exhibit complexity, they also interact with one another across landscapes, highlighting the importance of where livelihoods are interlocking. This is where change in one type of livelihood, for example enclosing land by settled agriculturalists, may impact another, for example pastoralists using the same landscape who then face a restriction on grazing area (Mohammed and Beyene, 2016). The networking relationships amongst people with similar types of livelihoods (say, a group of settled agriculturalists) is referred to as bonding social capital, while that between different groups (for example the ability to negotiate amongst agriculturalists and pastoralists) as bridging social capital (Putnam, 2000). These relationships amongst people and how they are mediated affect both the performance of agricultural systems and what approaches to agricultural development are appropriate. For example, farmers can produce valuable oil from the marula tree, *Sclerocarya birrea*, which is widely but sparsely dispersed in crop fields in southern Africa, but only female trees produce fruit so male trees tend to get removed by farmers, resulting in female trees not getting pollinated, and so also not producing fruit (Nghitoolwa et al., 2003). Social capital is required at a village level to regulate cutting of male trees and thereby maintain the capacity for females to be pollinated. Traditional societies often did regulate use and cutting of the trees (Hall et al., 2002). A systems approach to address the problem today would combine research on the effective pollination distance amongst trees (it is bee pollinated, but travel distances in savannah landscapes are not well understood) with addressing the need to re-establish the social capital necessary to agree upon and enforce regulations.

It is also quite common for livestock to roam over large distances, and collect and concentrate nutrients from across a landscape in key fields where crops are grown. Similarly, the strategic placement of perennials or other water flow regulation structures across landscapes may affect water infiltration, soil erosion control and groundwater recharge, feeding back to field level productivity (Ilstedt et al., 2016). Concerns about environmental degradation associated with agriculture have led to the emergence of a broader framework for considering a fuller range of ecosystem services provided by agricultural landscapes (Pagella and Sinclair, 2014). These include things like clean water, flood risk reduction, soil fertility maintenance and biodiversity conservation, as well as food production. Often farmers' decisions about land use result in key trade-offs amongst their impact on the provision of various ecosystem services that manifest at landscape scales and affect a range of people (Jackson et al., 2013). So establishing a eucalyptus woodlot in Ambo Ethiopia might provide an important timber provisioning service with the key product flowing to Addis Ababa, while the use of water by the woodlot reduces groundwater recharge and the water table depth, affecting crops growing locally in Ambo.

It is clear then that when considering smallholder farmers, we need to move from the industrial agriculture model of the crop field as a system in its own right, to understanding the crop field within the context of the livelihood system and the landscape. A long held maxim in systems research has been that it is necessary to take three scale levels into account, that above and that below the system of focal interest, and it is usually intractable to cope with more. For smallholder agriculture, this makes the livelihood system the key focus with the landscape above and the field (and other livelihood components) below. There are many examples that show how adopting such a multiple scale systems approach leads to more effective research, when evaluated in terms of development impact (Box 3.1).

Box 3.1   Systems research increases maize yield by 30%

A well-documented example of systems research breaking a log jam in smallholder agriculture comes from the mid-hills of Nepal where maize yields had been stagnant before a systems perspective was used to inform varietal selection. The first step was to understand how the maize fitted into the farmers' livelihood system by acquiring local knowledge from farmers about how they produced maize (Tiwari et al., 2004). Landscapes in the mid-hills included both individually and commonly held land. Individually utilized cropland could be divided into upper slope, rainfed 'bari' land, where maize was grown, and lower slope, irrigated 'khet' land, where rice was grown. Communal land included forest and grazing areas. A key element in maintaining the fertility of crop fields was through application of compost comprising crop residues and livestock manure. With reducing access to tree fodder from forest areas, as they came under community forest regulation, farmers allowed regeneration of fodder trees on their crop terrace risers to provide fodder in the dry winter period. Farmers did not follow agronomic recommendations for maize. Instead they planted initially at far higher densities than recommended, thinning down to far lower densities than recommended at harvest, using the thinnings as livestock fodder. They relay cropped with millet, and did this all on crop terraces with fodder trees on the risers that were competing with the crop. They tended to apply fertilizer purposively to patches of fields, where they thought it was most needed, rather than uniformly, in a local form of precision agriculture. It was clear that the farmer's objective was not to simply maximize grain yield of maize. Instead the farm household was interested in the total factor productivity of the farm, which rested on soil fertility from dung, as well other products from livestock and the yield of the relay cropped millet. Perhaps not surprisingly, screening maize varieties against farmer criteria and then allowing them to test different varieties themselves led to the identification and later to the release of varieties that outperformed those used previously by up to 30% (Tiwari et al., 2009). The new varieties replaced varieties that had been selected in 'ideal' conditions on researcher-managed plots that were quite different from typical farm conditions. The varieties that farmers preferred yielded better under farm conditions because they had longer roots and were thus able to capture resources competitively in the field conditions pertaining on farms (Tiwari et al., 2012).

Markets present a further complication to the consideration of the scale at which systems research needs to operate to improve smallholder agriculture, because they connect people across huge distances. For example, consumers in Europe or North America may pay higher prices for certified coffee or cocoa, linking them to famers in Africa, Asia or Latin America (Lyngbaek et al., 2001; Smith Dumont et al., 2014). In this case, consumer behaviour in one part of the world, via certification tracked through the value chain, affects the farming practices somewhere else. Agricultural research is often formulated around value chains, which has required systems research to embrace the whole food chain from farm to fork, including efficiency and waste at each stage (Jones and Street, 1990), leading to the emergence of a more holistic agri-food systems perspective (Thompson et al., 2007).

It is important to note that whether or not the predominant narrative of international agricultural development should continue to focus on supporting smallholder farmers have been hotly contested. It is argued that maintaining smallholder farming effectively consigns people to poverty (Collier, 2008) and that improvements in agricultural technology can rarely generate sufficient income on small areas for people to exit poverty (Harris and Orr, 2014). As is often the case, the way forward probably needs to embrace both these perspectives, supporting some smallholders to intensify within a largely commercial context referred to as 'stepping up', some to exit from agriculture as their major source of livelihood (stepping out) and to provide social protection for some of the poorest rural households with few assets or opportunities (hanging in) (Dorward, 2009). This categorization of livelihood trajectories emphasizes the importance of rural–urban linkages, a key element of which is migration from rural to urban areas that is set to increase hand in hand with population. A multiple scale systems understanding of rural livelihoods is relevant to guiding research to support smallholders in stepping up, stepping out and hanging in, and it is likely that smallholder farming will remain an important part of global food security for at least the time horizon of most current research and development initiatives.

Scale of impact

A key issue that has emerged with participatory research is how to scale up the innovations that arise from it. Participatory methods have undoubtedly proved hugely effective in co-developing agricultural improvements with local communities (Douthwaite et al., 2009), but the donors who fund international agricultural research are increasingly requiring candidate research to show how it can contribute to large scale impact. The problem, and indeed the value, of participatory research is that working intensively with particular communities will inevitably produce innovations that suit their local needs, but these will not necessarily then spread to other communities, even those close by, where conditions are different. The challenge is that many of the factors that affect the suitability of agricultural innovations, such as soils, climate, farming practices, household characteristics, markets, social capital and policy, vary at a fine scale, and this means that appropriate options for farmers to adopt to improve their livelihood systems also vary at a fine scale (Coe et al., 2016).

One approach to dealing with this problem is to posit that it is not innovations that can be scaled up but the whole innovation process, a notion that has considerable merit, and has fostered productive research on how innovation systems work and can be supported (Kilelu et al., 2013). But we know that innovations do spread in some circumstances, and it is likely that innovations suitable in one locality have relevance elsewhere. It would be highly inefficient to suggest that we need to support re-invention of the wheel in many different places, rather than understanding the contexts for which wheels are relevant, and then offering the wheel as an option for adoption in those contexts. While the wheel has fairly universal applicability, most agricultural innovations are suitable only in specific contexts. This requires us to move away from the notion that there is a fundamental dichotomy between bottom up and top down approaches, where bottom up refers to where innovation is locally driven, and top down to where innovations are compared across locations. These can be complementary where results from comparing performance of options across different contexts is used to inform local innovation (Coe et al., 2014).

Options

Large scale impact requires innovations to be widely adopted, for which it is necessary to generate innovations suitable for the range of contexts across large scaling domains and to understand which innovations are suitable for which contexts. In practice, innovations are often promoted collectively rather than individually and are referred to as options. The notion of informing farmers and other stakeholders along food chains about options that they may consider adopting and adapting to their circumstances is replacing prescriptive approaches. This is commensurate with the sophisticated local agro-ecological knowledge underpinning farmer practice that has been documented across the developing world (Sinclair and Walker, 1998; Sinclair and Walker, 1999; Lamond et al., 2016). Most options developed in the early days of farming systems research were technological at the field and farm level, relating to components and their management, such as crop varieties, cultivation methods, and pest, disease and weed control (Collinson, 2000). More recently the concept of what an option may consist of has broadened to embrace interventions to improve the enabling environment (Coe et al., 2014). This involves market interventions or changes to agricultural extension systems, as well as policy or institutional reform. This has largely come about because it was realized that they often interact. For example, a prerequisite for adoption of a technology like farmer-managed natural regeneration of trees on farms might be institutional reform to control livestock grazing that requires development of social capital. Adoption of a new crop might require market development for its product. This is not always the case, but a multiple scale systems understanding, that is of farm, landscape, market and governance systems, at different scales and their interconnections, facilitates identifying when innovations need to be promoted together with one another or when they are likely to be effective alone.

Context

The contextual factors that condition the suitability of agricultural innovations vary with location and with the type of innovation. When considering which contextual factors are important in a particular situation there are two key issues. The first is how important the factor is in determining the suitability of one or more options under consideration, where we are usually considering the applicability of a range of options across a large scaling domain; and the second is how widely this factor varies across the domain. For example, in considering how context affected adoption of a range of agronomic options developed by the Dryland Systems CGIAR Research Programme, ethnicity was important in the Sahel, and varied significantly across the east–west transect, but not along the north–south transect (Traore et al., 2012). These transects were constructed deliberately to encompass a range of contexts to facilitate efficient testing of options in relation to context. The north–south transect ran along an aridity gradient and the east–west transect across a gradient of population density and infrastructural development.

We know increasing amounts about the variability of common contextual factors that affect the suitability of agricultural innovations. For example, a large proportion of Africa's soils are now non-responsive, in the sense that even if fertilizer is applied, crop yield does not increase. There is a very fine scale distribution of these non-responsive soils, they even occur only in parts of fields, as well as patterned in landscapes through differences in management of fields as you move further away from homesteads (Vanlauwe et al., 2015). Reasons for non-responsiveness include combinations of chemical, physical and biological factors. Advances in remote sensing are giving us increasingly fine scale and rapid means to characterize soil and climatic variation across large scaling domains (Vagen et al., 2016). Less mappable contextual variables include extant farming practices and household characteristics such as labour availability, resource endowments and human capital. The broad categories used to map farming practices at continental scales (Dixon et al., 2001) are of little value in understanding variation at the scale required to understand adoption, and tend to focus on broad cropping patterns rather than other livelihood system components (Kmoch, 2014). Access to markets varies not only in relation to physical distance but also in relation to an individual farm household's agency and access to information and transport. Social capital also varies within, as well as amongst, communities and can be both developed and eroded (Pretty and Ward, 2001). While policies often operate over large areas, their implementation often varies locally and their impact acts differentially on households depending on their agency, creating fine scale variation in how policies impact rural people (Chomba et al., 2015). What is considered to be an option and what is considered as context can vary with the size and scope of a project or programme. A small research or development project may consider current land tenure and other policy issues as fixed elements of context because they cannot address them, even though they may have identified the requirement for policy reform, whereas a larger initiative might target policy reform as a key option for change (Smith Dumont et al., 2016).

The options by context approach

It is clear that to make impact at scale, research needs to be conducted at scale, so that understanding can be developed about which options are relevant to what contexts. The problem is that research projects and institutions rarely have sufficient resources to do research across large scaling domains, while the development agencies that do have resources to operate at large scale rarely have access to much evidence to target innovations to context (Coe et al., 2014). In fact, it is common practice to promote best-bet options over large areas, which means both that the options promoted are not suitable for many of the people in the area and that there is little opportunity to learn about which options suit different contexts. This has led to a recent paradigm shift from research for development (R4D) to research in development (RinD), where research perspectives and methods are embedded within development initiatives, to accelerate their impact through improving the speed and efficiency of learning about the suitability of different interventions for different people and places (Figure 3.3).

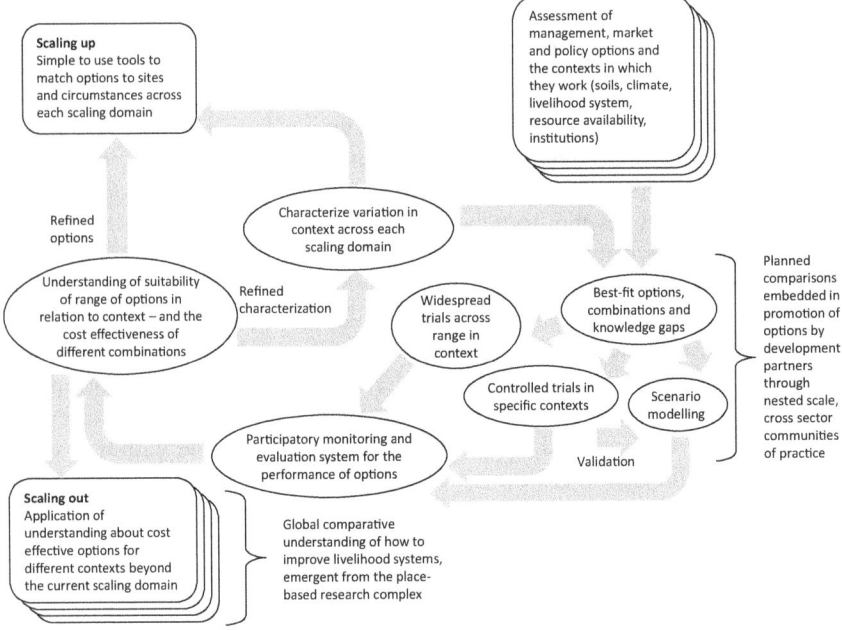

Figure 3.3 The research 'in' development co-learning cycle

Source: Adapted and re-drawn from Coe et al., (2014).

Generically RinD refers to where research is carried out within the complex social-ecological system that the research is aiming to impact (Douthwaite et al., 2015). Within the Forests Trees and Agroforestry CGIAR research programme, a pragmatic RinD framework was developed to ensure that livelihood systems research would make impact (Coe et al., 2014). This framework starts with developing an 'options by context' matrix to facilitate analysis of what options are available to address key issues in a scaling domain and for which contexts in the domain these are suitable (Smith Dumont et al., 2016). This identifies two key gaps. Firstly, contexts for which there may be no viable options, for example no options may be suitable for farmers with very small land areas but they may constitute a critical group that a development initiative seeks to target. Secondly, gaps in knowledge about which contexts some options are suitable for, resulting from options only having been tried out in a limited set of contexts in previous research. The first gap requires going back to the drawing board and looking for new options to suit the specified context. The second gap can be addressed through persuading development partners to operate so that ranges of options are tested across ranges in context and performance monitored (Figure 3.3).

The sustainability of a RinD approach requires that research and development partners, as well as farmers and other stakeholders in communities of practice, remain engaged (Figure 3.3). This makes it important during implementation to ensure that RinD supports the development process. Over time, development impact is accelerated but it is also important that development partners don't perceive any need to delay the promotion of options to farmers at the outset. Options are scaled out to farmers, as they would be without RinD, but the scaling out is done so that planned comparisons of the performance of options across contexts can be made as scaling up progresses (Coe et al., 2014). For example, there may be a knowledge gap relating to the contexts for which zai pits, micro-catchments dug within fields to collect water originating in the Sahel, are relevant in dryland Kenya, with different soil types, labour availability, availability of digging tools and crops being grown, expected to matter (Danjuma and Mohammed, 2015). A randomized control trial (RCT) can then be embedded in how zai pits are promoted to farmers through development programmes (White, 2013).

With advances in information and communications technology (ICT) it is becoming increasingly possible for farmers to report on the performance of options using simple recording formats including mobile phones. While trials across large numbers of farmers spanning a range of contexts (often referred to as large n trials, because the sample size is large) are often the most appropriate way to close knowledge gaps, some questions are more amenable to more controlled trials over smaller sample frames or through simulation modelling (Luedeling et al., 2016). It is imperative that the learning gleaned from testing options over ranges in context is then made available in a form that can be readily used by grassroots development workers to match options to local circumstances and so promote more suitable options. This reduces the risks that farmers face in adopting innovations (Coe et al., 2016). There are a growing body of examples of how large scale trialling of options with farmers is proving successful in making impact (Coe and Sinclair, 2016), from micro-dosing of nutrients in southern Africa (Twomlow et al., 2010) to increasing climbing bean productivity in Rwanda (Franke et al., 2016).

Conclusion

There have been debates over many years about the relationship between research and development in the context of international agricultural research. What is proposed here is a systems framework for bringing them together. This will ensure that, on the one hand, research outputs are relevant for, and can be adopted by, development partners, and on the other, that development initiatives operate efficiently and effectively in delivering suitable options for agricultural improvement to farmers and other stakeholders along the food chain. This requires profound changes in the behaviour of both research and development partners. Researchers can no longer view their responsibility as ending once they have produced research outputs. They need to take joint responsibility with development partners for ensuring that these outputs are adopted, and in doing so, their research priorities and ways of working will change.

Individual researchers, projects and institutions are increasingly being evaluated by donors in terms of outcomes and impact, rather than only output, which reinforce this transition. Similarly, development partners need to acknowledge the partial state of knowledge that exists globally, about how particular innovations will perform in specific situations, and base decisions about what to promote to farmers, and how to dialogue with them, on the basis of evaluating what evidence there is. In implementing RinD, research and development partners come together in joint ventures, generating broad based learning alliances that can deliver greater impact, but this does not confuse the distinctive expertise and contributions that they bring to the joint process. Research centres are working with development partners; they are not becoming development agencies. It is making full use of the complementarity of roles through collaboration that generates impact. The 2030 Agenda for Sustainable Development is ambitious in setting out to end poverty and hunger globally, and this can only be achieved if innovations are adopted locally. RinD can make an important contribution to this through enabling systems science to operate at the scale at which impact needs to be made, shifting our focus from empowering the farmer to empowering millions of farmers.

References

Allen, W. (1965) *The African Husbandman*, Oliver and Boyd, Edinburgh, 505 pp.

Antle, J.M. and Capalbo, S.M. (1988) 'An Introduction to Recent Developments in Production Theory and Productivity Measurement', in Capalbo, S.M. and Antle, J.M. (Eds), *Agricultural Productivity, Measurement and Explanation*, Resources for the Future, Washington, DC, 17–95 pp.

Carney, D. (2002) *Sustainable Livelihoods Approaches: Progress and Possibilities for Change*, DFID, London.

Chomba, S.W., Nathan, I., Minang, P.A. and Sinclair, F. (2015) 'Illusions of empowerment? Questioning policy and practice of community forestry in Kenya', Ecology *and Society*, 20 (3), http://dx.doi.org/10.5751/es-07741–200302.

Coe, R., Njoloma, J. and Sinclair, F. (2016) 'Loading the dice in favour of the farmer: reducing the risk of adopting agronomic innovations', *Experimental Agriculture*, doi:10.1017/S0014479716000181.

Coe, R. and Sinclair, F. (2016) 'The options by context approach: a paradigm shift in agronomy', *Experimental Agriculture* (Special Issue, in press).

Coe, R., Sinclair, F. and Barrios, E. (2014) 'Scaling up agroforestry requires research "in" rather than "for" development', *Current Opinion in Environmental Sustainability*, 6, pp. 73–77. Collier, P. (2008) 'The Politics of Hunger: How Illusion and Greed Fan the Food Crisis', *Foreign Affairs*, Nov/Dec 2008, http://www.foreignaffairs.com/articles/64607/paul-collier/the-politics-of-hunger.

Collinson, M. (2000) *A History of Farming Systems Research*, FAO and CABI Publishing, Wallingford, UK.

Conway, G. (2012) *One Billion Hungry: Can We Feed the World?* Cornell University Press, Ithaca, NY, 439 pp.

Danjuma, M.N. and Mohammed, S. (2015) 'Zai pits system: a catalyst for restoration in the dry lands', *Journal of Agriculture and Veterinary Science*, 8 (2), pp. 1–4.

Dixon, J., Gulliver, A. and Gibbon, D. (2001) *Farming Systems and Poverty: Improving Farmers' Livelihoods in a Changing World*, Food and Agriculture Organization of the United Nations and The World Bank, Rome and Washington, DC, 420 pp.

Dorward, A. (2009) 'Integrating contested aspirations, processes and policy: development as hanging in, stepping up and stepping out', *Development Policy Review*, 27 (2), pp. 131–146.

Douthwaite, B., Apgar, J.M., Schwarz, A., McDougall, C., Attwood, S., Senaratna Sellamuttu, S. and ClayTon, T. (Eds) (2015) Research in development: Learning from the CGIAR Research Program on Aquatic Agricultural Systems. Penang, Malaysia: CGIAR Research Program on Aquatic Agricultural Systems. *Working Paper*: AAS-2015–16.

Douthwaite, B., Beaulieu, N., Lundy, M. and Peters, D. (2009) 'Understanding how participatory approaches foster innovation', *International Journal of Agricultural Sustainability*, 7, pp. 42–60.

Franke, A.C., Baijukya, F., Kantengwa, S., Reckling, M., Vanlauwe, B. and Giller, K.E. (2016) 'Poor farmers – poor yields: socio-economic, soil fertility and crop management indicators affecting climbing bean productivity in northern Rwanda', *Experimental Agriculture*, doi:10.1017/S0014479716000028.

Grigg, D.B. (1974) *The Agricultural Systems of the World*, Cambridge University Press, Cambridge.

Hall, J.B., O'Brien, E.M. and Sinclair, F.L. (2002) *Sclerocarya Birrea: A Monograph*, School of Agricultural and Forest Sciences Publication 19, University of Wales, Bangor, UK, 157 pp.

Harris, D. and Orr, A. (2014) 'Is rainfed agriculture really a pathway from poverty?' *Agricultural Systems*, 123, pp. 84–96.

HarvestChoice (2015) *New Global Crop Data Aid in Food Policy Decisions*, International Food Policy Research Institute, Washington, DC, and University of Minnesota, St. Paul, MN, http://harvestchoice.org/node/9983.

Ilstedt, U., Bargues Tobella, A., Bazie, H.R., Bayala, J.H., Verbeeten, E., Nyberg, G., Sanou, J., Benegas, L., Murdiyarso, D., Laudon, H., Sheil, D. and Malmer, A. (2016) 'Intermediate tree cover can maximize groundwater recharge in the seasonally dry tropics', Nature Scientific Reports 6, Article number: 21930.

Jackson, B., Pagella, T., Sinclair, F., Orellana, B., Henshaw, A., Reynolds, B., Mcintyre, N., Wheater, H. and Eycott, A. (2013) 'Polyscape: a GIS mapping toolbox providing efficient and spatially explicit landscape-scale evaluation of multiple ecosystem services', *Landscape and Urban Planning*, 112, pp. 74–78.

Jones, J.G.W. and Street, P.R. (Eds) (1990) *Systems Theory Applied to Agriculture and the Food Chain*, Elsevier Applied Science, London, 365 pp.

Keating, B.A., Carberry, P.S., Bindraban, P.S., Asseng, S., Meinke, H. and Dixon, J. (2010) 'Eco-efficient agriculture: concepts, challenges, and opportunities', *Crop Science*, 50, pp. 109–119.

Kilelu, C.W., Klerkx, L. and Leeuwis, C. (2013) 'Unravelling the role of innovation platforms in supporting co-evolution of innovation: contributions and tensions in a smallholder dairy development programme', *Agricultural Systems*, 118, pp. 65–77.

Kmoch, L. (2014) *Utilising local knowledge for improved resilience of smallholder farming systems in northern Morocco*, MSc Dissertation, School of the Environment, Natural Resources and Geography, Bangor University, Wales, UK, 82 pp.

Lamond, G., Sandbrook, L., Gassner, A. and Sinclair, F.L. (2016) 'Local knowledge of tree attributes underpins species selection on coffee farms', *Experimental Agriculture* (online), doi: 10.1017/S0014479716000168.

Luedeling, E., Smethurst, P.J., Baudron, F., Bayala, J., Huth, N.I., van Noordwijk, M. and Sinclair, F.L. (2016) 'Field-scale modeling of tree–crop interactions: challenges and development needs', *Agricultural Systems*, 142, pp. 51–69.

Lygnbæk, A.E, Muschler, R.G. and Sinclair, F.L. (2001) 'Productivity and profitability of multistrata organic versus conventional coffee farms in Costa Rica', *Agroforestry Systems*, 53, pp. 205–213.

MA (2005) *Millennium Ecosystem Assessment Synthesis Report*, 1st edn, Island Press, Washington, DC.

Mohammed, A. and Beyene, F. (2016) 'Social capital and pastoral institutions in conflict management: evidence from eastern Ethiopia', *Journal of International Development*, 28, pp. 74–88.

Mueller, N.D., Gerber, J.S., Johnston, M., Ray, D.K., Ramankutty, N. and Foley, J.A. (2012) 'Closing yield gaps: nutrient and water management to boost crop production', *Nature*, 490, pp. 254–257.

Nghitoolwa, E., Hall, J.B. and Sinclair, F. (2003) 'Population status and gender imbalance of the marula tree, Sclerocarya birrea subsp. caffra in northern Namibia', *Agroforestry Systems*, 59, pp. 289–294.

Pagella, T.F. and Sinclair, F.L. (2014) 'Development and use of a new typology of mapping tools to assess their fitness for supporting management of ecosystem service provision', *Landscape Ecology*, 29 (3), pp. 383–399.

Pretty, J. and Ward, H. (2001) 'Social capital and the environment', *World Development*, 29 (2), pp. 209–227.

Putnam, R.D. (2000) *Bowling Alone: The Collapse and Revival of American Community*, Simon & Schuster, New York.

Ruthenberg, H. (1971) *Farming Systems in the Tropics*, Clarendon Press, Oxford, 286 pp.

Sinclair, F.L. and Walker, D.H. (1998) 'Qualitative knowledge about complex agroecosystems. Part 1: a natural language approach to representation', *Agricultural Systems*, 56, pp. 341–363.

Sinclair, F.L. and Walker, D.H. (1999) 'A utilitarian approach to the incorporation of local knowledge in agroforestry research and extension', in Buck, L.E., Lassoie, J.P. and Fernandes, E.C.M. (Eds) *Agroforestry in Sustainable Agricultural Systems*, Lewis Publishers, New York, 245–275 pp.

Smith Dumont, E., Bonhomme, S., Pagella, T. and Sinclair, F.L. (2016) 'Structured stakeholder engagement leads to development of more diverse and inclusive agroforestry options', *Experimental Agriculture* (in press).

Smith Dumont, E., Gnahou, G.M., Ohouo, L., Sinclair, F.L. and Vaast, P. (2014) 'Farmers in Co^te d'Ivoire value integrating tree diversity in cocoa for the provision of ecosystem services', *Agroforestry Systems*, 88 (6), pp. 1047–1066.

Spedding, C.R.W. (1996). *Agriculture and the citizen*. Chapman and Hall, London, 282p.

Tiwari, T.P., Brook, R.M. and Sinclair, F.L. (2004) 'Implications of hill farmers' agronomic practices in Nepal for crop improvement in maize', *Experimental Agriculture*, 40, pp. 1–21.

Tiwari, T.P., Brook, R.M., Wagstaff, P. and Sinclair, F.L. (2012) 'Effects of light environment on maize in hillside agroforestry systems of Nepal', *Food Security*, 4, pp. 103–114.

Tiwari, T.P., Virk, D.S. and Sinclair, F.L. (2009) 'Rapid gains in yield and adoption of new maize varieties for complex hillside environments through farmer participation. Improving options through participatory varietal selection (PVS)', *Field Crops Research*, 111, pp. 137–143.

Thompson, J., Millstone, E., Scoones, I., Ely, A., Marshall, F., Shah, E. and Stagl, S. (2007) Agri-food system dynamics: pathways to sustainability in an era of uncertainty, *STEPS Working Paper* 4, Brighton: STEPS Centre.

Traore, P.C.S., Lamien, N., Ayantunde, A.A., Bayala, J., Kalinganire, A, Binam, J.N., Carey, E., Emechebe, A. Namara, R., Tondoh, J.E. and Vodouhe, E. (2012) Sampling the vulnerability reduction – sustainable intensification continuum: a West African paradigm for selection of Dryland Systems sites. *11th International Conference on Drylands Development*, Beijing.

Twomlow, S., Rohrbach, D., Dimes, J., Rusike, J., Mupangwa, W., Ncube, B., Hove, L., Moyo, M., Mashingaidze, N. and Mahposa, P. (2010) 'Micro-dosing as a pathway to Africa's green revolution: evidence from broad-scale on-farm trials', *Nutrient Cycling in Agroecosystems*, 88 (1), pp. 3–15.

United Nations (2015) *Transforming our world: the 2030 agenda for sustainable development*, United Nations General Assembly Resolution 70/1.

Vågen, T.G., Winowiecki, L., Tondoh, J.E., Desta, L.T. and Gumbricht, T. (2016) 'Mapping of soil properties and land degradation risk in Africa using MODIS reflectance', *Geoderma*, 263, pp. 216–225.

van Ginkel, M., Sayer, J., Sinclair, F. Aw-Hassan, A., Bossio, D., Craufurd, P., El Mourid, M., Haddad, N., Hoisington, D., Johnson, N., León Velarde, C., Mares, V. Mude, A. Nefzaoui, A., Noble, A., Rao, K.P.C., Serraj, R., Tarawali, S., Vodouhe, R. and Ortiz Rios, R.O. (2013) 'An integrated agro-ecosystem and livelihood systems approach for the poor and vulnerable in dry areas', *Food Security*, 5 (6), pp. 751–767.

Vanlauwe, B., Descheemaeker, K., Giller, K.E., Huising, J., Merckx, R., Nziguheba, G., Wendt, J. and Zinger, S. (2015) 'Integrated soil fertility management in sub-Saharan Africa: unravelling local adaptation', *SOIL*, 1, pp. 491–508.

Von Bertalanffy, L. (1968) *General System Theory: Foundations, Development, Applications*, George Braziller, New York.

White, H. (2013) 'An introduction to the use of randomised control trials to evaluate development interventions', *Journal of Development Effectiveness*, 5 (1), pp. 30–49.

4   Understanding African farming systems as a basis for sustainable intensification

Dennis Garrity, John Dixon and Jean-Marc Boffa

Introduction

Africa has witnessed an extraordinary rebound in economic growth over the past decade. This has inspired more confidence in the continent's future. But economic growth has not significantly reduced extreme poverty or food insecurity for the third or more of the population experiencing chronic or crisis-driven hunger. Most of Africa's poor are rural, and most rely largely on agriculture for their livelihoods. Sustainably intensifying smallholder agriculture is fundamental to overcoming the seemingly intractable problem of African poverty, but how?

The African farming context is immensely diverse. This requires new ways of organizing and governing the innovation process, from upstream research to downstream implementation. The farming systems framework is especially helpful to aggregate locations with similar constraints and investment opportunities, identify common issues and provide options for managing risk and enhancing productivity. Such an approach requires cross- or trans-disciplinary thinking, bringing the best of socioeconomic and biophysical analysis together. We assembled 15 multi-disciplinary teams from across the continent to analyze each major farming system (Garrity et al., 2012). This chapter examines five of these farming systems as the basis for sustainable intensification. It also examines the key cross-cutting drivers to provide a stimulus for the rebuilding of the farming systems perspective as a critical input to both policy and practice.

The aim of this chapter is to explore how deeper knowledge of African farming systems structure and function can identify strategic interventions for sustainable agricultural intensification and contribute to poverty reduction and livelihood improvement. We consider farming systems to be substantial populations of individual farm households with broadly similar patterns of livelihood and consumption patterns, constraints and opportunities, and for which similar development strategies and interventions would be appropriate. Sustainable intensification emphasizes greater production from existing resources without loss or degradation of the resource base. The farming systems approach was inspired by the words of William Allen in his monumental study of African farming systems a half century ago (*The African Husbandman*, 1965), who said, *'We must try to see the situation through the eyes of the farmer, and put aside for the time being our own preconceived ideas, prejudices, and conceptions of good land-use, which derive from very different societies and environments'*. This notion is no less cogent today than it was then.

Effective sustainable intensification leads to productive diversified farming systems that maintain the resource base and respond to external opportunities and pressures from changing markets, population pressure and policies. National and regional decision makers face the challenge of identifying and investing in specific agricultural and rural development opportunities where the greatest impact on food security and poverty will be achieved. Experience has shown that policy-making must be better grounded in context-specific evidence and analysis, complemented by innovative ways of thinking about future pathways for agricultural development.

The chapter presents elements of a farming systems framework, notably the cross-cutting drivers and challenges, to support decision makers in research and development endeavours, both public and private, drawing, inter alia, upon ongoing work led by the World Agroforestry Centre (ICRAF) and supported by the Australian Centre for Agricultural Research (ACIAR) (Garrity et al., 2012). FAO and the World Bank published an analysis that examined agricultural development issues and priorities across six developing regions, based on the contrasting perspectives of farm households in different types of farming (Dixon et al., 2001; www.fao.org/farmingsystems/). The classifications proved to be a valuable tool in targeting and prioritizing agricultural research and development in recent years. The African classification proved particularly useful in studies and planning to support many international research and development efforts (InterAcademy Council, 2004).

Leading scientists called for an update of the analysis for the period 2000–2015, noting substantial new dimensions in the drivers of change in African agriculture. A book-length update of the year 2000 analysis is underway with much greater detail (the analysis has been elaborated from a single chapter into a 600-page book) with inputs from more than 70 African and international scientists. The effort began with a validation and fine-tuning of the African farming systems classification, and the establishment of 15 multi-disciplinary teams, one for each major farming system across the region. The analytical framework was anchored in characterizing broad groups of farm households with broadly similar patterns of livelihood and consumption, incorporating inherent local variation. A wealth of spatial data on production, marketing, nutritional and natural resources were assembled in the course of the work.

More than 70% of the rural poor in sub-Saharan Africa (SSA) reside in five broad regional farming systems being the focus of this chapter: (1) the highland perennial, (2) maize-mixed, (3) cereal-root crop mixed, (4) agro-pastoral and (5) highland mixed farming systems (Figure 4.1). Our analysis shows that these five major farming systems provide the principal livelihoods of more than two-thirds of Africa's poor. This chapter focuses on an analysis of the five systems. The sustainable intensification constraints and potentials of each of the five systems are evaluated, and this is followed by a summation of the cross-cutting drivers impinging upon their sustainable intensification.

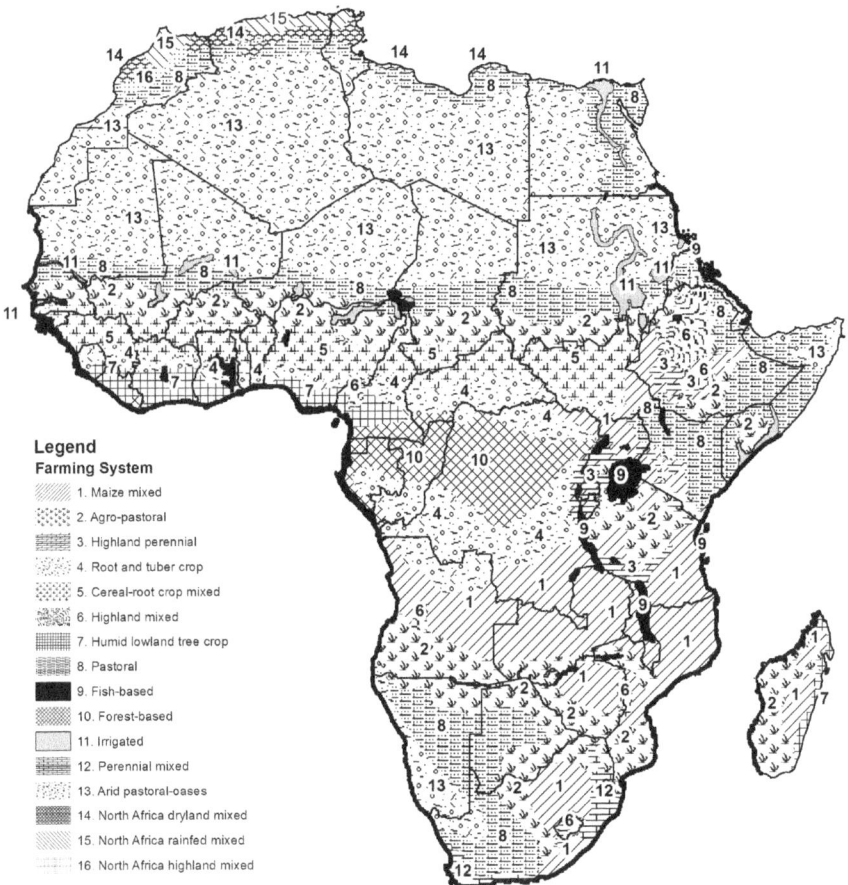

Legend
Farming System

1. Maize mixed
2. Agro-pastoral
3. Highland perennial
4. Root and tuber crop
5. Cereal-root crop mixed
6. Highland mixed
7. Humid lowland tree crop
8. Pastoral
9. Fish-based
10. Forest-based
11. Irrigated
12. Perennial mixed
13. Arid pastoral-oases
14. North Africa dryland mixed
15. North Africa rainfed mixed
16. North Africa highland mixed

Figure 4.1 Sub-Saharan African farming systems

Highland perennial farming systems: Intensification against the minimum limits of farm size

With extremely high population densities, but high development potential (soils, rainfall, markets), the highland perennial farming system has been a natural experiment in the interaction between population growth, declining farm sizes and the intensification of farming systems. The system has the highest productivity among rainfed farming systems and would further benefit from improved land tenure, better market access and improved labour mobility. These challenges will be faced by many other farming systems in the coming years, as sustainable intensification encounters increasing agricultural population density and declining farm sizes, and the positive consequences and limits of farming systems commercialization.

Highland perennial farming systems occupy a highly favourable agroecology in a mountainous terrain. They are based on a diverse range of enterprises anchored by perennial crops, particularly East African Highland Bananas, coffee and agroforestry, which are suited to hillside farming. Cattle are also an integral part of the system, with a recent shift to higher value dairy production and a corresponding increase in forage production. These favourable ecologies are limited in area while their average rural population densities are the highest in sub-Saharan Africa. The average farm size in many parts of the East African highlands has now reached critical thresholds that limit their capacity to support livelihoods purely through agricultural production. The density of poor households in these areas is also the highest on the continent.

The liberalization of agricultural markets in the 1990s reinforced market integration. Where market infrastructure is now sufficiently well-developed, highland perennial farming systems have diversified into higher value cash crops, particularly horticulture, smallholder dairy, high quality coffee and smallholder tea. However, such market-led intensification has not been universal and has depended on integration into national markets. Central Kenya and Northern Tanzania have been the principal beneficiaries, while Western Kenya, Southern Tanzania and the Albertine Rift are only beginning such diversification pathways. Southern Ethiopia has the highest population densities and is the least integrated into national markets. Farming is becoming unsustainable in this area.

Highland perennial farming systems thus have the potential of being the most productive and at the same time the largest source of rural–urban migration in East Africa. But due to the critically small farm sizes, agriculture will only be a share of a diversified livelihood strategy, with an increasing participation in the growing rural non-farm economy, and reliance on remittances from migration of household members.

Maize-mixed farming systems: Engine for rural growth?

The maize-mixed farming system is a crop-livestock system lying largely in the sub-humid zone of Southern and Eastern Africa, with a growing season of around six months. Subsistence-oriented, rainfed staple food production is the predominant agricultural activity. The widespread adoption of maize was driven by its lower requirement for labour than for the earlier small grain staples such as millet and sorghum. Legumes are grown as intercrops, or in rotation with maize. Cash crops include coffee, tobacco, cotton, groundnuts and sunflower. Household income is typically further supported by trading and small businesses.

A large proportion of smallholder farmers in the system are net purchasers of food grains, although they will typically sell staples at harvest time to bring in needed cash. Many farmers typically use slow, inefficient and physically demanding manual cultivation techniques (complemented with animal draught where available). The agricultural population is aging and has an increasing number of female-headed households. Biotic and abiotic constraints limit the productivity of agriculture in general, and maize and legumes in particular. Despite improved market access, institutional and socioeconomic constraints make it difficult for smallholder farmers to access seeds, inputs and output markets in order to respond to market price signals. Crop pests reduce yields by as much as half.

There are significant constraints to effective value chain operation. Weak rural institutions for delivery of services and inadequate farmer organization contribute to poor capacity for remedying market imperfections in the supply of key inputs and marketing surplus produce. There is a significant private sector seed industry, but the seed production and deployment environment is poor and dominated by hybrid maize seed. There are signs of serious fertility decline over much of the system. Mineral fertilizers provide only a partial solution as there is good evidence of increasing soil acidity where there has been prolonged use of inorganic fertilizers without the use of manures or retention of crop residues to build soil organic matter. Climate change and the associated progressive increase in temperatures and decline in precipitation are expected to reduce maize crop yields substantially, especially in Southern Africa.

The maize-mixed farming system has a greater agricultural population and more poor farm families in aggregate than any of the other farming systems. It is a potential food basket, with good potential for intensification and diversification, and it can be a driver of agricultural growth and food security in the Eastern and Southern African region. But sustainable intensification in this system demands strategic interlinked initiatives aimed at improving access to agricultural resources, smallholder competitiveness and household risk management.

Agro-pastoral farming systems: Achieving resilience under duress?

The agro-pastoral farming system is found throughout Sahelian West Africa stretching from Senegal to Sudan, as well as in parts of Eastern and Southern Africa. A relatively short growing period underpins the importance of millet (*Pennisetum glaucum*) and sorghum (*Sorghum bicolor*) as the dominant cereal crops and grass pasture as the rangeland component of the system. For generations, populations have adapted their farming systems and way of life to the dramatic rainfall variability and its uncertain effects on the production of crops, trees and grazing resources. Drought is a regular phenomenon. Not only are farmers used to it, it is a central feature in their economic planning. Thus, the primary concern of agro-pastoral farm households is to ensure their survival and to minimize the risk of failure to produce their means of subsistence.

Households in this mixed crop–livestock system typically integrate the growing of food or cash crops with trees in their crop fields (parkland agroforestry), combined with an extensive and pastoral-type livestock production system. Livestock activities involve cattle, sheep, goats, donkeys, camels and poultry. In places affected by long cycles of drought, there has been a shift from cattle to small ruminants as they are less costly, better adapted to drought, easier to feed and reproduce faster than cattle. Sedentary farmers and nomadic herders have traditionally sustained functional links through the exchange of grain and crop residues for manure. Livestock, rangeland and cropland productivities are closely linked. During the dry season the livestock graze on crop residues and their manure enhances soil fertility for crop production. Rangelands and fallow lands provide livestock forage and nutrients are transferred to cropland through manure.

The agro-pastoral system has the second largest population of extremely poor people among the farming systems of Africa, and the population is growing at the rate of 2.8% per year. Demographic pressure on land and stagnating revenues from agricultural production severely limit prospects for further increases. The system is rapidly transforming to more sedentary forms of mixed production. Labour migration is expanding rapidly as a response to the limitations of the environment. Household dependence on off-farm income and remittances is growing.

Current trends of urbanization and increasing disposable income in urban areas are predicted to more than triple urban demand for foodstuffs during the next 40 years, especially high value foods including dairy and meat. This provides the major opportunity for intensification. It implies optimizing crop–livestock interactions, integrated soil fertility replenishment, greater reliance on livestock feeds and veterinary products, and intensifying the agroforestry component.

One of the bright spots observed in this farming system during the last two decades has been the growing expansion of regreening practices, and the widespread farmer-to-farmer diffusion in this regard. Over five million hectares of farmer-managed natural regeneration of indigenous trees on croplands has been mapped in Niger and Mali (Reij et al., 2009). Trees are naturally regenerated and cultured in the crop fields to provide biofertilizers for increased cereal yields, enhanced fodder production, fuel wood, timber and other environmental services. This has been called the most dramatic positive environmental transformation recently seen in Africa.

Local livelihoods in the agro-pastoral farming system have adapted to rainfall variability, low ecosystem productivity and climatic and economic risks. Strategies to cope have included labour mobility, diversification of activities and income, intensification and collective resource management. The past decade of reasonable rainfall has led to re-integration of tree cover into the farmlands, and a great expansion of millet and sorghum growing areas. Sustainable intensification priorities should aim to enhance adaptation capacities and food security, focusing on integrated, multi-scale participatory approaches, flexible tenure regimes, locally adapted information systems and government support for the supply of agricultural services.

Cereal-root crop mixed farming systems: West Africa's future breadbasket?

The cereal-root crop mixed farming system is considered to have one of the highest agricultural growth potentials in Africa, through the expansion of cropping area and through sustainable intensification, including mechanization and higher crop and livestock productivity. The crops include sorghum, millet, maize and rice on over half the area, root and tuber crops (cassava, sweet potato and yam) on about one-tenth with annual leguminous crops or pulses (cowpeas, pigeon peas, dry beans) on 6%, and oilseed crops (groundnut, soybean, sesame) on about 10%. Cotton occupies just under 5%, and other crops about 15% of the cropped area.

This farming system has long been seen as a major source of agricultural growth for Africa (Dixon et al., 2001). But the variable annual rainfall and generally poor soil quality make this a challenging agroecological environment. This region has the potential to become a food production powerhouse that could feed Africa, and eventually create a booming export business. However, the immediate realities within the cereal-root crop mixed system are still showing high poverty rates. In 2005, about 47% of the rural population had a per capita daily income of less than US $1.25. The average annual increase in the agricultural population in this particular system is relatively low (1.1% per year during the 2005–2010 period). In much of this farming system, labour (not land) is the limiting constraint to the expansion of production. Thus, expanding the cultivated area of the small farm by increasing the efficiency and returns to labour is critical. There is ample opportunity for growth through expansion of the cropped area as well as through higher yields per hectare. There is substantial tree cover on croplands in major parts of this farming system, particularly the cultivation of shea nut for oil and cosmetics. Further expansion and intensification of the agroforestry parklands is a key pathway to higher incomes and the regeneration of soil health.

Although the agroecological conditions are similar to the maize-mixed system, the population density is lower and the farm sizes are somewhat larger. The development of the system will benefit from sustainable and efficient labour-saving patterns of resource management, e.g. through conservation agriculture to address current land degradation (nutrient mining and soil erosion), and promotion of smallholder-led commercialization, along with the reduction of deficiencies in transport, processing and storage infrastructure.

Highland mixed farming systems

The highland mixed farming system occurs in the Eastern, Southern and Southwestern African highlands, hosting 67 million people on about 48 million hectares of land, of which about 99% is agricultural. These are cool highlands > 1700m above sea level. Mountainous and undulating terrain predominates. The challenges experienced in the highland mixed farming system, the largest part of which occurs in Ethiopia's highlands, include high population density and declining land per capita, fragmented and eroded farms, insecure land tenure, and poor market infrastructure. Yet, this system represents an agricultural growth pole for the country. It is supported by a strong policy environment and by the availability of improved crop and livestock technologies. An important investment priority lies in the development of private sector-supported smallholder commercial agriculture including specialized high altitude temperate enterprises such as seed production, dairy production and temperate vegetables supported by improved road connectivity, and input markets.

There are tight interactions between crop and livestock production. The three main subsystems are livestock-cereal based, mixed wheat-pulse based and mixed maize based. The system is constrained by limited human and financial capital, and poor market access with at least seven hours travel to reach the nearest town of 50,000 people. High population density and rapid growth, severe soil erosion, farm fragmentation, a declining land holding per capita, declining soil fertility and overstocking are common. Only about 2% of the rural population has access to electrical power. The majority depend on biomass energy.

The system has distinct mountain products and services. Sustainable intensification priorities include the facilitation of mountain niche markets (e.g. wool and mohair, highland flowers, organic honey), commercialization of livestock systems, diversifying livelihood options (including on-farm and out-migration) and increased access to technological information and knowledge.

Common challenges and policy implications across farming systems

This analysis has highlighted the enormous variation in the underlying potential for particular areas to participate in market-driven sustainable intensification opportunities. If the variation among the farming subsystems is compared by two factors, the land available per household and the accessibility of agricultural services (particularly markets), virtually all of them suffer from severe inadequacy in either one or both factors (Table 4.1).

By and large, changes within farming systems tend to be evolutionary and are path dependent. They build upon the structure of farming by households that are working smallholdings, often held under communal systems of tenure. In response to the forces of population pressure and market demand, farmers change their cropping patterns, redeploy household labour and make small capital investments in inputs, draught animals and tools. New techniques are generally adopted by making small changes to existing systems. Over time, the accumulation of successive changes can transform farming, landscapes and society. But such transformations are generally seen in the medium to long term. In this respect, the African experience may not be so very different from that seen in much of Asia, where the apparent quantum leaps of the Green Revolution were actually, on closer inspection, the cumulative effect of a series of modest improvements for any given crop or locality.

Each farming system has distinct pathways toward food and nutrition security requiring different types of public support. However, some common cross-cutting requirements also emerge. Ultimately, the key is how the drivers of change affect farm household decisions to change practices, adopt new crops or livestock, and market their produce in different ways.

Table 4.1 Farming systems by land availability and access to agricultural services

Agricultural services (marketing, credit, insurance)	Very small (<1 ha)	Small (1–3 ha)	Medium (3–10 ha)	Very large (>10 ha)
Very good / **Fair**	**Highland perennial**: Central Kenya			
Poor	**Highland perennial**: Mount Kilimanjaro			
Very poor	**Highland perennial**: Albertine Rift			
	Maize-mixed: Malawi	**Maize-mixed**: Zambia		
	Agro-pastoral: Burkina Faso, SC Niger	**Cereal, root & tuber crops**: West Africa		
	Highland mixed: Ethiopia	**Agro-pastoral**: Mali	**Maize-mixed**: DRC, Mozambique, Angola	
	Land per household			

Source: Garrity et al., 2012.

The population explosion: Food security, poverty and land

Africa's population growth rates have accelerated tremendously during the past 50 years, and have become the highest in the world. The big change that has recently disrupted the rural society in many farming systems across the continent has been the abrupt closure of the land frontier. Urban populations have expanded due to rural displacement. The highland perennial farming system of Eastern Africa is a classic case of the buildup of extreme pressure on the land under comparatively favourable agroecological and market conditions. But the pressures are also prevalent in the highland mixed farming systems of Ethiopia, the vast maize-mixed farming systems and the agro-pastoral systems of the severely climate-constrained drylands. This process has happened so rapidly that public policy has been unable to cope effectively. The new path to family security is limited to educating a small number of children. The majority of these children will then find their own way in an extremely competitive job market. The future narrative of Africa's farming systems will be influenced enormously by how fast rural population growth rates can adjust to the new realities of extreme land scarcity.

Natural resources management and climate change challenges

Biomass productivity has been declining on a huge scale in Africa (Bai et al., 2008). Large swaths of Southern Africa are affected, particularly in the maize-mixed farming systems in Zambia, Angola, DRC, Mozambique and Tanzania. Much of this loss of biomass productivity is due to forest clearing for agriculture, in addition to reductions in the productivity of the land previously cleared for agriculture, aggravated more recently by climate change especially in Southern Africa. But there has also been a widespread farmer-managed regreening in parts of the Sahel (Reij et al., 2009) and in the Ethiopian drylands (highland mixed farming systems) during this period, managed by farmers as low-cost land improvement for multiple benefits.

Soil fertility replenishment

Reversing the trend of soil fertility depletion, which governs agricultural productivity, in all African farming systems has become a major development policy issue (Vanlauwe, 2010). As rural populations have grown, farm sizes have decreased and fallowing has diminished in most farming systems. Farmyard manure supplies are also declining in many areas. Soil impoverishment has become a primary concern for smallholder farmers across a range of countries (Bunch, 2011). Crop-livestock-tree interactions are of great significance for maintaining soil nutrients. Culturing fertilizer trees in crop fields is also becoming increasingly popular as a component of integrated soil fertility management (Garrity et al., 2010).

Tree dynamics

The decline in natural forest area is expected to continue, particularly in the Congo Basin and Southern Africa. Of the farming systems described in this chapter, the cereal-root crop mixed system is most closely linked to deforestation. Currently, the maize-mixed and the highland perennial are experiencing particularly acute fuel wood shortages. However, there is evidence of a turnaround in tree cover. Two-thirds of the cultivated lands of Kenya contain at least 10% tree cover, often as trees on field boundaries (Kenya Landscape Restoration Technical Working Group, 2015).

In Africa, farmers historically sustained medium to high densities of trees in their cropping systems. Trees serve many purposes, but particularly to provide a source of livestock fodder as community grazing lands are depleted, as a source of fuel wood and timber for home consumption and sale, and as a source of biofertilizers to sustain soil fertility (particularly leguminous trees such as *Faidherbia albida*), as well as the production fruits, leaf vegetables, medicinals and other products, and the enhancement of local environmental services (Garrity et al., 2010). The recognition of these multiple co-benefits has stimulated increasing interest across the continent in the upscaling of these evergreen agriculture systems. They have been particularly successful in the agro-pastoral systems of the Sahel where young agroforestry parklands are expanding (Reij et al., 2009).

Water management

Irrigation holds great potential for agricultural growth, food security and poverty alleviation in SSA, but its contribution to date has been constrained by lack of investment and the poor performance of existing public sector-managed large-scale irrigation schemes (Kizito et al., 2012). With few exceptions, however, large-scale public investments in irrigation have often failed, despite many millions of dollars invested in large public schemes for surface irrigation. There are, however, other options to improve the capture and utilization of water in agricultural systems that deserve more attention, particularly smaller-scale activities in field water management. Many smallholder farmers procure irrigation equipment (buckets, pumps, drips, pipes and sprinklers) either individually or in small groups. In addition, rainwater harvesting technology can be applied at the farm or community level across wide areas, particularly in agro-pastoral areas with little or no groundwater or surface water potential.

Markets and trade

Alongside the growth in consumption requirements, demand from the export of agricultural commodities (cocoa, coffee, cotton, sugar, tea, tobacco, etc.) was, for much of the last century, the major stimulus to agriculture. Intra-regional trade in food and cash commodities is now growing and offers substantial promise, for example livestock in West Africa and maize in Eastern and Southern Africa. The rate of urbanization in Africa is proceeding exceptionally fast. Consequently, the urban demand for agricultural products is growing very rapidly, providing market opportunities for a wide diversity of farm products. More than a decade ago, Diao et al. (2003) estimated that the potential demand for agricultural products in Africa far exceeded supply. This is a major stimulus for both the intensification and diversification of farming systems. Throughout West and East Africa there are thriving belts of agriculture surrounding cities, supplying all manner of produce including the vegetables, fruit, dairy and other livestock produce that command higher than average returns.

Science and technology

A majority of poor smallholders operate mixed crop–livestock farming systems and so research on crop–livestock must interface among fodder, cash flow and risk management. Perhaps the most important research challenge is organization and delivery of mission-directed systems research. A related challenge is the effective integration of socioeconomic and biophysical research. There are also issues related to addressing gender inequity in agriculture. The most basic of these relate to the control and management of agricultural resources, especially land tenure and access to credit though banks (micro-finance opportunities not withstanding).

Major advances in communications technology have brought information and knowledge much closer to small farm households. Dramatic expansion in mobile phones has occurred in most countries of the region, reducing the gender inequity in access to agricultural information, including market prices and 'mobile banking'. The functionality of mobile phones is likely to grow dramatically in the coming decades for many forms of information provision, e.g. disease identification, and decision support tools. Beyond the mobile phone, there are also advances in a number of new ICTs and new variants on old communications methods, e.g. local community radio. These tools will reinforce and increase the effectiveness of innovation platforms to accelerate the diffusion of knowledge more effectively than in the past.

Conclusion

Bold initiatives are needed to drive sustainable intensification in African farming systems, underpinned by new ways of organizing and governing the innovation process. Mission-directed research within a farming systems framework is needed to better guide policy and investment decisions. Such a framework is a practical means to target science and public policy to the contrasting needs of different farming systems.

Our analysis shows that the five major farming systems covered in this chapter provide the principal livelihoods of more than two-thirds of sub-Saharan Africa's poor. They account for a major part of African crop, livestock, fish and forestry production and exports. With appropriately targeted strategic interventions, they will underpin national economic growth and sustainable development. Success requires investments in local, national and regional innovation systems and in targeted policy analysis and implementation. The establishment of such frameworks and the building of capacity of researchers and policymakers should attract substantial investment from national, regional and international bodies.

References

Allan, W. (1965) *The African Husbandman*. Oliver and Boyd, London. Reprinted by International African Institute, Munster.

Amanor, K.S. (2005) 'Agricultural Markets in West Africa: Frontiers, Agribusiness and Social Differentiation'. *IDS Bulletin* 36(2): 58–62.

Bai, Z., Dent, D., Olsson, L. and Schaepman, M.E. (2008) *Global Assessment of Land Degradation and Improvement. 1. Identification by Remote Sensing*. GLADA Report 5. FAO, Rome.

Bunch, R. (2011) *Africa's Soil Fertility Crisis and Coming Famine*. State of the World Report. WorldWatch, Washington DC.

Diao, X., Dorosh, P.A. and Rahman, S.M. (2003) Market opportunities for African agriculture: an examination of demand-side constraints on agricultural growth. Development Strategy and Governance Division Discussion Paper 1, International Food Policy Research Institute.

Dixon, J., Gulliver, A. and Gibbon, D. (2001) *Farming Systems and Poverty*. Improving Farmer's Livelihoods in a Changing World, FAO & World Bank, Rome, Italy & Washington DC.

Garrity, D., Dixon, J. and Boffa, J. (2012) *Understanding African Farming Systems*. Australian Centre for International Agricultural Research, 55 p.

Garrity, D.P., Akinnifesi, F.K., Ajayi, O.C., Weldesemayat, S.G., Mowo, J.G., Kalinganire, A., Larwanou, M. and Bayala, J. (2010) 'Evergreen Agriculture: A Robust Approach to Sustainable Food Security in Africa'. *Food Security* 2(3): 197–214.

InterAcademy Council (2004) *Realizing the Promise and Potential of African Agriculture*, InterAcademy Council, Amsterdam, 171 p.

Kenya Landscape Restoration Technical Working Group (2015) 'Mapping to Guide Planning and Coordination of Tree-Based Landscape Restoration in Kenya: Summary Report and Results'. Unpublished, 36 p.

Kizito, F., Williams, T.O., McCartney, M. and Erkossa, T. (2012). *Green and Blue Water Dimensions of Foreign Direct Investment in Biofuel and Food Production in West Africa: The Case of Ghana and Mali*, Routledge, London.

Reij, C., Tappan, G. and Smale, M. (2009) Agricultural transformation in the Sahel: another kind of "green revolution", IFPRI Discussion Paper no. 00914.

Vanlauwe, B., Bationo, A., Chianu, J., Giller, K.E., Merckx, R., Mokwunye, U., Ohiokpehai, O., Pypers, P., Tabo, R., Shepherd, K.D., Smaling, E.M.A., Woomer, P.L. and Sanginga, N. (2010) 'Integrated Soil Fertility Management: Operational Definition and Consequences for Implementation and Dissemination'. *Outlook on Agriculture* 39(1): 17–24.

5 Does sustainable intensification offer a pathway to improved food security for aquatic agricultural system-dependent communities?

Simon J. Attwood, Sarah Park, Jacqueline Loos, Michael Phillips, David Mills and Cynthia McDougall

Introduction

Aquatic agricultural systems (AAS) are diverse production and livelihood systems that occur along inland lakes and rivers, freshwater floodplains, estuarine deltas and coasts. These diverse production systems are typically characterised by seasonal changes in productivity and water availability, driven by periodic variation in rainfall, river flow and/or coastal and marine processes (WorldFish, 2011). Globally, AAS are highly significant; first, they cover vast areas of the non-OECD[1] world, with approximately 2.5 million km^2 of inland AAS coverage and a further 2 million km^2 of coastal system coverage (Béné and Teoh, 2014). This represents approximately 27% of the non-OECD cultivated area. Second, an estimated 500 million people are dependent on AAS in the non-OECD world, with approximately three-quarters of this number dependent on inland systems. This represents approximately 16% of the total estimated rural population in non-OECD countries (Béné and Teoh, 2014). Within AAS, food production and livelihoods depend on diverse activities and resources. Interdependent terrestrial and aquatic resource uses include agricultural practices (e.g. fish farming, crops, livestock), considerably supplemented by wild harvested foods (e.g. caught fish) from native ecosystems. This diversity of production and resource-use options is coupled with or based on high agricultural and wild biodiversity and associated ecosystem services. Community well-being within AAS heavily depends on the various ecosystem services these systems provide, as well as processes and institutions which mediate access to services. However, AAS and their beneficial characteristics are extremely sensitive to environmental changes, thus rendering communities vulnerable to ecosystem shocks and disturbances (Halwart, 2006; WorldFish, 2011; Weeratunge et al., 2012; Castine et al., 2013). Implicit in this are issues relating to rights and equitable access to the benefits of this productivity (Birch et al., 2014; Loos et al., 2014), where 'distribution gaps' may be more important than 'yield gaps' and 'nutrition gaps' (Pinstrup-Andersen, 2009; Tscharntke et al., 2012). Consequently, the productive intensification of these systems through approaches that are both ecologically meaningful and socially and economically equitable is a high priority. Box 5.1 provides an example of the complex interaction of components in the AAS of the Barotse floodplain in Zambia, and the ingenuity of those that manage and depend upon them.

Box 5.1 System diversity for livelihoods in the Barotse floodplain, Western Province, Zambia

The AAS in the Barotse region of Zambia contain a multitude of different land and waterscape components. These interact to support a wide diversity of crops, livestock and fish (Baidu-Forson et al., 2014). These, in turn, provide considerable potential to supply a nutritionally diverse diet throughout the year (Luckett et al., 2015). Whilst fish are abundant in many of the waterways, crops are also produced on the many land types spanning the floodplain. These areas differ greatly in terms of their productive potential, soil type and native vegetation associations (Estrada-Carmona, 2014). Seasonal flooding renders some of the most productive areas unavailable during the flooding period, when communities are forced to move to higher ground. However, in these seasonal refuges, soil structure, nutrient status and water holding capacity are much less suitable for cropping. Opportunities for growing food during the flood season thus become scarce, with an attendant hunger season and malnutrition occurring. Although the flooding cycle may appear a temporary threat to crop and livestock production, it provides essential services in maintaining the nutrient status and water holding capacity of cropping lands for the subsequent season, and enables fishing activities to occur. Traditional governance systems mediate the use and management of natural resources in this dynamic and complex social-ecological system.

Aquatic agricultural systems as multifunctional landscapes

A unifying factor amongst many AAS is the considerable diversity of land-use and water-use types. These land and water uses include differing spatial extents, arrangements and community-utilisation of capture fisheries, aquaculture, cultivated and wild plant species, and livestock grazing. However, the dominance of water-based features in these systems drives much of the ecosystem service generation and landscape multi-functionality, e.g. managing water for fisheries or aquaculture, irrigation water, hydropower, flood protection, drinking water, biodiversity conservation (Gordon et al., 2010). Implicit within these dynamics are synergies and trade-offs among a wide range of ecosystem services that underpin community livelihoods and well-being (Power, 2010). Trajectories of change that commonly occur in AAS include both expansion and intensification of terrestrial agriculture (Boserup, 1965; Laurance et al., 2014), decline in condition and productivity of natural fisheries (Hall, 2013), increased irrigation of crops and consequent water abstraction and diversion (Brummett et al., 2012), encroachment and loss of wetland habitats such as mangroves, and expanding and intensifying aquaculture (Troell et al., 2014). Figure 5.1 depicts a number of common synergistic and antagonistic interactions among different land and water uses in an AAS in Tonle Sap, Cambodia. The interactions depicted are: (1) water availability for irrigation of crops and watering of livestock, and pollution of water (e.g. eutrophication, turbidity, agrochemical residues) from adjacent terrestrial production; (2) and (3) soil erosion from livestock access to watering points; (4) integration of livestock and cropping – cattle feeding on rice stubble *and* maintaining soil fertility; (5) habitat provisioning for pollinators (e.g. native bees, hoverflies), beneficial natural enemies such as predators (e.g. spiders, Coccinellidae beetles) and parasitoids (e.g. native wasps) through mixed cropping and perennial vegetation; (6) more nuanced understanding and effective management of floodwaters – floodwaters can be a risk to cropping, but also are essential for maintaining and replenishing soil nutrients

Figure 5.1 Potential interactions among different land and water elements in an aquatic agricultural landscape in Tonle Sap, Cambodia

Source: Photo credit: Eric Baran.

and moisture, and are fundamental in underpinning productivity in AAS; (7) integration of fish farming and capture fisheries into the cropping and livestock grazing landscape; (8) retention and restoration of native vegetation. Fringing vegetation can filter particulate and agrochemical pollutants from cropland into waterways, reduce soil erosion, provide building materials, fodder for livestock and habitat for biodiversity.

Inherent within such a system are complex arrays of ecosystem services that underpin the many production and livelihood activities undertaken by communities. Figure 5.2 displays a range of ecosystem services and how they might be used by farmers and fishers.

The complexity of the landscape and the juxtaposition of terrestrial and aquatic elements in AAS provide a diverse range of ecosystem services (e.g. Figure 5.2). From a nutrition perspective, there is potential to deliver high dietary diversity, including fish, aquatic plants, cereal crops, vegetables, fruit trees, livestock products and wild harvested animals and plants (Halwart, 2008; Bharucha and Pretty, 2010). Furthermore, the retention or regrowth of elements of native vegetation is likely to provide a wide range of regulating services (e.g. maintenance of soil fertility, soil formation, water filtration), and

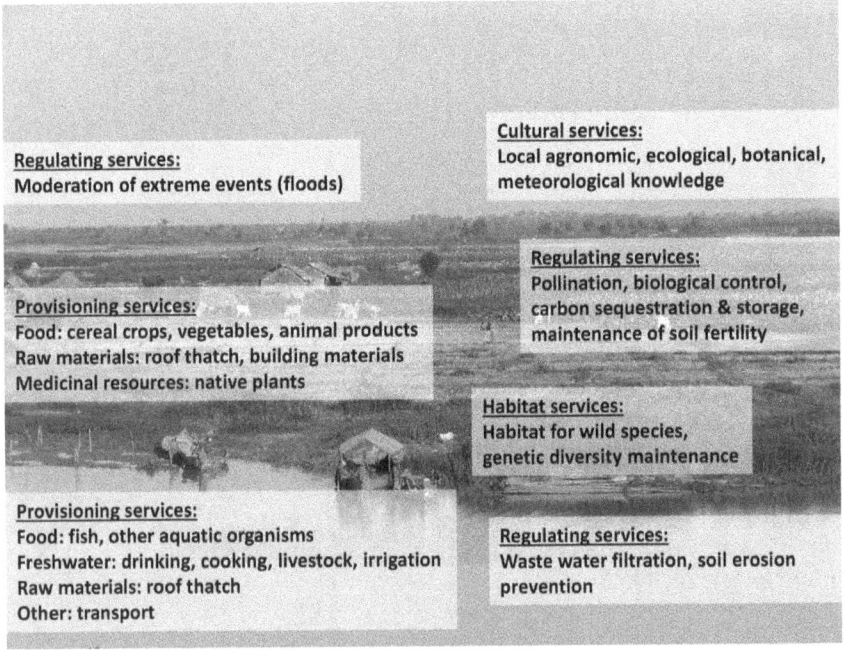

Figure 5.2 Potential ecosystem services and applications from various land and water elements in an aquatic agricultural landscape in Tonle Sap, Cambodia

Source: Photo credit: Eric Baran.

habitat services for wild biodiversity, including pollinators and natural enemies of agricultural pests (Attwood et al., 2008). This latter point may include species of conservation concern (Wright et al., 2012), species which can be harvested, species which may perform regulatory services useful to communities, and crop wild relatives that enhance domesticated crops through gene flow or adaptive trait availability.

Two approaches to intensification: Conventional and sustainable

Conventional intensification

A powerful and largely unchallenged paradigm for humanity's future is the contention that we need to produce more food (Godfray et al., 2010; Bajželj et al., 2014; Tilman and Clark, 2014) and set about closing so-called 'yield gaps' (Mueller et al., 2012). The drivers of this demand for increased production include an increasing global human population (Gerland et al., 2014) and altered patterns of food consumption, particularly increased levels of meat and dairy products (Kastner et al., 2012). However, land that is suitable for crop production is becoming scarce, with the vast majority of this land already being used for agricultural production; furthermore, land clearing comes with significant environmental and social costs (Lambin et al., 2013). Consequently, producing more food from available agricultural land is now often considered the most feasible means by which to address food security, and intensification is outstripping expansion as a means to increase production (Foley et al., 2011). The shortcomings of an approach that focuses ostensibly on increasing yields has, however, been well documented. For instance, whilst few deny that intensification is likely to be needed for future food security, other issues such as nutritional content, health values and greenhouse gas emissions of different production systems are critical (Tilman and Clark, 2014). Similarly, by reducing food waste, we may be able to address food security in tandem with intensification (Bajželj et al., 2014).

Conventional agricultural intensification has generally been achieved through processes such as improved crop varieties and animal breeds, NPK fertiliser application, synthetic chemical pesticide and herbicide use, increased mechanisation, wetland drainage, increased livestock stocking rates and specialisation of production (itself leading to loss of heterogeneity at many scales) (Benton et al., 2003; Tscharntke et al., 2012). This impacts on- and off-farm biodiversity, environmental function and ecosystem services in many systems globally (Brodie et al., 2005; Attwood et al., 2008). AAS are frequently ecologically highly significant, with a considerable overlap with areas of high biodiversity (e.g. many AAS are Ramsar sites and Important Bird and Biodiversity Areas, or may include biodiverse marine ecotypes such seagrass and coral reefs). Consequently, there is a need for intensification that takes account of environmental impacts and the need to conserve critical ecosystem services.

Sustainable intensification

In recent years, the concept of sustainable intensification has gained increasing popularity as a means to address the need for productive intensification without the attendant environmental (and social) costs often associated with conventional intensification (The Royal Society, 2009). A general understanding of sustainable intensification is an increase of outputs on existing agricultural land while minimising the negative environmental impacts and sparing the need for appropriation of additional land for food and commodity production (Garnett and Godfray, 2012). This framing of sustainability is implicitly linked to the concepts of efficient or even optimal resource use and primarily refers to environmental sustainability. A broader and more holistic understanding of sustainability aims at ensuring human well-being over the long term. Several aspects need to be considered in order to achieve this goal, for example by integrating inter- and intra-generational distributive and procedural justice in the long term (Loos et al., 2014). Especially in recent years, the notion of sustainable intensification has received growing attention in research and policy cycles, using the rationale that a growing and wealthier global population will demand increases in food production (Godfray et al., 2010). Sustainable intensification promises to meet these demands without further jeopardising essential ecosystem functions and services on which humanity depends.

Thus, in order to improve human well-being, an isolated focus on production will not necessarily improve the poor life conditions and the food insecurity of many residents in AAS. If anything, increases in production that are ecologically meaningful need to be combined with fair procedures, for example by truly allowing stakeholders to be involved in the process of decision making on what, how and where to intensify (Rosegrant and Cline, 2003). The benefits of these measures, moreover, need to be fairly distributed among members of the communities in these systems, while a conservative handling of limited resources can help ensure that future generations will still have access to these resources and build their livelihoods upon them.

Whilst the individual approaches and management actions to actually deliver sustainable intensification are likely to be production system and site specific, there are already a wide range of actions available to intensify terrestrial cropping and livestock systems, aquaculture systems and capture fisheries in a more integrated manner. In many cases, these are likely to be less external input dependent or demanding, and hence more suitable for poor economic groups (e.g. integrated pest management through retention or encouragement of structurally complex vegetation, versus the sourcing, purchase and use of synthetic chemical pesticides). A selection of examples for both conventional and integrated/sustainable approaches are depicted in Table 5.1, with some of the potential advantages of each approach (conventional versus integrated/sustainable) described.

The potentially negative impacts of unfettered agricultural intensification, and the opportunities for sustainable intensification, are relatively well-researched in

Table 5.1 Examples of technologies and techniques that integrate aspects of sustainability into intensification as an alternative to conventional methods of intensification

Production type	Conventional intensification	Sustainable/integrated intensification	Benefits of conv. approach	Benefits of sustainable/integrated intensification
Terrestrial cropping	Chemical pest control	IPM, biological control, push-pull planting, natural enemy habitat, cropping abutting native vegetation	• Generally effective; spatially and temporally controllable • Science of pesticides well understood • Incremental technology improvement	• Reduced agrochemical costs • Reduced on-site and off-site pollution and impacts • Improved water quality and fishery health • Improved habitat and ecosystem services • Reduced human health risks
	NPK fertiliser	Integration of livestock manure Composting crop residues and mulching	• Consistently increases yields • Combined with precision agriculture, can be very efficiently applied	• Reduced agrochemical costs • Reduced erosion • Reduced on-site and off-site pollution and impacts • Improved water quality • Waste minimisation • Reduced GHG emissions
	Focus on fewer crops (specialisation)	Increase/maintain crop diversity Intercropping and rotations	• Simplified management • Simplified business model and value chains • More efficient storage, processing and transport	• Production integration • Reduced waste • Reduced off-site impacts • Increased dietary diversity • Improved varietal matching to farmer needs
	Landscape simplification, removal of non-cropped elements	Retention of native vegetation Agroforestry Restoration	• Farm mechanisation improved • Greater area under intensive agriculture	• Improved pest control • Filtration of runoff • Improved water quality • Wild biodiversity conservation • Reduced soil erosion • Crop shelter • Increased ecosystem services

(Continued)

Table 5.1 (Continued)

Production type	Conventional intensification	Sustainable/integrated intensification	Benefits of conv. approach	Benefits of sustainable/integrated intensification
	Stubble burning	Livestock grazing or stubble retention	• Pest management	• Permanent soil cover – reduced erosion • Reduced GHG emissions • Increased soil organic matter (SOM) and soil carbon (C) • Improved nutrient cycling
	Tillage	Reduced or no-till methods	• Good for poorly drained soils • Well-tilled seedbed	• Reduced soil erosion • Reduced soil compaction • Improved soil water holding capacity • Increased water infiltration • Improved soil C sequestration • Reduced labour/fuel costs
	High levels of water use (e.g. irrigation)	Water conservation/collection management	• High yields compared to rainfed (e.g. rice) • Pest control	• Improved water quality and quantity • Reduced flooding • Reduced dryland salinity • Reduced depletion/pollution of ground water • Reduced soil erosion
Aquaculture	Chemicals for fish disease prevention and control Intensified use of supplementary feeds and feeding practices	Emphasis on improving pond management and husbandry Pond systems optimise use of natural fertility and feeding practices	• Short-term fix approach • Easily followed instructions • Formulated feeds increasingly available in many countries	• Reduces cost and improved product quality • Fewer downstream environmental impacts • Fewer human health risks • Reduced costs • Reduced risk for farmers • More efficient use of nutrients • Less effluent and waste • Reduced reliance on external feed inputs for pond fertility
	Monoculture over polyculture (specialisation in fewer species)	Polyculture systems, integrated farming, efficient use of pond niches and resources Multiple harvest of different crops	• Economic forecasting and cropping patterns more easily managed	• Increased risks associated with specialisation (e.g. disease risks, economic risks for small farmers)

Fisheries	High levels of water use	Water conservation, integration of fish culture into existing water uses	Water quality maintenance easier with flow through systems	• Improved water-use efficiencies
	Small number of premium target species managed for maximum yield	Broad spectrum of species targeted with emphasis on lower food chain, higher productivity species	Managers comfortable with traditional management systems; Science for single species management simpler and tested	• Uses properties and productivity of system to optimise yield and food supply • Redistributes pressure away from vulnerable target species
	Quotas, catch shares, and management systems targeted at maximising efficiency	Participatory management systems prioritise food security and livelihoods for the most vulnerable	Ownership with a few larger operators, simplifying inspection and management; Benefits easier to account for	• Fisheries serve broader societal objectives of resilient food systems and well-being • Stakeholder participation allows effective feedback on management objectives
	Technology targeted at increasing selectivity and catch efficiency	Technologies such as FADs that promote resilience among small-scale fishers (SSF) and equity in access to benefits	Attainable for larger operators with higher capital capacity	• More equitable distribution of ecosystem benefits • Greater contribution of fish resources to well-being
	Management that mandates or incentivises dead bycatch discards	Management that mandates retention and use of bycatch	Simplifies management, enforcement and handling of catch	• Optimises use of fish resources
	Government subsidies for fuel and inputs	Subsidies only considered if improve governance or post-harvest quality of catch	Politically palatable as seen to be improving the lives of fishers	• Maximises benefits from subsidies to improve sustainable management and food supply/safety

terrestrial cropping systems. Much less work has been conducted in aquatic production systems, despite a general agreement that impacts are likely to be severe (Jackson et al., 2001; Gessner et al., 2004). Yet, there are an increasing number of approaches where efforts are being made to intensify production in ways that are more environmentally responsible and aimed at harnessing, rather than depleting, the natural capital of the system. Boxes 5.2 and 5.3 describe technologies used in AAS in Timor-Leste and Bangladesh, respectively, aimed at sustainably intensifying productivity in systems where water is the dominant feature.

Box 5.2   Intensification through governance and technology innovation in marine aquatic agricultural systems (AAS)

Developing coastal aquaculture in cages or ponds, or enhancing wild stocks (stock enhancement or sea ranching) through release of juveniles from hatchery systems can be considered marine correlates of terrestrial agriculture. While these systems offer promise for increased production, the barriers of high establishment costs and technical complexity in many cases exclude the most vulnerable people from engaging. More accessible, and often not requiring new investments or technologically complex solutions at the operator level, are opportunities to intensify the flow of ecosystem services to the poor through changes in governance and technologies in coastal fisheries. Rebuilding overharvested stocks through investments by governments or external actors in improved resource management can, on varying scales, substantially increase harvestable resources (Worm et al., 2009). Examples of successful rebuilding in the developing world are, however, rare. Providing exclusive access rights to small-scale fishers over nearshore areas can increase yield by redressing issues of competition with larger, more efficient commercial vessels. While such zones exist on paper in a number of developing countries (e.g. Vietnam, Philippines), ineffective enforcement has limited their impact. There is growing evidence that management promoting less-targeted harvesting, and instead focusing on balancing harvest according to productivity, could improve sustainability outcomes while substantially increasing yields from marine fisheries (Zhou et al., 2015).

Among technological innovations, nearshore fish aggregating devices (FADs) show promise in increasing yields, diversity and sustainability of catch for small-scale and subsistence fishers (see Figure 5.3). Nearshore FADs are simple-technology anchored buoy systems designed to attract and aggregate pelagic fish close to shore, increasing catchability. While FADs do not increase biological productivity of the system, under effective management they have the potential to sustainably improve yields for small-scale fishers, and improve resilience of food systems for the coastal poor.

Figure 5.3 Fishers in Timor-Leste inspect a nearshore fish aggregating device
Source: Photo courtesy of David Mills.

Box 5.3   Intensification in Bangladesh through ghers

The coastal southwest of Bangladesh is one of the country's poorest areas. During the 1970s it experienced a growth in the export-oriented pro-duction of black tiger shrimp (*Penaeus monodon*) and giant freshwater prawn (*Macrobrachium rosenbergii*). However, the resulting expansion of salinisation of large areas of former rice land has negatively impacted the physical, social and natural environment. In response, many paddy fields have been converted to ghers to enable increased productivity of shrimp or prawn farming, and provide other ecosystem services. Ghers are produced by constructing peripheral trenches around an aquaculture pond to provide shelter for stocked aquatic animals and to prevent flood-ing or escape (see Figure 5.4). Whilst in high salinity regions agricultural biodiversity is low and generally characterised by shrimp-fish farming, in intermediate salinities integrated prawn-shrimp-rice-vegetable farming is practiced, and in low saline areas integrated giant freshwater prawn-fish-rice-vegetable farming systems are also common. While these three systems fall roughly along an increasing salinity gradient that runs from inland to coastal areas, a great deal of heterogeneity can be observed among nearby villages, and even within the same village, depending on the conditions of local or 'micro' agro-ecologies. Consequently, the approach is highly adaptable to a range of conditions and enables integration of different production elements.

Perhaps inevitably, due to the technological focus of many production solutions to global food security, sustainable intensification discourse often focuses upon ameliorating environmental impacts. However, in order to meet demands for food security, increases in food production through intensification oblige coactions with social aspects of food systems. So far, little attention has been given to the impacts of production changes on various stakeholders. Similarly under-studied are the institutional arrangements for managing such change, particularly given the often sectoral nature of many management bodies associated with AAS.

In addition to the diversity of biophysical elements in the system, and the implications of this for production, the often complex and community-specific

Figure 5.4 A gher system in southwest Bangladesh

Source: Photo courtesy of Michael Phillips.

social interactions and governance arrangements play an important role. For instance, in many communities women may be marginalised in terms of land ownership (Antwi-Agyei et al., 2015), access to resources for production or access to benefits of production (e.g. Gladwin et al., 2001). Such inequitable access, even at intra-household levels, can lead to other implications, such as child malnutrition. Therefore, the intensification of production that results in increasing yields may not necessarily bring equal benefits to different members of communities. Accordingly, any intensification of production needs to ensure that social equity, distribution, rights and access to food are all accounted for (Loos et al., 2014). This remains a critical area of research, in order to understand how sustainable intensification can be promoted in a more socially equitable manner.

The need for a systems approach to aquatic agricultural systems management

Due to the complex array of ecological, social and production element interactions and cascades, there is an urgent need to adopt a systems approach to research, engagement, development and management actions in AAS. Conventional approaches to intensification, which largely focus on increasing yields, carry a considerable risk of negatively impacting on other system elements (Tscharntke et al., 2012; Hochman et al., 2013). Integrated approaches that explicitly consider and address the many synergies and trade-offs among different land and water uses in aquatic agricultural systems are therefore required (Kremen and Miles, 2012).

Any approach to broadscale intensification of production in AAS must be sensitive to their ecological fragility, and the highly interdependent nature of different elements of the system, both biophysically and socially. For instance, Table 5.1 offers a simple comparison between a conventional approach to intensifying production in three broad activities in AAS, and integrated approaches to intensification, that seek to minimise environmental impacts and draw upon ecological processes and functions to sustainably increase yields. Whilst this analysis treats the three production systems as largely independent, there are clearly many interactions operating across land and water elements that support sustainable livelihoods. These interactions, in turn, provide opportunities for integrated management among different production elements. For instance, NPK fertiliser has a range of negative environmental impacts commonly associated with its application; in a more integrated system, manure from livestock or waste from aquaculture ponds could be utilised as a complete or partial substitute for inorganic chemical fertilisation.

In a more specific landscape example (Figures 5.1 and 5.2), integration of components could occur by harvesting fringing native vegetation for roof thatch. However, sustainable management of the resource requires that extraction rates are kept in balance with production so as to maintain not only thatch yields, but also the other ecosystem services associated with fringing vegetation, such as the natural filtration of waste water draining from agricultural land into adjacent waterways (Figures 5.1 and 5.2). Such an ecosystem service can prevent chronic sedimentation and turbidity in areas that support fish nurseries and feeding grounds. The ecological interactions among multiple livelihood activities in complex landscapes require that any attempt to increase production in one element of the system needs to be considered in respect to likely impacts on other system components. Such a holistic perspective becomes extremely important in multi-use and multi-user landscapes were trade-offs and synergies in the supply and use of ecosystem services may be widely distributed across many actors and livelihood activities. Whilst a conventional approach to intensification often does not incorporate this systems perspective of natural resources management, it is clearly needed in attempts to sustainably intensify production systems.

Conclusion

This chapter has shown how AAS consist of myriad biotic and abiotic components interacting across scales from the landscape to the individual farm or field. Managing these systems consequently presents a wide range of opportunities (e.g. ecosystem service synergies, land- and water-use efficiencies) and challenges (e.g. trade-offs and negative impacts of the management of one system element on another). An integrated systems research approach is essential for promoting effective and holistic management of these systems if they are to be sustainably intensified in ways that simultaneously deliver increased agricultural productivity, enhanced functioning of ecological processes and an equitable distribution of the benefits to those who use, manage and rely upon them for their livelihoods. The strong dependence of the poor and vulnerable upon ecosystem services in many AAS also argues for an approach to intensification that is both sensitive to the needs and opportunities of the poor, and to their participation in research and development processes.

Note

1 Organisation for Economic Development and Cooperation

References

Antwi-Agyei, P., Dougill, A.J., and Stringer, L.C. (2015) 'Impacts of land tenure arrangements on the adaptive capacity of marginalized groups: The case of Ghana's Ejura Sekyedumase and Bongo districts', *Land Use Policy*, vol 49, pp. 203–212.

Attwood, S.J., Maron, M., House, A.P.N., and Zammit, C. (2008) 'Do arthropod assemblages display globally consistent responses to intensified agricultural land use and management?' *Global Ecology and Biogeography*, vol 17, no 5, pp. 585–599.

Baidu-Forson, J.J., Phiri, N., Ngu'ni, D., Mulele, S., Simainga, S., Situmo, J., Ndiyoi, M., Wahl, C., Gambone, F., Mulanda, A., and Syatwinda, G. (2014) 'Assessment of agrobiodiversity resources in the Borotse flood plain, Zambia', CGIAR Research Program on Aquatic Agricultural Systems, Penang, Malaysia. *Working Paper: AAS-2014–12.*

Bajželj, B., Richards, K.S., Allwood, J.M., Smith, P., Dennis, J.S., Curmi, E., and Gilligan, C.A. (2014) 'Importance of food-demand management for climate mitigation', *Nature Climate Change*, vol 4, no 10, pp. 924–929.

Béné, C., and Teoh, S.J. (2014) 'Estimating the numbers of poor living in aquatic agricultural systems', WorldFish, Penang, Malaysia, unpublished.

Benton, T.G., Vickery, J.A., and Wilson, J.D. (2003) 'Farmland biodiversity: is habitat heterogeneity the key?' *Trends in Ecology and Evolution*, vol 18, no 4, pp. 182–188.

Bharucha, Z., and Pretty, J. (2010) 'The roles and values of wild foods in agricultural systems', *Philosophical Transactions of the Royal Society B: Biological Sciences*, vol 365, no 1554, pp. 2913–2926.

Birch, J.C., Thapa, I., Balmford, A., Bradbury, R.B., Brown, C., Butchart, S.H., Gurung, H., Hughes, F.M., Mulligan, M., Pandeya, B., and Peh, K.S.H. (2014) 'What benefits do community forests provide, and to whom? A rapid assessment of ecosystem services from a Himalayan forest, Nepal', *Ecosystem Services*, vol 8, pp. 118–127.

Boserup, E. (1965) *The conditions of agricultural growth: the economics of agrarian change under population pressure*, George Allen and Unwin, London.

Brodie, J., Fabricius, K., De'ath, G., and Okaji, K. (2005) 'Are increased nutrient inputs responsible for more outbreaks of crown-of-thorns starfish? An appraisal of the evidence', *Marine Pollution Bulletin*, vol 51, no 1, pp. 266–278.

Brummett, R.E., Beveridge, M., and Cowx, I.G. (2012) 'Functional aquatic ecosystems, inland fisheries and the millennium development goals', *Fish and Fisheries*, vol 14, no 3, pp. 312–324.

Castine, S.A., Sellamuttu, S.S., Cohen, P., Chandrabalan, D., and Phillips, M. (2013) 'Increasing productivity and improving livelihoods in aquatic agricultural systems: a review of interventions', CGIAR Research Program on Aquatic Agricultural Systems, Penang, Malaysia. *Working Paper: AAS-2013–30.*

Estrada-Carmona, N. (2014) 'Using participatory mapping with a gender lens to understand how landscapes are used for nutrition'. [Online] Available from http://www.a4nh.cgiar.org/2014/11/06/participatory-mapping-with-a-gender-lens/ [Accessed: 04/12/2015].

Foley, J.A., Ramankutty, N., Brauman, K.A., Cassidy, E.S., Gerber, J.S., Johnston, M., Mueller, N.D., O'Connell, C., Ray, D.K., West, P.C., and Balzer, C. (2011) 'Solutions for a cultivated planet', *Nature*, vol 478, no 7369, pp. 337–342.

Garnett, T., and Godfray, C. (2012) 'Sustainable intensification in agriculture. Navigating a course through competing food system priorities', *Food Climate Research Network and the Oxford Martin Programme on the Future of Food*, University of Oxford, UK, p. 51.

Gerland, P., Raftery, A.E., Ševčíková, H., Li, N., Gu, D., Spoorenberg, T., Alkema, L., Fosdick, B.K., Chunn, J., Lalic, N., and Bay, G. (2014) 'World population stabilization unlikely this century', *Science*, vol 346, no 6206, pp. 234–237.

Gessner, M.O., Inchausti, P., Persson, L., Raffaelli, D.G., and Giller, P.S. (2004) 'Biodiversity effects on ecosystem functioning: insights from aquatic systems', *Oikos*, vol 104, no 3, pp. 419–422.

Gladwin, C.H., Thomson, A.M., Peterson, J.S., and Anderson, A.S. (2001) 'Addressing food security in Africa via multiple livelihood strategies of women farmers', *Food Policy*, vol 26, no 2, pp. 177–207.

Godfray, H.C., Beddington, J.R., Crute, I.R., Haddad, L., Lawrence, D., Muir, J.F., Pretty, J., Robinson, S., Thomas, S.M., and Toulmin, C. (2010) 'Food security: the challenge of feeding 9 billion people', *Science*, vol 327, no 5967, pp. 812–818.

Gordon, L.J., Finlayson, C.M., and Falkenmark, M. (2010). 'Managing water in agriculture for food production and other ecosystem services', *Agricultural Water Management*, vol 97, no 4, pp. 512–519.

Hall, S. (2013) *Blue frontiers: managing the environmental costs of aquaculture*, WorldFish, Penang Malaysia.

Halwart, M. (2006) 'Biodiversity and nutrition in rice-based aquatic ecosystems', *Journal of Food Composition and Analysis*, vol 19, no 6, pp. 747–751.

Halwart, M. (2008) 'Biodiversity, nutrition and livelihoods in aquatic rice-based ecosystems', *Biodiversity*, vol 9, pp. 36–40.

Hochman, Z., Carberry, P.S., Robertson, M.J., Gaydon, D.S., Bell, L.W., and McIntosh, P.C. (2013) 'Prospects for ecological intensification of Australian agriculture', *European Journal of Agronomy*, vol 44, pp. 109–123.

Jackson, J.B., Kirby, M.X., Berger, W.H., Bjorndal, K.A., Botsford, L.W., Bourque, B.J., Bradbury, R.H., Cooke, R., Erlandson, J., Estes, J.A., and Hughes, T.P. (2001) 'Historical overfishing and the recent collapse of coastal ecosystems', *Science*, vol 293, no 5530, pp. 629–637.

Kastner, T., Rivas, M.J.I., Koch, W., and Nonhebel, S. (2012) 'Global changes in diets and the consequences for land requirements for food', *Proceedings of the National Academy of Sciences*, vol 109, no 18, pp. 6868–6872.

Kremen, C., and Miles, A. (2012) 'Ecosystem services in biologically diversified versus conventional farming systems: benefits, externalities, and trade-offs', *Ecology and Society*, vol 17, no 4, p. 40.

Lambin, E.F., Gibbs, H.K., Ferreira, L., Grau, R., Mayaux, P., Meyfroidt, P., Morton, D.C., Rudel, T.K., Gasparri, I., and Munger, J. (2013) 'Estimating the world's potentially available cropland using a bottom-up approach', *Global Environmental Change*, vol 23, no 5, pp. 892–901.

Laurance, W.F., Sayer, J., and Cassman, K.G. (2014) 'Agricultural expansion and its impacts on tropical nature', *Trends in Ecology and Evolution*, vol 29, no 2, pp. 107–116.

Loos, J., Abson, D.J., Chappell, M.J., Hanspach, J., Mikulcak, F., Tichit, M., and Fischer, J. (2014) 'Putting meaning back into "sustainable intensification"', *Frontiers in Ecology and the Environment*, vol 12, no 6, pp. 356–361.

Luckett, B.G., DeClerck, F.A., Fanzo, J., Mundorf, A.R., and Rose, D. (2015) 'Application of the nutrition functional diversity indicator to assess food system contributions to dietary diversity and sustainable diets of Malawian households', *Public Health Nutrition*, vol 18, no 13, pp. 2479–2487.

Mueller, N.D., Gerber, J.S., Johnston, M., Ray, D.K., Ramankutty, N., and Foley, J.A. (2012) 'Closing yield gaps through nutrient and water management', *Nature*, vol 490, no 7419, pp. 254–257.

Pinstrup-Andersen, P. (2009) 'Food security: definition and measurement', *Food Security*, vol 1, pp. 5–7.

Power, A.G. (2010) 'Ecosystem services and agriculture: tradeoffs and synergies', *Philosophical Transactions of the Royal Society B*, vol 365, pp. 2959–2971.

Rosegrant, M.W., and Cline, S.A. (2003) 'Global food security: challenges and policies', *Science*, vol 302, no 5652, pp. 1917–1919.

The Royal Society (2009) 'Reaping the benefits: science and the sustainable intensification of global agriculture'. Ref 11/09. [Online] Available from http://royalsociety.org/Reapingthebenefits/ [Accessed: 25/11/2015].

Tilman, D., and Clark, M. (2014) 'Global diets link environmental sustainability and human health', *Nature*, vol 515, no 7528, pp. 518–522.

Troell, M., Naylor, R.L., Metian, M., Beveridge, M., Tyedmers, P.H., Folke, C., Arrow, K.J., Barrett, S., Crépin, A.S., Ehrlich, P.R., and Gren, Å. (2014) 'Does aquaculture add resilience to the global food system?' *Proceedings of the National Academy of Sciences*, vol 111, no 37, pp. 13257–13263.

Tscharntke, T., Clough, Y., Wanger, T.C., Jackson, L., Motzke, I., Perfecto, I., Vandermeer, J., and Whitbread, A. (2012) 'Global food security, biodiversity conservation and the future of agricultural intensification', *Biological Conservation*, vol 151, no 1, pp. 53–59.

Weeratunge, N., Chiuta, T.M., Choudhury, A., Ferrer, A., Husken, S.M.C., Kura, Y., Kusakabe, K., Madzudzo, E., Maetala, R., Naved, R., Schwarz, A., and Kantor, P. (2012) 'Transforming aquatic agricultural systems towards gender equality: A five country review', World-Fish, Penang, Malaysia, *Working Paper: AAS-2012–21.*

WorldFish (2011) 'CGIAR Research Program: Aquatic Agricultural Systems. *Program Brief.' WorldFish, Penang, Malaysia.*

Worm, B., Hilborn, R., Baum, J.K., Branch, T.A., Collie, J.S., Costello, C., Fogarty, M.J., Fulton, E.A., Hutchings, J.A., Jennings, S., and Jensen, O.P. (2009) 'Rebuilding global fisheries', *Science*, vol 325, no 5940, pp. 578–585.

Wright, H.L., Lake, I.R., and Dolman, P.M. (2012) 'Agriculture – a key element for conservation in the developing world', *Conservation Letters*, vol 5, no 1, pp. 11–19.

Zhou, S., Smith, A.D., and Knudsen, E.E., (2015) 'Ending overfishing while catching more fish', *Fish and Fisheries*, vol 16, no 4, pp. 716–722.

6 What works where for which farm household?

Rapid approaches to food availability analysis

Randall S. Ritzema, Romain Frelat, James Hammond and Mark T. van Wijk

Introduction

Despite decades of development investment, pervasive poverty, food insecurity and malnutrition are still evident in many areas throughout the developing world (UNDP, 2014), particularly in rural areas. Lack of access to food of adequate quantity or quality is both manifested and often measured in the context of the smallholder household. Although these rural households are characterized by high degrees of poverty, they are also generally viewed as possessing the potential to meet both rural and urban food needs in the future, in the face of continuing population growth, climate change and resource degradation (Vanlauwe et al., 2014).

Researchers and development planners thus often seek to identify on-farm and off-farm interventions that effect change at the household level, and for cost efficiency reasons try to do so in a way that produces wide-scale impact. However, formulating comprehensive recommendations that decrease household food insecurity or poverty at wide scales is not an easy task (Chikowo et al., 2014; Giller et al., 2011). The heterogeneity of household food security status, composition and livelihood strategies, as well as the agroecological and socioeconomic contexts in which these households are situated (Tittonell, 2014; Tittonell et al., 2010) complicate strategy formulation (Vanlauwe et al., 2014). Options that work well for some farm households may not work well for others, for many reasons. Thus, the challenge facing development planning is to 'customize' intervention strategies to fit local situations while extracting lessons that are applicable at broader scales, thereby increasing both the effectiveness and efficiency of the planning process.

Quantitative analysis plays a crucial role in both characterizing household food and nutritional status, and assessing the potential benefits of proposed interventions. Ex ante analysis using household models (van Wijk et al., 2014) is a common approach. These models attempt to represent complex and integrated household processes, and are useful for in-depth exploration of household decision making and resource allocation in response to resource constraints (Herrero et al., 2014; van Wijk et al., 2014). Limited research resources often force modelling analyses to focus on a small number of farm 'types' extracted from household typology analysis (Cortez-Arriola et al., 2014; Klapwijk, Bucagu et al., 2014). The effects of agricultural and other livelihood interventions can be tested by contrasting modelled household welfare with the implemented intervention against the 'baseline' household condition. Extrapolation of results is considered appropriate to the extent that the households in the study population resemble the analyzed household types.

However, though commonly applied, household models have high data demands, and are time-consuming and costly to develop (Thornton and Herrero, 2001). Furthermore, comprehensive datasets are needed to drive household models, and thus modelling analyses are often site-specific and yield results that are difficult to extrapolate (van Wijk et al., 2014). Large-scale analyses, in particular, require sizeable input datasets, but few are adequately comprehensive to provide informative results at regional scales or above (van Wijk, 2014; van Wijk et al., 2014). Combining datasets and subsequently deriving household typologies to reach higher scales is often awkward, as differing datasets tend to vary widely in sample sizes, content, assumptions and data quality. Alternatively, analysis performed on collectively comprehensive but disaggregated datasets requires adaptation of the household model to each dataset, further increasing model development demands.

The use of farm typologies can also be problematic. Household livelihood strategies, even within a small sample of households, can be highly diverse. Analysis of representative household types thus results in a loss of nuanced information on individual farm households (Herrero et al., 2014). The use of household types does not consider or highlight successful strategies already being implemented by insightful and progressive individual farm households. Secondary issues arise when testing interventions: it is difficult to estimate intervention effects on the study population (such as estimated shifts in aggregate household poverty) by using a household model, and uncertainties arise when especially successful interventions could potentially change membership in the defined household types.

This chapter describes a different quantitative approach for characterizing livelihoods and assessing potential wide-scale impact of interventions that complements detailed household modelling. This approach answers a different set of research questions than household modelling and can be applied across diverse agroecological and socioeconomic contexts. Two linked studies on East and West Africa farming systems illustrate the utility of the approach (Frelat et al., 2015; Ritzema et al., 2016). The discussion then considers how results inform intervention strategy, and how use of this approach has catalyzed the development of advanced data collection tools and associated analysis software that enhances the utility of the method.

A 'complementary approach' for household-level analyses

The methodology we forward in this chapter can be considered an 'inverse' of the typical approach of applying (1) complex household modelling to (2) a limited number of household types. Rather, it applies (1) simple calculation schemes that generate indicator values for (2) every household contained within one or more household datasets.

Simple calculation schemes do not fully capture household processes and are thus not intended to accurately portray household behaviour in the way that more complex household models attempt. However, the computational simplicity of the method enables rapid development of analytical frameworks and eases model adaptation requirements between differing datasets. It can also be rapidly deployed to test indicator responses under proposed interventions. Because of reduced development demands, results from this simple scheme can therefore provide guidance in early stages of intervention strategy formulation, as well as identify key areas for deeper analysis using more complex modelling approaches. In this way, the simple scheme described here complements complex household modelling, and the two approaches can be used in tandem to enhance household analysis.

This approach uses parsimonious sets of indicators and associated input variables, limiting the number of required parameters to those commonly gathered by household surveys. Thus, data typically included in otherwise dissimilar datasets can often be aggregated to increase household representation at higher scales and enhance the ability to formulate wide-scale recommendations.

Because indicator results are calculated for each household, detailed information on 'what works where for which farm household' is readily obtained, and households of particular interest can be identified for further detailed exploration. Alternatively, results can be presented using descriptors of the sample population of households, including statistics or distributions of indicator values and household characteristics.

Indicator values can be used descriptively, e.g. to establish a baseline characterization of a sample of households. Furthermore, the effects of proposed interventions can be assessed by adjusting appropriate parameter values, recalculating the indicators of interest and contrasting indicator values under the interventions with baseline values.

Examples: Food availability analyses in sub-Saharan Africa

Two example studies explore the utility of this approach to household analysis. Both studies focus on 'food availability' (FA) as a component of the broader and more complex concept of food security (Coates, 2013; Headey and Ecker, 2013) and encompass large geographic areas with diverse agroecological and socioeconomic characteristics. Ritzema et al. (2016) studied 1,800 households in 9 sites in 7 countries across East and West Africa, and Frelat et al. (2015) combined 6 differing datasets containing data on more than 13,500 households in 77 sites in 15 countries in sub-Saharan Africa.

The studies applied the method outlined in this chapter and built on previous research by Hengsdijk et al. (2014) by calculating a similarly-defined FA indicator for every household within the studies' respective datasets. The indicators essentially estimated the average daily food energy potentially available for each farm household member. Available food energy was calculated as the sum of food crop and livestock production (with food energy content differentiated by crop and livestock type) for consumption or the market, and cash crop production and off-farm income that contribute to cash reserves. Cash reserves did not include expenses, and were assumed to be available to purchase the staple crop identified for each research site. Household size was quantified in terms of Male Adult Equivalents (MAE). The indicator is thus an analytical simplification, as it does not consider dietary or nutritional diversity, but is restricted to food energy content. Both studies then estimated the fraction of households that have inadequate or sufficient food availability to meet household needs.

Though using similar FA indicators at the household level, the studies applied the indicators differently. In Ritzema et al. (2016), indicator values, as well as the relative contribution of the six livelihood components to those values, were calculated for each household. Figure 6.1 displays an example for one study site: Lushoto, Tanzania.

A sensitivity analysis tested the effect of two categories of agricultural productivity interventions (a 50% increase in crop production and a 50% increase in the production of livestock products) alongside an off-farm intervention (a USD 200 increase in off-farm income per year) on FAI values (Table 6.1, with Figure 6.1[b] showing the crop production boost results as an example).

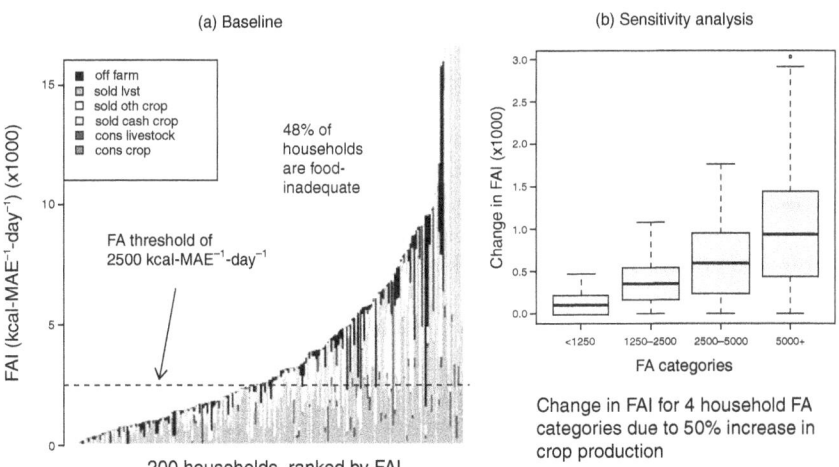

Figure 6.1 Lushoto food availability index (FAI) distribution and intervention analysis. Results are shown for baseline and sensitivity analysis for 200 households in Lushoto, Tanzania. Food Availability is abbreviated as 'FA', and the Food Availability Index as 'FAI'. Figure (a) shows FAI baseline results, as well as potential food sources contributing to FAI values. Households with FAI values less than the FA threshold of 2,500 kcal-MAE^{-1}-day^{-1} are considered 'food-inadequate'. Figure (b) shows potential shifts in FAI for four categories of households, grouped according to baseline FAI ranges, from a hypothetical 50% increase in staple crop production

Source: Adapted from Ritzema et al., 2016.

Table 6.1 Sensitivity analysis results. Results show the percentage of households with Food Availability Index baseline values below the threshold of 2,500 kcal-MAE^{-1}-day^{-1}. Three options are considered in the sensitivity analysis: increasing staple crop production by 50% (Crop Boost), increasing production of all livestock products by 50% (Livestock Boost), and increasing off-farm income by USD 200 (Off-Farm Boost). (Details on study sites can be found in Rufino et al., 2012)

Site	% of households that are food-inadequate			
	Baseline	*Crop Boost: 50%*	*Livestock Boost: 50%*	*Off-Farm Boost: USD 200*
Yatenga, Burkina Faso	23.5	22.0	23.0	10.0
Borana, Ethiopia	46.5	42.0	43.2	19.0
Lawra, Ghana	48.2	41.7	47.7	18.6
Nyando, Kenya	35.0	30.5	31.5	18.5
Wote, Kenya	40.2	35.7	35.7	14.6
Kaffrine, Senegal	24.0	18.5	23.5	14.0
Lushoto, Tanzania	47.5	40.2	46.7	16.1
Hoima, Uganda	28.0	24.0	27.5	2.5
Rakai, Uganda	19.0	14.5	19.0	8.5

Source: Adapted from Ritzema et al., 2016.

Livelihood strategies across the FAI distribution in the Lushoto study site were highly diverse and shifted with increasing FA (Figure 6.1[a]). Distribution characteristics provide key insights into the correlation between livelihood sources and food availability: households indicated a relatively stronger reliance on crop production for consumption at the lower end of the FA spectrum, and shifted toward market-oriented crop and livestock production and off-farm sources at higher levels of food adequacy. The key finding in this study was that interventions focused on raising crop and livestock productivity, while benefitting food-adequate households to some degree, only minimally benefit the most food-inadequate households (Figure 6.1 and Table 6.1). Other types of interventions, such as increasing off-farm income opportunities, are required to reach these households.

Frelat et al. (2015) did not focus on interventions, but on the correlations between household food availability and farm characteristics. The study quantified the contribution of different household activities on the overall FA across sub-Saharan Africa. Crop production was the major source of food energy, comprising 60% of FA. The off-farm income contribution to FA ranged from 12% for food-inadequate households to 27% of income for the 58% of food-adequate households.

As reported in Frelat et al. (2015), the correlation between FA and self-assessed food security status, calculated on more than 8,000 households, was statistically significant with a correlation coefficient of 0.29 across the datasets ($p < 0.05$). These results support the conclusion that despite simplified underlying assumptions, FA gave reasonable insight regarding the overall food security status of individual farm households. This was confirmed in more recent farm household survey work in Guatemala, Tanzania and Vietnam, where FA had a significant positive relationship with household-level diet diversity and the USAID Household Food Insecurity Access Scale.

The food availability status for 72% of the households could be correctly predicted with an Artificial Neural Network (ANN) mini-model using only three explanatory variables: household size, number of livestock and land area. The ANN response model was used to predict the food availability 'frontier' using a livestock–cropland threshold curve that shifted upwards with increasing family size (Figure 6.2). Based on the number of livestock and household size, a land size threshold value could thus be defined, above which a smallholder farm is likely to be able to produce enough food and cash to feed the family. This mini-model is particularly useful as these three variables can be easily and rapidly collected for large numbers of households, in contrast to variables like productivity, consumption and sales, which require implementation of detailed survey instruments.

The relationships in Figure 6.2 were strongly affected by market access: when farmers have good market access, the size of the farm needed to produce and purchase enough food to feed the family can be small. Similar to Ritzema et al. (2016), this analysis also suggested that targeting poverty through improving market access and off-farm opportunities is a better strategy to improve food security than focusing solely on agricultural production.

Findings in both studies offer an alternative perspective to the premise that agricultural intensification and productivity increases, e.g. closing yield gaps (Loos et al., 2014; Sumberg, 2012), will be the primary means of improving

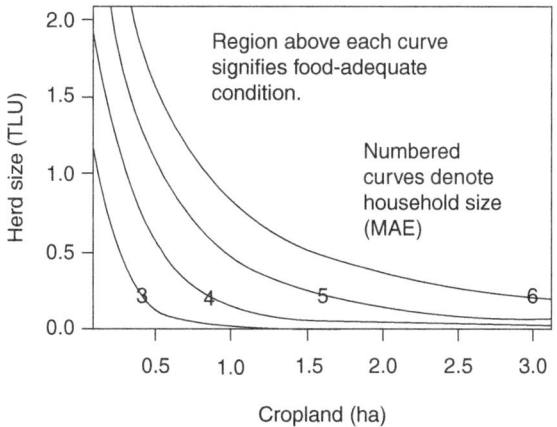

Figure 6.2 Predicted food availability threshold curves. Curves indicate family size, ranging from 3 to 6 Male Adult Equivalents. In general, households with a herd size and cropland that placed them above their respective threshold curve were found to be 'food-adequate'

Source: Adapted from Frelat et al., 2015.

food security in sub-Saharan Africa. While increased production is important from a societal point of view, the studies suggest that other non-agricultural and off-farm options, with associated policy support, are needed to reach the segments of rural society that are most food-insecure.

Back to the beginning: From analysis to data collection

As outlined in the example studies on sub-Saharan Africa above, the use of simple calculation schemes on individual households, though applied with necessary simplifications, can provide key insights for intervention strategy and policy formulation. The studies demonstrated the utility of the approach in assessing the 'performance' of farm households in terms of food availability, and in elucidating the links between productivity and production strategies vs. food security and nutrition.

However, successful analysis depends upon data of sufficient quantity and quality. Effective baseline characterization of the current capacity of farming households to meet their nutritional and livelihood needs, for example, is essential to identify drivers of farm household performance and to assess how different intervention options might improve that performance. These characterizations aspire to capture how households balance multiple factors and objectives, not only farm productivity. Typically, trade-offs and causal links exist between these indicators and objectives (Klapwijk, van Wijk et al., 2014), and these trade-offs might differ according to the local context and by the scale at which they are considered. At the farm system level, links between e.g. agricultural production and food security, or agricultural production and nutrition, might appear straightforward: higher production leads to improved food security and nutrition. However, analysis considering the perspective of the smallholder household often reveals the relationship to be more complex. For example, increased production of food crops through improved varieties (e.g. maize) may lead to declines in the overall nutritional values of diets of the rural poor if the new crop replaces original varieties with higher nutritional content (Welch and Graham, 1999). The measurement of poverty and food security, and their links to agriculture, face similar challenges, most notably data quality issues (Tiffen, 2003) and the absence of comparable indicators and monitoring systems. The analyses by Frelat et al. (2015) showed large differences among different survey instruments, with lack of standardization of indicators, thereby leading to an analysis in which only a small amount of the information collected during lengthy surveys could actually be used for cross-site comparisons.

These challenges in systems characterization, in combination with the findings of the FA indicators in Frelat et al. (2015) and Ritzema et al. (2016), catalyzed the development of a new flexible and modular multi-indicator survey framework designed for use at the farm household level called Rural Household Multi-Indicator Survey (RHoMIS) (Hammond et al., 2016). RHoMIS uses the information needed for an FA indicator, here labelled Food Availability Ratio (FAR), as a basis and puts these in the context of a series of other farm household performance indicators. The survey framework makes use of recent developments in on-line survey and statistical analysis software to expedite data collection processes.

The design of the RHoMIS tool followed five principles:

i) The survey had to be rapid enough to avoid participant fatigue and accompanying declines in data quality.
ii) The survey had to be utilitarian, i.e. only questions that contributed to a subsequent pre-defined analysis were included, in order to minimize superfluous data collection.
iii) The survey had to be user-friendly, so that all participants in the process of collecting and analyzing data could perform the tasks with minimum hassle, leading to higher data quality.
iv) The survey had to be flexible, so that it could be modified easily to suit the local context of the farming systems and farm households where it would be deployed.
v) The survey had to be robust and reliable, relying as much as possible on knowledge that is inherent in respondents' work routines, and that respondents are ready to share.

The survey was trialled in two contrasting agro-ecosystems, in the Lushoto district of Tanzania (n = 151) and in the Guatemalan Trifinio region (n = 285), giving quantitative insight into the fundamental relationships between livelihood strategies, agricultural production, food security, nutrition, gender equity and greenhouse gas emissions among smallholders. Findings showed that the Guatemalan site had more households in poverty, greater experience of food insecurity and lower dietary diversity than the Tanzanian site. Despite this, the households in Tanzania had access to, on average, fewer calories per person per day than the Guatemalan site. The major differences in the farming systems were more significant livestock ownership and market orientation in the Tanzanian site. Detailed analysis of the Tanzania site identified 'positive deviants' within the smallholder farm household population with successful coping strategies who scored highly in dietary diversity despite low food availability (less than 2,500 kcal per day). Male-headed households in which the male was usually away scored significantly lower on various performance indicators than female-headed households, suggesting the former may benefit from targeted agricultural interventions. A key finding is that in both locations, the FAR strongly covaried with the USAID FANTA Hunger and Food Insecurity Access Scale (HFIAS) indicator (Coates et al., 2007) and household level dietary diversity (Coates, 2013), showing that despite the assumptions and simplifications used in developing the FAR, it gives a fairly accurate indication of the food security status of smallholder farm households. Both HFIAS and diet diversity exhibited a linear relationship to FAR values of up to 4,000–5,000 kcal per MAE per day in both locations, suggesting that the FAR can be used for cross-site comparisons of food security status. Above FAR values of 4,000–5,000 kcal per MAE per day, FAR dissociates from the other food security indicators, showing that above this threshold other factors like education, access to food, and personal preferences and incentives become more important in driving food security and nutrition rather than the availability of agricultural produce and cash.

Conclusion

Quantitative analysis of farm households informs the intervention strategy formulation process by characterizing household livelihoods, as well as by assessing how proposed interventions might improve household food availability. However, smallholder households exhibit high variability and diversity in food availability levels, livelihood strategies and agricultural system characteristics. Datasets describing these systems and supporting food security analysis at higher scales are likewise of variable quality and availability. These realities heighten the need for analytical approaches that address both complexities.

This chapter has described a method of applying simple indicator-based analysis to every household in large household datasets. The method allows for analysis to be developed and applied rapidly across diverse contexts and datasets, forming a strong analytical base for large-scale studies and potentially providing guidance in early stages of intervention and policy development. The approach furthermore retains the information content in household distributions, thereby clarifying the importance of various livelihood sources across households and sites, providing initial indications concerning how proposed interventions will affect households across those distributions, elucidating cross-site and intra-site patterns, and identifying areas for further in-depth analysis using household models.

Household modelling can yield detailed understanding of how farm households behave, but the resource-intensive model development process and site-specific nature of household modelling places constraints on the ability to apply those results at higher scales. This research suggests that the method proposed here can be coupled with both household modelling and advanced data collection tools such as RHoMIS to form a robust analytical platform to support large-scale intervention strategy formulation that is both effective and efficient.

As outlined in Ritzema et al. (2016), further advancements are envisioned for the method described here. First, the incorporation of simple optimization routines will improve the explanatory power of indicator values by better incorporating household resource allocation processes into indicator valuations. Second, risk analysis capabilities will be enhanced through explicit inclusion of factor uncertainty and variability.

Further developments are envisioned for the RHoMIS technology as well, including broadening the scope of possible farm household level analyses by: (i) incorporating more indicators in the survey framework; (ii) applying RHoMIS in more sites in Central America, sub-Saharan Africa and Southeast Asia, and thereby building a library of standardized surveys and databases; (iii) expanding the existing farm household database used in the FAR analyses by including results from more projects as well as the World Bank's Living Standards Measurement Study – Integrated Surveys on Agriculture (LSMS-ISA) initiative; and

(iv) continuing to develop standardized data analysis procedures such as those in Hammond et al. (2016). Another priority is to embed farm household characterization exercises like the RHoMIS framework into ongoing efforts to outscale agricultural intervention options and into monitoring and evaluation frameworks. Rapid and automated tools like RHoMIS and FA indicators can furthermore accelerate the monitoring and evaluation cycle, thereby ensuring that key performance indicators for productivity, food security and nutrition are quantified on a regular basis and fed into project adaptation processes.

References

Chikowo, R., Zingore, S., Snapp, S. and Johnston, A. (2014) 'Farm typologies, soil fertility variability and nutrient management in smallholder farming in Sub-Saharan Africa', *Nutrient Cycling in Agroecosystems*, 100(1), pp. 1–18.

Coates, J. (2013) 'Build it back better: Deconstructing food security for improved measurement and action', *Global Food Security*, 2(3), pp. 188–194.

Coates, J., Swindale, A. and Bilinsky, P. (2007) *Household Food Insecurity Access Scale (HFIAS) for Measurement of Food Access: Indicator Guide*, Washington, DC: USAID.

Cortez-Arriola, J., Groot, J.C.J., Massiotti, R.D.A., Scholberg, J.M.S., Aguayo, D.V.M., Tittonell, P. and Rossing, W.A.H. (2014) 'Resource use efficiency and farm productivity gaps of smallholder dairy farming in North-west Michoacan, Mexico', *Agricultural Systems*, 126, pp. 15–24.

Frelat, R., Lopez-Ridaura, S., Giller, K.E., Herrero, M., Douxchamps, S., Djurfeldt, A., Erenstein, O., Henderson, B., Kassie, M., Paul, B., Rigolot, C., Ritzema, R.S., Rodriguez, D., van Asten, P. and van Wijk, M.T. (2015) 'Drivers of household food availability in subSaharan Africa based on big data from small farms', *Proceedings of the National Academy of Sciences of the United States of America*, 113(2), pp. 458–463.

Giller, K.E., Tittonell, P., Rufino, M.C., van Wijk, M.T., Zingore, S., Mapfumo, P., Adjei-Nsiah, S., Herrero, M., Chikowo, R., Corbeels, M., Rowe, E.C., Baijukya, F., Mwijage, A., Smith, J., Yeboah, E., van der Burg, W.J., Sanogo, O.M., Misiko, M., de Ridder, N., Karanja, S., Kaizzi, C., K'Ungu, J., Mwale, M., Nwaga, D., Pacini, C. and Vanlauwe, B. (2011) 'Communicating complexity: integrated assessment of trade-offs concerning soil fertility management within African farming systems to support innovation and development', *Agricultural Systems*, 104(2), pp. 191–203.

Hammond, J., Fraval, S., van Etten, J., Suchini, J.G., Mercado, L., Pagella, T., Frelat, R., Lannerstad, M., Douxchamps, S., Teufel, N., Valbuena, D. and van Wijk, M.T. (2016) 'The Rural Household Multi-Indicator Survey (RHoMIS) for rapid characterisation of households to inform climate smart agriculture interventions: description and applications in East Africa and Central America', *Agricultural Systems* (in press). Available online at: http://www.sciencedirect.com/science/article/pii/S0308521X16301172.

Headey, D. and Ecker, O. (2013) 'Rethinking the measurement of food security: from first principles to best practice', *Food Security*, 5(3), pp. 327–343.

Hengsdijk, H., Franke, A.C., van Wijk, M.T. and Giller, K.E. (2014) *How Small Is Beautiful? Food Self-Sufficiency and Land Gap Analysis of Smallholders in Humid and Semi-Arid Sub Saharan Africa*, Wageningen, The Netharlands: Wageningen UR.

Herrero, M., Thornton, P.K., Bernués, A., Baltenweck, I., Vervoort, J., van de Steeg, J., Makokha, S., van Wijk, M.T., Karanja, S., Rufino, M.C. and Staal, S.J. (2014) 'Exploring future changes in smallholder farming systems by linking socio-economic scenarios with regional and household models', *Global Environmental Change*, 24(1), pp. 165–182.

Klapwijk, C.J., Bucagu, C., van Wijk, M.T., Udo, H.M.J., Vanlauwe, B., Munyanziza, E. and Giller, K.E. (2014) 'The 'one cow per poor family' programme: current and potential fodder availability within smallholder farming systems in southwest Rwanda', *Agricultural Systems*, 131, pp. 11–22.

Klapwijk, C.J., van Wijk, M.T., Rosenstock, T.S., van Asten, P.J.A., Thornton, P.K. and Giller, K.E. (2014) 'Analysis of trade-offs in agricultural systems: current status and way forward', *Current Opinion in Environmental Sustainability*, 6, pp. 110–115.

Loos, J., Abson, D.J., Chappell, M.J., Hanspach, J., Mikulcak, F., Tichit, M. and Fischer, J. (2014) 'Putting meaning back into "sustainable intensification"', *Frontiers in Ecology and the Environment*, 12(6), pp. 356–361.

Ritzema, R.S., Frelat, R., Douxchamps, S., Silvestri, S., Rufino, M.C., Herrero, M., Giller, K.E., Lopez-Ridaura, S., Teufel, N., Paul, B. and van Wijk, M.T. (2016) 'Is production intensification likely to make farm households food-adequate? A simple food availability analysis across smallholder farming systems from East and West Africa', submitted to *Food Security*.

Rufino, M.C., Quiros, C., Boureima, M., Desta, S., Douxchamps, S., Herrero, M., Kiplimo, J., Lamissa, D., Joash, M., Moussa, A.S., Naab, J., Ndour, Y., Sayula, G., Silvestri, S., Singh, D., Teufel, N. and Wanyama, I. (2012) *Developing generic tools for characterizing agricultural systems for climate and global change studies (IMPACTlite-phase 2)*, CGIAR Research Program on Climate Change, Agriculture and Food Security (CCAFS).

Sumberg, J. (2012) 'Mind the (yield) gap(s)', *Food Security*, 4(4), pp. 509–518.

Thornton, P.K. and Herrero, M. (2001) 'Integrated crop-livestock simulation models for scenario analysis and impact assessment', *Agricultural Systems*, 70(2–3), pp. 581–602.

Tiffen, M. (2003) 'Transition in sub-Saharan Africa: agriculture, urbanization and income growth', *World Development*, 31(8), pp. 1343–1366.

Tittonell, P. (2014) 'Livelihood strategies, resilience and transformability in African agroecosystems', *Agricultural Systems*, 126, pp. 3–14.

Tittonell, P., Muriuki, A., Shepherd, K.D., Mugendi, D., Kaizzi, K.C., Okeyo, J., Verchot, L., Coe, R. and Vanlauwe, B. (2010) 'The diversity of rural livelihoods and their influence on soil fertility in agricultural systems of East Africa – a typology of smallholder farms', *Agricultural Systems*, 103(2), pp. 83–97.

UNDP (2014) *Human Development Report 2014. Sustaining Human Progress: Reducing Vulnerabilities and Building Resilience*, New York: United Nations Development Programme.

van Wijk, M.T. (2014) 'From global economic modelling to household level analyses of food security and sustainability: How big is the gap and can we bridge it?', *Food Policy*, 49(P2), pp. 378–388.

van Wijk, M.T., Rufino, M.C., Enahoro, D., Parsons, D., Silvestri, S., Valdivia, R.O. and Herrero, M. (2014) 'Farm household models to analyse food security in a changing climate: a review', *Global Food Security*, 3(2), pp. 77–84.

Vanlauwe, B., Coyne, D., Gockowski, J., Hauser, S., Huising, J., Masso, C., Nziguheba, G., Schut, M. and Van Asten, P. (2014) 'Sustainable intensification and the African smallholder farmer', *Current Opinion in Environmental Sustainability*, 8, pp. 15–22.

Welch, R.M. and Graham, R.D. (1999) 'A new paradigm for world agriculture: meeting human needs – Productive, sustainable, nutritious', *Field Crops Research*, 60(1–2), pp. 1–10.

7 Approaches to operationalizing integrated systems research

Hervé D. Bisseleua and Ann Degrande

Introduction

Today's rapidly changing world faces greater challenges than ever before, such as climate change, hunger, malnutrition and natural resources degradation (IPCC, 2014; FAO, 2015). Increasing demand for food, expected to double worldwide by 2050, calls for sustainable intensification of agricultural production (Garnett et al., 2013;Vanlauwe et al., 2014). Food security and food autonomy are critical where the poor live, which is often within landscapes that paradoxically are rich in biodiversity (Henry et al., 2009; Maas et al., 2013). However, 90% of farmers in the developing world, who farm less than 2 ha of land, strongly rely on this biodiversity and associated ecological processes for agricultural production (beneficial trophic interactions, soil food webs, stress-adapted crop genotypes) (Barrios, 2007; Bisseleua et al., 2009). Hence, ensuring sustainable and affordable food production in different sites with diverse production systems and livelihood strategies remains a huge challenge. Nevertheless, some options, considered 'sustainable', are put forward, from sustainable intensification and climate-smart agriculture (CSA) to agro-ecological intensification (Tittonell, 2013). *Sustainable intensification* focuses on optimizing yields from a defined area of land while reducing environmental impacts and enhancing environmental services (Vanlauwe et al., 2014). *Climate-smart agriculture* aims to sustainably increase agricultural productivity and incomes, adapt and build resilience to climate change and reduce or remove greenhouse gas emissions, when and where appropriate (Tittonell, 2013). *Ecological intensification* is an alternative concept to achieve higher yields by enhancing ecological processes such as biological pest control or pollination with known beneficial effects on crop yields rather than relying on chemical inputs (Bommarco et al., 2013).

While some of the above approaches may address inequality and foster empowerment of smallholder farmers as the most decisive actor in the search for solutions, the debate about which approach is most appropriate often puts alternatives against each other. In reality, a combination of approaches is necessary in addressing the complexity of smallholder livelihood systems and food production. Unfortunately, the needs of farmers living in such complex environments are obviously not adequately taken into account in mainstream research and extension systems, whose agendas are mainly determined by researchers' perspectives, and often target one commodity or specific components of the system, rather than looking at the diverse and integrated nature of the prevalent farming systems and their interconnections to smallholders' livelihoods. There is thus need to redefine the research agenda, so that greater focus is put on the emerging role and importance of smallholders in food production and food autonomy and the multifunctionality of small-scale farms, recognizing that there exist overlapping environmental services and related stakes, such as cultural ecosystem services, planned and associated biodiversity, ecological benefits and payments for ecosystems services (Tscharnkte et al., 2012).

We therefore argue that the 'agricultural research for development' paradigm, aiming at enhancing the livelihoods of smallholder farmers, must emphasize an increase in agro-ecological capacity to support resilient and productive smallholder systems through integrated systems approaches. Moving to more integrated systems research also requires adapting the research agenda in order to address increasing economic disparities, social segregation and resulting high poverty levels. In this respect, there is a need to better understand the conditions and processes that lead to disparities and social segregation and the role of interventions (services and markets) to counter these negative developments. Finally, a redefined research agenda is needed to promote rural development through endogenous solutions. This will entail giving serious and adequate attention to local cases (context-driven) and meso-scale studies, building collective action through participatory action research and developing novel partnerships using systems thinking. Both interdisciplinary and transdisciplinary research are required in an integrated systems research model. *Interdisciplinarity* involves the combining of two or more academic disciplines into one activity (e.g. a research project). It is about creating something new by crossing boundaries, and thinking across them. *Transdisciplinarity* connotes a research strategy that crosses many disciplinary boundaries to create a holistic approach. It applies to research efforts focused on problems that cross the boundaries of two or more disciplines.

Research institutions at national and international level, policy makers, agro-industry, non-governmental organizations and civil society now have a strong will to collaborate and jointly come up with solutions, which marks a clear shift from 'agricultural research' or 'agricultural research and development' (ARD) to 'agricultural research for development' (AR4D) and 'integrated agricultural research for development' (IAR4D) (Hawkins et al., 2009). Within this changing context, since 2008 the CGIAR[1] is engaged in a restructuring process moving from the 'traditional' mandate to increase the productivity of crops, livestock, trees and fish and to improve the management of natural resources, to a renewed and expanded research effort. The new vision is not merely about obtaining higher yields from improved varieties and practices, but puts greater emphasis on new themes such as climate-smart[2] and nutrition-sensitive agriculture, faster adoption of new technologies, higher profitability from the small farm and food processing sectors, as well as on cross-cutting issues such as gender and youth,[3] policies and institutions[4] and capacity development[5] (Lundy et al., 2012; Leeuwis et al., 2014). The CGIAR restructuring process resulted in three systems CRPs where research adopts a systems approach, encompassing the full range of intervention points from soil–plant–water–livestock relationships to markets and value chains: i.e. Humidtropics, Dryland Systems and Aquatic Agricultural Systems.[6] These CRPs integrate social and biophysical sciences with the use of both local and expert knowledge to understand and solve complex problems affecting lives and livelihoods in these systems. This is being achieved through integrated systems approaches in *Research for Development* (R4D), with a focus on sustainable intensification and capacity to innovate through partnerships and broad stakeholder participation. In these programs, a number of CGIAR Centres are brought together to jointly develop integrated systems research approaches and engage with partners at various levels towards their implementation. These research activities implemented through strategic platforms (R4D) and operational innovation platforms (IPs) go beyond individual research action and single component focus. They rather use a new mode of operation that brings groups of partners together to work on commonly identified challenges in a way that goes beyond individual partners' capacities. This calls for a whole-system (holistic) perspective and analysis, integrated and inter-/transdisciplinary research approaches and new institutional arrangements (at landscape level and beyond). In these processes, farmers and their communities are at the centre to plan, analyze, test and implement – in collaboration with development partners – feasible combinations of technical, market, governance and policy options capable of improving livelihoods. By improving the understanding of place-based social, financial, technical and environmental contexts, integrated systems research provides a knowledge resource to enhance the targeting and relevance of potential systems interventions, with an aim to scale these out to similar extrapolation domains. The approach also entails further development of monitoring and evaluation methods with indicators that show whether systems approaches are working, for whom, where and to what extent.[7]

Systems research is not the exclusive 'right' of Systems CRPs. Within the international research systems paradigm, systems research should be seen as functioning at varying levels throughout the entire agri-food system, with built-in interconnections. This, however, requires a massive mind-set change, including competency assessment and enhancement, and strengthened strategic partnerships related to areas of integrated systems research and analysis along the impact pathway.

This chapter presents a framework for this integrated systems research and then demonstrates its utility through case studies where the approach has been applied. From this, the chapter draws lessons for practitioners of research for development, in particular for those working in integrated systems research for agricultural transformation.

What do we understand by integrated systems approach?

From the above, it is obvious that the complexity of smallholder agricultural systems calls for inter-/transdisciplinary and integrated approaches to tackle the diversity of socio-economic and biophysical constraints to farming and livelihood systems. Integrated approaches are expected to identify, quantify and address a complex set of interactions[8] that shape and constrain farming systems and natural resources integrity (Bawden, 1996; Röling and Jiggins, 1998; Barrios, 2007; Bellon and Hemptinne, 2012; Berkhout et al., 2015). They should also be able to foster an enabling environment through institutions and delivery mechanisms that support the scaling of successful innovations, and influence the policy environment in which the system operates (Hounkonnou et al., 2012; Schut, Klerkx, Sartas et al., 2015). Policy related constraints such as limited access to markets, state control over production technologies and marketing facilities (roads, storage infrastructures), price volatility, poor service delivery, lack of frameworks for producer associations and inadequate finance, can affect farmers' ability to benefit from opportunities created by integrated systems research, and should be given due attention through continuous interactions and dialogue with policy makers (Renkow and Byerlee, 2014).

The application of innovative system thinking is imperative in systems research to cope with productivity trade-offs and synergies, climate change, land degradation, gender inequities and youth unemployment at the expected scale of impact (i.e. millions of farmers across millions of hectares of land). Integrated systems research for development is expected to directly improve the effectiveness of development spending at local scales, at the same time as producing generalizable knowledge and forging new partnerships that will impact livelihoods. Research on foresight, synergies and trade-offs, for example, are core components that help prioritize interventions and predict possible early successes (Darnhofer et al., 2012). These approaches may be used to assess and refine a pre-selected innovation or to compare alternative innovations and inform choices; or when used in combination, can help to integrate, synthesize, evaluate and pre-select alternatives.

In contrast to the traditional linear 'research for development' approach,[9] integrated systems research (Table 7.1) puts research within the context of development practices and engages with farmers, their communities and a range of development partners, in a medium to long term co-learning agenda using multi-stakeholder processes (Coe et al., 2014). This involves working with all interested stakeholders to define what impacts are needed. This is then followed by defining what outcomes will deliver these impacts (indicators and targets), what outputs (if adopted) will produce the desired outcomes and finally what research will lead to these outputs (i.e. following the impact pathway backwards) (Van Mierlo et al., 2010; Neef and Neubert, 2011). Such iterative and interactive learning, and action-based participatory research and development processes, should continue to improve options available to farmers and to better localize and contextualize their needs, making research more demand-driven, focused and result-oriented (Bawden, 2010). The iteration helps to develop options targeting users' needs with greater impacts on poverty reduction, food security and the resource integrity.

Table 7.1 Comparison of the systems approach with conventional approaches

Conventional approaches	Integrated systems research approach
Focus is on single commodities and single livelihood components	Focus is on farming *systems* and livelihood *portfolios*
Aimed at improving productivity and closing yield gaps, regardless of risk	Explicit consideration of trade-offs among multiple aims – improving productivity, reducing risk, and social, economic, and environmental sustainability. Aimed at multiple wins where possible, or balance among trade-offs where not
Focus is on discrete value chains, overlooking externalities	Attention given to interactions between value chains, explicitly considering externalities
Focus is on innovations and investments responding to specific drivers of change within sectors at discrete scales	Focus is on interactions between multiple drivers of change, and innovation and investment options across sectors and scales
Linear, research-*for*-development approaches	Iterative research-*in*-development approach
Mono- or multi-disciplinary	Inter- or trans-disciplinary

Source: CGIAR Research Program on Dryland Systems (2015), available at http://mel.cgiar.org/xmlui/handle/20.500.11766/4505 (accessed October 2015).

The principles, trade-offs and challenges of integrated systems research

Partnerships dimension

Changing individual and collective decision-making in agricultural systems requires new incentives and forms of information, but also conducive processes of engagement, learning and interaction with partners. Approaches that produce research outputs and support scaling of innovation are proliferous (Schut, Klerkx, Sartas et al., 2015), and range from classical public extension and private engagement, to multi-stakeholder IPs and learning alliances, as well as innovative forms of public–private partnerships, engagement with civil society organizations and media, and innovative funding models (e.g. impact investments). Moreover, within such approaches, different strategies, methodologies and mechanisms are used. Each of these approaches uses different models and mechanisms with their specific strengths and weaknesses, making them more or less appropriate in a specific problem and/or governance context. What is particularly important though is a clear definition of the roles of each and every partner, and the new and conducive institutional arrangements in different stages of the impact pathway (van Huis, 2012). For instance, Humidtropics systems CRP recognizes different types of partners: (1) science partners providing key science expertise on livelihood systems in the broadest sense; (2) national R4D actors co-leading multi-stakeholder network processes (e.g.

platforms) where investment priorities, implementation, tracking and learning will take place; (3) locally operating next users with interest in scaling up and out innovations; and (4) advocacy, media and ICT partners for knowledge-sharing (communication), awareness and policy engagement. Humidtropics employs structured multi-stakeholder platforms to drive research for development programs and agricultural innovation systems with active participation of key stakeholders in program delivery to ensure sustainability and scalability of successful innovations. This entails R4D platforms at the country level, composed of different stakeholder groups (farmers, donors and development organizations, private sector, government, researchers, media) which assist in guiding research for development interventions, including removal of constraints for the scaling of technological and institutional innovation. The operational IPs on the ground (at field site level), designed by the R4D platforms, are mechanisms for joint exploration of prioritized constraints and opportunities, and joint implementation of research for development activities (Figure 7.1). Such new institutional arrangements allow farmers and other stakeholders to become experts, instead of 'users' or 'adopters', of scientific recommendations.

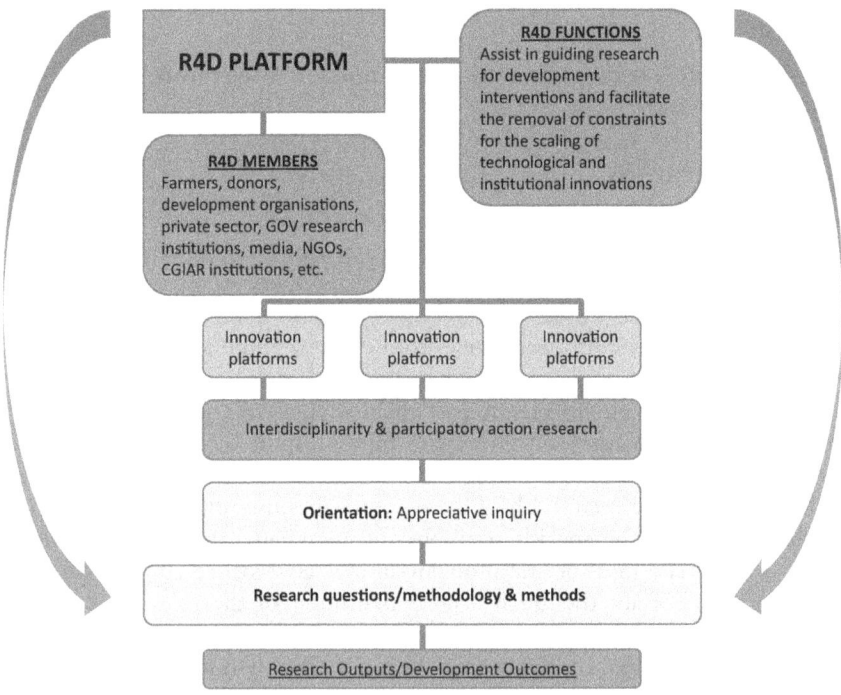

Figure 7.1 Humidtropics multi-stakeholders' process applied in action areas for integrated systems research for agricultural development

Gender dimension

Gender plays a significant role in integrated systems research because gender norms and divisions of labour can be considered social institutions that pose both opportunities and constraints to addressing challenges in the agricultural transformation. Women and youth are more vulnerable and worst affected by land- and labour-related issues. Also, women and youth are often net suppliers of labour into production systems, but seldom control resources and benefits from these systems (Filmer and Fox, 2014). Even when young females are able to move up the ladder in the educational system, unemployment rates are often higher than those of their male counterparts. Therefore, in integrated systems research, special attention should be given to women and youth through:

- open recognition of social and gender differentiation as an integral component of a new theory of scaling;
- particular attention to specific gender and youth issues and information in the development and testing of decision support tools;
- development and testing of institutional innovations beneficial for women and youth;
- gender and youth sensitive assessment and evaluation of different scaling interventions and the development of decision frameworks.

Humidtropics systems CRP uses gender-sensitive systems analysis to explore the similarities and the differences between men's, women's and youth's use of resources, access to benefits and experiences at the R4D and IPs. This aims to effectively mainstream gender dimensions into every component of research and development activities and to prioritize appropriate gender-sensitive solutions that increase benefits from innovations. The gender dimension is mainstreamed into every component of systems improvement research targeting social and technical systems innovations through the following: engage women and youth in on-farm experimentation; integrate gender issues using gender-sensitive qualitative and quantitative data collection and analysis methods; and take into account the gender dimension in results-based monitoring and evaluation.

Institutional arrangements dimension

Institutions are both formal and informal arrangements that orient human behaviour and interactions (Brouwer et al., 2015). Where existing interaction and decision patterns tend to (re)produce poverty, low productivity, degradation, malnutrition and inequality, institutional innovations are needed to modify incentive patterns in livelihood systems. Agricultural sciences, including in international agricultural research organizations, have a strong reputation in developing and testing new technologies and technical management practices. In comparison, the capacity to develop and test new institutional arrangements and options (linked to the broader integrated systems research agenda) is much weaker, even though diagnostic research often points to a primacy of institutional constraints (Schut, Rodenburg et al., 2015). In the social sciences, too, there is still limited understanding of the interplay between technical opportunities and institutional change (Leeuwis, 2013). Although appreciation for the systemic approach is growing within the scientific community, it is clear that disciplinary approaches focusing on transfer of technology still dominate. Integrated systems research arose as a reaction to the inadequacies of discipline research based on narrow disciplinary perspectives, and the increased recognition that farmers and societal actors (e.g. local people) are inside the system boundary (Darnhofer et al., 2012). The aggregated potential benefits of disciplinary approaches remain to some extent mitigated because of the wrong measures of performance (for researchers and influential stakeholders). In addition, weak institutional arrangements, coupled with professionals lacking the appropriate mix of capabilities in system enquiry and inter- and transdisciplinary research modes, failed to secure an ongoing learning and adaptation paradigm. This situation should be remedied without losing the benefits of disciplinary expertise, as long as future practice is primarily context (rather than discipline) focused. In addition, globalization of agricultural research for development is leading to pervasive institutional arrangements, such as universal metrics, that create or have the potential to create a form of systemic failure (Ison, 2012: pp. 154). Integrated systems research, as currently practiced by most research and development institutions, still lacks conducive institutional settings for ongoing success and must therefore be re-framed. Applied systems thinking should be extended to both the object of study (farms, rural territories, civic food networks, etc.) and the research setting. Conducive institutional settings should be implemented to incentivize researchers who engage in integrated systems research, or any such practice, to comply with the academic merit system and thus further their career possibilities. However, this is seldom the case as institutions continue to support and reward disciplinary more than interdisciplinary scientific outputs. For integrated systems research to be effective and efficient there must be new institutional innovations central and not peripheral. This must be supported by new capacities for trans- and interdisciplinary systemic action research.

The Humidtropics experience and role in integrated systems research

Humidtropics – the CGIAR Research Programme (CRP) on Integrated Systems for the Humid Tropics – was initiated in June 2012, with research activities beginning in 2013. The goal of Humidtropics was to improve overall agricultural productivity of poor smallholder farmers within the humid tropics, and transform lives of rural poor in target regions, while ensuring that environmental and natural resources integrity are respected. This was achieved through integrated systems approaches in R4D, with a focus on sustainable intensification and capacity to innovate through partnerships and broad stakeholder participation. Within the context of the CGIAR, the Programme aims at bringing together a number of Centres to develop these integrated systems research approaches and engage with partners at various levels towards their implementation. The intention is to go beyond individual research action and single component focus in research, to a new mode of operation that brings groups of partners together to work on commonly identified challenges in a way that goes beyond individual partner capacities. The logic of the Programme, how it has been implemented in action areas in Africa, Asia and Latin America and experiences from multi-stakeholder platform processes are further described by Hiwasaki et al. (this volume). To realize its objectives, Humidtropics adopted three main Strategic Research Themes (SRTs) to underpin the research process:

• *Systems Analysis and Global Synthesis* that explores the baseline situation and synthesizes progress towards the expected outcome situation;
• *Integrated Systems Improvement* that involves researching and mainstreaming promising systems interventions related to productivity, natural resources management, markets and institutions;
• *Scaling and Institutional Innovation* that aims to improve stakeholders' capacity to innovate and to support the scaling of interventions developed under the theme described above at farm, national and global levels.

Research and development actors need each other to achieve impact at scale. However, collaboration between different stakeholders in agricultural research for development (AR4D) has been insufficient so far. Humidtropics has built on the Forum for Agricultural Research in Africa (FARA) innovation platform initiatives and successes (Adekunle et al., 2012) to pilot a two-level Multi-Stakeholder Process (MSPr), i.e. R4D Platforms and Innovation Platforms (IP), to develop joint action and science-based solutions through integrated agricultural systems research. The implementation of this approach starts with the establishment of a R4D platform, composed of representatives from farmer organizations, private sector, government, research institutions and civil society. This is followed by the identification of entry themes for R4D interventions in agriculture. Humidtropics MSPr then used a Rapid Appraisal of Agricultural Innovation Systems (RAAIS) to identify entry points, research topics and research questions in sustainable agriculture and natural resources management (Schut, Klerkx, Rodenburg et al., 2015). Stakeholders involved in the different platforms (R4D and IP) agree on ways to work together while implementing integrated systems research. IPs are the operational bodies where the different stakeholders jointly carry out R4D activities. The experience of Humidtropics using MSPr shows that they could play a strategic role in the transformation of Africa agriculture and are potentially solid mechanisms to operationalize the Comprehensive Africa Agriculture Development Programme (CAADP) and the National Agriculture and Food Security Investment Plan (NAFSIP) process and priorities at country level.

Humidtropics design: From plot to landscape

Agroecosystem design and the development of appropriate methods are core field activities of Humidtropics. In practical terms, the research on 'whole farm productivity improvements' includes diversification of existing farming systems and integrating legumes, livestock, fruit trees and vegetables for increased income and more diverse diets. Platform-based research activities also require collaboration with other CRPs (RTB[10] and Maize, for example) on social science challenges such as a 'gender norms' study. Research on scaling and institutional innovation is based on various dimensions. For example, in Nicaragua research is carried out to understand incentive systems for scaling organic farming practices through in-depth monitoring of IPs and social processes across geographies, with a focus on social networks, decision-making, participation, learning, power and ownership, and investments. Elsewhere, tools to facilitate, track and scale participatory multi-stakeholder networks and policy engagement are developed. For example, the bilateral project on policy action on sustainable intensification of cropping systems (PASIC – www.pasic.ug) in Uganda sees IITA, CIP and IFPRI scientists, in collaboration with the Ministry of Agriculture Animal Industry and Fisheries in Uganda (MAAIF), developing scientific evidence to develop Zonal Investment Plans (ZIP) for the potato-based farming systems of the South-west Uganda highlands.

In the Cote d'Ivoire and Cameroon Action Site, a participatory action research method was developed that includes 'Participatory Trial Design' (ParTriDes) workshops and 'mother and baby trials' (Snapp, 2002) to facilitate the large-scale testing of best-bet sustainable intensification and diversification options, co-developed with farmers and relevant stakeholders across heterogeneous sites. In Cote d'Ivoire, ParTriDes focused on farmer–researcher collaboration through multi-stakeholder process, with emphasis on developing innovations for women farmers. The objective was to test and evaluate improved soil fertility management techniques for sustainable food crop production in the South-West of Côte d'Ivoire. Experimentation was based on the 'mother and baby' trial design, which makes it possible to collect quantitative data from mother trials (managed by researchers) and farmers' perception and evaluation of the technology from the baby trials (managed by farmers). The 'mother' plots tested technologies to intensify cassava and maize cropping (comparing local and improved planting material of cassava/maize in different soil fertilization methods and association with legumes or leguminous trees). The 'mother' trial was a collective learning farm, set up by women groups and the research facilitator (R4D platform) on land belonging to the group. The 'baby' plots tested a subset of technologies against a control (farmers' normal practice), with each baby plot being a replicate of the mother trial. Women's perceptions were systematically collected together with agronomic and biological performances of the technologies. Methodical cross-checking of performance evaluation by researchers and farmers provided complementary rather than competing information. With this experience, the 'mother and baby' trial concept was found to be an efficient way to test the potential for widespread adoption of improved soil fertility management in cocoa-based systems in Cote d'Ivoire.

In Ethiopia, intensive fodder production on terraces, in flood plains or valleys is a common feature of the agricultural landscape. The association of fodder species such as *Chloris gayana*, *Brachia humidicola* and *Pennisetum pedicellatum* with teff (*Eragrostis tef*), maize, beans, sometimes groundnuts, potatoes or beans is a system that proves sustainable. Participatory research performed in collaboration with universities (Ambo University), local governments, national and international research institutions, the civil society, and local NGOs is aimed at optimizing such systems. Results showed an increasing complexity in the system, moving from sole fodder to maize intercropped with beans + soil bund + fodder + tree + legumes + sweet potatoes + potatoes, resulting in more than doubling crop and fodder yields. This leads to providing substantial amounts of protein and income to farmers, as well as environmental benefits in term of soil erosion control and incentives to invest in long term natural resources management.

Another example of integrated systems research for the design of complex adaptive systems is the integration of small ruminants and agroforestry practices in potato production systems in Rwanda to enhance soil conservation practices that will impact human nutrition and agricultural productivity which is performed in collaboration with national and international research institutions, local government, the civil society and universities. The design of integrated crops and free-range chicken production systems, using mobile chicken housing in Cote d'Ivoire, is another illustration of integrated systems research. Yet, trade-offs may emerge around the implementation of such complex management systems at the scales of the farm or the landscape, especially in situations of resource scarcity.

Humidtropics efforts to transit to 'results-based management' (RBM) was supported through pilot-funding by the CGIAR Consortium and helped to experiment with a new mode of financing implementation of integrated systems research directly through operational innovation and R4D platforms. These platform research projects are small projects designed based on RBM principles and with wider participation of stakeholders. They were building on systems approaches, including systems analysis, integrated systems improvement and development of a monitoring and evaluation (M&E) framework for improved monitoring of results, budgets and enhanced decision-making by different levels of management. About 24 such livelihood systems research projects were established by multi-stakeholder platforms under Humidtropics, with many addressing key challenges, such as: (1) developing interventions in North Nicaragua to make production systems more diverse and resilient; (2) improving 'whole farm' productivity by optimal enterprise combinations in the cocoa-based farming systems in West Africa (Ghana, Cote d'Ivoire, Cameroon and Nigeria); (3) improving agroforestry and Irish potato-based cropping system productivity as well as dietary diversity in tropical highlands of Northern Rwanda; (4) improving root/tuber/banana–legume–livestock integration in Mushinga, South-Kivu, in the Democratic Republic of Congo (DRC) for improved income and human and animal nutrition; and (5) enhancing livelihoods and better natural resources management through appropriate integration and diversification on smallholder farms in the Central Highlands of Vietnam.

Within Humidtropics, capacity development is driven by clearly defined objectives. Successful implementation of the strategy relies on the active and ongoing engagement of partners, both amongst the program partners and, more critically, with boundary partners and members of the multi-stakeholder networks/platforms. In line with its mandate, Humidtropics puts a strong emphasis on building capacity for integration of women and youth in the IAR4D process, and the mainstreaming of gender and youth strategies as part of interventions. Capacity development activities involve many partners spread across a wide range of geographical and subject areas. Appropriate learning and communication approaches are leveraged to ensure co-learning, information sharing and coordination across Humidtropics in the following areas:

- systems thinking and methodologies amongst partners and stakeholders;
- trade-offs and synergy analysis within systems research;
- innovation systems and institutional innovation analysis;
- facilitation for multi-stakeholder engagement, including R4D and IPs;
- action research and behavioural change methodologies focused on gender and youth;
- scaling approaches and processes, including key communication and engagement skills;
- support graduate training aligned with the research in action areas, through a postgraduate research fellowship scheme, preferably with local and international student 'blends';
- monitoring and evaluation for adaptive management and RBM.

These capacities are implemented at all levels and comprise many forms and methodologies, including workshops and training-of-trainers approach, e-learning and blended learning, coaching, mentoring and formal training. These activities also capitalize on ICT tools (e.g. Digital decision support tools) to develop novel ways for learning at scale.

Conclusion

Integrated system approaches with a livelihoods orientation to address the productivity and sustainability needs of smallholder farmers should include a broader engagement using a combination of interdisciplinary and farmer participatory approaches revolving around multi-stakeholder, multi-sector processes through new institutional arrangements such as R4D platforms and IPs. Capacity development should embrace all stakeholder categories from farmers to the private sector, research scientists, development partners, advisory services and policy makers. This implies greater emphasis on systems research and component research when prioritizing problems and also potential entry points for targeted interventions. Although there is a strong call to operationalize integrated systems research and support innovation in smallholder agricultural systems, the enforcement of these concepts is still dominated by disciplinary approaches focused on transfer of technology. This tension is having a negative impact on the capacity to fully operationalize integrated systems research for development as a coherent whole. As Cees Leeuwis states,

> Stakeholders involved in integrated systems research indirectly start from conflicting assumptions and views about how research may contribute to development, what kinds of public goods research may deliver, in which contexts place-based research should take place, and how (and by whom) research agendas should be set.[11]

Arguably, the enactment of integrated systems research from farm to landscape and beyond should strengthen the following areas:

1   Development of a strategy for achieving impact at scale for systems research by engaging integrated systems research processes with regional bodies and processes such as CAADP, the Common Market for East and Southern Africa (COMESA) and the United Nations Convention to Combat Desertification (UNCCD), as well as with government and donor development programs/projects and so forth.
2   Capacity building and competency development for the implementation of integrated systems research, including assessment of systems interactions, trade-off analysis and synergy development. Such capacity needs to be further strengthened, recognized and rewarded in National Agricultural Research and Education Systems (NARES), as well as in international agricultural research organizations.
3   Clearer guidelines and instruments on 'how to conduct and assess systems research' are needed to guide research groups and teams. This implies developing aggregate indices that demonstrate changes in performance of dominant systems, such as:

    a   Systems Productivity Index ($ and/or food)
    b   Systems Innovation Index (% knowledge gain)
    c   Systems Resilience Index (%)

4   Development of a toolbox such as the tools for systems analysis (TOSA) toolbox developed by Humidtropics in collaboration with partners and database on systems methodologies and tools, with examples of how they have been used in different situations, will contribute towards capacity development and conduct of systems research.
5   Increased emphasis is needed in institutional arrangements for conducting systems research, and in institutional innovation and scaling strategies.

Acknowledgments

This study forms part of the CGIAR Research Programme on Integrated Systems for the Humid Tropics (Humidtropics). Two anonymous reviewers gave valuable feedback that enabled us to improve the quality of the manuscript.

Notes

1 CGIAR, the Consultative Group on International Agricultural Research, composed of 15 Centres, is the only worldwide partnership addressing agricultural research for development, whose work contributes to the global effort to tackle poverty, hunger and major nutrition imbalances, and environmental degradation.
2 All research and development activities need to build in resilience to climate shocks and a focus on adaptation to and mitigation of climate change.
3 The main challenge here is to ensure that all research conducted by CGIAR and its partners is gender-sensitive and promotes gender equity – that is, it is adapted to both the needs and the aspirations of poor women.
4 This concerns the need to reform the policies and institutions that affect agri-food systems so that these become more conducive to pro-poor development and to the protection of natural resources.
5 There is a great need to strengthen capacity, both in the research and development organizations that are CGIAR's partners and also at a grassroots level.
6 http://www.cgiar.org/our-strategy/cgiar-research-programs/
7 http://www.managingforimpact.org/sites/default/files/case/wageningen_19_march_2015_eric_koper.pdf
8 Interactions between cropping and animal husbandry, between on- and off-farm work, between technologies and agro-ecosystems, or interactions between production methods and cultural landscapes, between economic incentives and farm diversity, between farmers and other rural actors
9 The impact pathway includes four steps: research activity, output, outcomes and impact
10 Roots, Tubers and Bananas
11 http://www.wageningenur.nl/en/activity/Cees-Leeuwis-KTI-Systems-research-in-the-CGIAR-as-a-multi-dimensional-arena-of-struggle.htm

References

Adekunle, A.A., Ellis-Jones, J., Ajibefun, I., Nyikal, R.A., Bangali, S., Fatunbi, O. and Ange, A. (2012) Agricultural innovation in sub-Saharan Africa: experiences from multiple-stakeholder approaches. Forum for Agricultural Research in Africa (FARA), Accra, Ghana.

Barrios, E. (2007) 'Soil biota, ecosystem services and land productivity', *Ecological Economics*, 64, pp. 269–285.

Bawden, R.J. (1996) 'On the systems dimension of FSR', *Journal of Farming Systems Research and Extension*, 5 (2), pp. 1–18.

Bawden, R.J. (2010) 'The community challenge: The learning response', in Blackmore, C. (ed.) *Social learning systems and communities of practice*. London, Springer, pp. 39–56.

Bellon, S. and Hemptinne, J.L. (2012) 'Redefining frontiers between farming systems and the environment', in Darnhofer, I., Gibbon, D. and Dedieu, B. (eds) *Farming systems research into the 21st century: the new dynamic*. Dordrecht, Springer, pp. 307–333.

Berkhout, E., Glover, D. and Kuyvenhoven, A. (2015) 'On-farm impact of the system of rice intensification (SRI): evidence and knowledge gaps', *Agricultural Systems*, 132, pp. 157–166.

Bisseleua, D.H.B., Missoup, A.D. and Vidal, S. (2009) 'Biodiversity conservation, ecosystem functioning and economic incentives under cocoa agroforestry intensification', *Conservation Biology*, 23 (5), pp. 1176–1184.

Bommarco, R., Kleijn, D. and Potts, S.G. (2013). 'Ecological intensification: harnessing ecosystem services for food security', *Trends in Ecology and Evolution*, 28 (4), pp. 230–238.

Brouwer, H., Woodhill, J., Hemmati, M., Verhoosel, K. and van Vugt, S. (2015) *The MSP guide: how to design and facilitate multi-stakeholder partnership*. Wageningen, Wageningen University, p. 188.

CGIAR Research program on drylands (2015) http://drylandsystems.cgiar.org/content/integrated-systems-approach (accessed on October 2015).

Coe, R., Sinclair, F. and Barrios, E. (2014) 'Scaling up agroforestry requires research 'in' rather than 'for' development', *Current Opinion in Environmental Sustainability*, 6, pp. 73–77.

Darnhofer, I., Gibbon, D. and Dedieu, B. (2012) 'Farming systems research: an approach to inquiry', in Darnhofer, I., Gibbon, D. and Dedieu, B. (eds) *Farming systems into the 21st century: the new dynamic*. Dordrecht, Springer, pp. 3–31.

Filmer, D. and Fox, L. (2014) *Youth employment in sub-Saharan Africa, Africa development series*. Washington, DC, World Bank.

Food and Agriculture Organization (FAO), International Fund for Agricultural Development (IFAD) and World Food Programme (WFP) (2015) The State of Food Insecurity in the World 2015, Meeting the 2015 international hunger targets: taking stock of uneven progress, Rome, FAO.

Garnett, T., Appleby, M.C., Balmford, I., Bateman, J., Benton, T.G., Bloomer, P., Burlingame, B., Dawkins, M., Dolan, L., Fraser, D., Herrero, M., Hoffman, P., Smith, P., Thornton, P.K., Toulmin, C., Vermeulen, S.J. and Godfray, H.C.J. (2013) 'Sustainable intensification in agriculture: premises and policies', *Science*, 314, pp. 33–34.

Hawkins, R., Heemskerk, W., Booth, R., Daane, J., Maatman, A. and Adekunle, A.A. (2009) 'Integrated Agricultural Research for Development (IAR4D)', A Concept Paper for the Forum for Agricultural Research in Africa (FARA) Sub-Saharan Africa Challenge Programme (SSA CP), FARA, Accra, Ghana.

Henry, M., Tittonell, P., Manlay, R.J., Bernoux, M., Albrecht, A. and Vanlauwe, B. (2009) 'Biodiversity, carbon stocks and sequestration potential in aboveground biomass in smallholder farming systems of western Kenya Agriculture', *Ecosystems and Environment*, 129, pp. 238–252.

Hiwasaki, L., Idrissou, L., Okafor, C. and van der Hoek, R. 'Constraints and opportunities in using multi-stakeholder processes to implement integrated agricultural systems research: the Humidtropics case', chapter 24 (this volume).

Hounkonnou, D., Kossou, D., Kuyper, T., Leeuwis, C., Nederlof, S., Röling, N., Sakyl-Dawson, O., Traore M. and van Huis A. (2012) 'An innovation systems approach to institutional change: Smallholder development in West Africa', *Agricultural Systems*, 108, pp. 74–84.

Intergovernmental Panel on Climate Change, 2014. Climate Change 2014: Synthesis Report. Contribution of Working Groups I, II and III to the Fifth Assessment Report of the Intergovernmental Panel on Climate Change [Core Writing Team, R.K. Pachauri and L.A. Meyer (eds.)], IPCC, Geneva, Switzerland, p. 151.

Ison, R.L. (2012) 'Systems practice: making the systems in farming systems research effective', in Darnhofer, I., Gibbon, D. and Dedieu, D. (eds.) *Farming systems research into the 21st century: the new dynamic*. Dordrecht, Springer, pp. 141–157.

Leeuwis, C. (2013) *Coupled Performance and Change in the Making*, Second inaugural lecture, Wageningen University.

Leeuwis, C., Schut, M., Waters-Bayer, A., Mur, R., Atta-Krah, K. and Douthwaite, B. (2014) *Capacity to innovate from a system CGIAR research program perspective*, Penang, Malaysia:

CGIAR Research Program on Aquatic Agricultural Systems. Program Brief AAS-2014–29, p. 5.

Lundy, M. et al. (2012) *LINK methodology: a participatory guide to business models that link small-holders to markets*, CIAT, Cali.

Maas, B., Clough, Y. and Tscharntke, T. (2013) 'Bats and birds increase crop yield in tropical agroforestry landscapes', *Ecology Letters*, 16, pp. 1480–1487.

Mierlo, B.C. van, Arkesteijn, M.C.M. and Leeuwis, C. (2010) 'Enhancing the reflexivity of system innovation projects with system analyses', *American Journal of Evaluation*, 31 (2), pp. 143–161.

Neef, A. and Neubert, D. (2011) 'Stakeholder participation in agricultural research projects: a conceptual framework for reflection and decision-making', *Agriculture and Human Values*, 28, pp. 179–194.

Renkow, M. and Byerlee, D. (2014) Assessing the Impact of Policy-Oriented Research: a Stocktaking, Paper prepared for a workshop on Assessing the Impact of Policy-Oriented Research, Washington, DC, November 11–12 2014.

Röling, N. and Jiggins, J. (1998) 'The ecological knowledge system', in Röling, N. and Wagermakers, M.A. (eds) *Facilitating sustainable agriculture: Participatory learning and adaptive management in times of environmental uncertainty*. Cambridge, Cambridge University Press, pp. 283–311.

Schut, M., Klerkx, L., Rodenburg, J., Kayeke, J., Raboanarielina, C., Hinnou, L.C., Adegbola, P.Y., van Ast, A. and Bastiaans, L. (2015) 'RAAIS: Rapid appraisal of agricultural innovation systems (Part I). A diagnostic tool for integrated analysis of complex problems and innovation capacity', *Agricultural Systems*, 132, pp. 1–11.

Schut, M., Klerkx, L., Sartas, M., Lamers, D., Mc Campbell, M., Obgonna, I., Kaushik, P., Atta-Krah, K. and Leeuwis, C. (2015) 'Innovation platforms: experiences with their institutional embedding in agricultural research for development', *Experimental Agriculture*, 44 (1), pp. 37–60. doi: 10.1017/S001447971500023X.

Schut, M., Rodenburg, J., Klerkx, L., Hinnou, L.C., Kayeke, J. and Bastiaans, L. (2015), 'Participatory appraisal of institutional and political constraints and opportunities for innovation to address parasitic weeds in rice', *Crop Protection*, 74, pp. 158–170.

Snapp, S. (2002). 'Quantifying farmer evaluation of technologies: the mother and baby trial design', in Bellon, M.R. and Reeves, J. (eds) *Quantitative analysis of data from participatory: methods in plant breeding*. Mexico, CIMMYT, pp. 9–16.

Tscharntke, T., Tylianakis, J.M., Rand, T.A., Didham, R.K., Fahrig, L., Batáry, P., Bengtsson, J., Clough, Y., Crist, O.T., Dormann, C.F., Ewers, R.W., Fründ, J., Holt, R.D., Holzschuh, A., Klein, A.M., Kleijn, D., Kremen, C., Landis, D.A., Laurance, W., Lindenmayer, D., Scherber, C., Sodhi, N., Steffan-Dewenter, I., Thies, C., van der Putten, W.H. and Westphal, C. (2012) 'Landscape moderation of biodiversity patterns and processes – eight hypotheses', *Biological Reviews*, 87, pp. 661–685.

Tittonell, P. (2013) *Farming Systems Ecology: Towards Ecological Intensification of World Agriculture*, inaugural lecture, Wageningen University Press, p. 44.

Van Huis, A., 2012 'An innovation systems approach to institutional change: Smallholder development in West Africa', *Agricultural Systems*, 108, pp. 74–83.

Vanlauwe, B., Coyne, D., Gockowski, J., Hauser, S., Huising, J., Masso, C., Nziguheba, G., Schut, M. and Van Asten, P. (2014) 'Sustainable intensification and the African smallholder farmer', *Current Opinion in Environmental Sustainability*, 8, pp. 15–22.

8 Systems thinking

Towards transformation in praxis and situations

Bernard Hubert and Ray Ison

Revisiting traditional paths of innovation

If the issue of innovation in agriculture is so difficult to address today, it is because there is no one clear objective, no road is signposted. If producing more remains a requirement, it is essential not only to produce more efficiently but also to produce other goods, tangible like yield improvement as well as intangible, e.g., cultural goods and symbols, like livestock in pastoral societies, gifts and exchanges of yam and tarots in the Pacific. Society's demands on agriculture are increasingly complex: from environmental services, inclusion of marginalized populations, and quality differentiation to revitalization of rural territories and energy production; i.e., demands are multifaceted, interdependent, and always historically and contextually situated. These growing demands challenge us to rethink the role and even the functions of the agricultural sector as well as the place of research in innovation. Human-induced climate change sits as a backdrop which demands a new beginning, the acknowledgment of uncertainty, rather than a clamouring for certainty.

Society's attitudes towards research are often contradictory: research is expected to provide the answers to humanity's problems, and society is disillusioned with research advances and rejects its value (see Abate et al., 2009; Beintema et al., 2009; Calvo et al., 2009; Kammili et al., 2009). For some people in society, the world of research has 'collaborated' with the productivist paradigm; for others, new scientific knowledge and technologies beget new dangers to the environment or health. Moreover, science now addresses subjects that are so complex that the results may seem confusing even to the informed public. Solutions are not self-evident or self-imposing. We must recreate the subtle link between the objects which research builds and the way they are taken up by society, by defining new relations between science and society. We name as an 'object' the outcome of the reification of the processes, issues, phenomena we are handling and investigating in order to produce knowledge in a complex world without reducing it to a bundle of simple cause/effect relationships (Hubert, 2004). However, these new objects are not only more and more complex but they are also sometimes new – often unknown until now because of the activities of research, and built by scientific knowledge. Thus, the role of research, data gathering, modelling, setting diagnosis, and building scenarios is critical for dealing with issues such as climate change, erosion of biodiversity, transgenics, and emerging diseases. For us, this contributes to a new paradox for research in society, of having to explore evidence-based questions which trouble society and which are set up by scientific data, e.g., where policy perversities and technologies produce unwanted, or unintended, consequences as the links between obesity, corn, and farm policy in the USA, or the issue of energy transition facing new fossil fuels resources. These new questions are often dealt with within the disciplinary fields which generated them, but shouldn't they also be debated within transdisciplinary groups, which would also include those who generate more than scientific knowledge? It corresponds to what is now known and developed as 'participatory research', 'translational research', 'research in partnership', 'intervention research', and so forth, relying on multi-stakeholder working groups experiencing cross learning processes.

Unfortunately, these arguments are not new; Hubert et al. (2000) characterized the interdependent set of issues in terms of a persistent and seemingly intractable *problematique*. However, the global response to this *problematique* has been weak as has also been the case for the International Assessment of Agricultural Knowledge, Science and Technology for Development (Scoones, 2009). The political economy of research for development continues to be occupied in the main by actors committed to what Donald Schön (1995) described as the 'high ground' of technical rationality. In the context of international assessments,[1] which is one way to understand changing CGIAR policy, Scoones (2009, p. 547) argues that '*the politics of knowledge needs to be made more explicit, and negotiations around politics and values, framings and perspectives, need to be put centre-stage in assessment design*'. The same could be said of the CGIAR policy. In the first instance, the 'problematique' requires a helpful framing and then a meaningful engagement over a long time frame. It is therefore important, we argue, for research to assume a different role in relation to society; research in which responsibilities are shared through the development of innovation networks among scientists, policymakers, and actors in civil society.

It should be clear to everyone by now, thinking beyond our own research specialties, that fully satisfying solutions to achieve shared goals do not yet exist and that we must now think outside the productivist and linear model, which has long been the main frame of reference. We must define together a new vision. Surely, we cannot respond to future issues using recipes from a model that is questioned by a changing world and new ways of perceiving it. In the 1950s, the vision seemed clear, and a consensus emerged that drove innovation towards the goal of increased productivity, as embodied by the Green Revolution. Research institutions, technical extension institutions, and cooperatives were all compatibly organized, working towards this common goal. Today, the path is far from being as clear; in fact, there is no longer a single path; instead, a whole diversity of paths is appearing leading to transition pathways (Hubert and Ison, 2011; Coudel et al., 2012).

Transitions have long been a subject of interest among historians, agronomists, and sociologists, but their inner process dynamics often remain obscured between contextual determinism and 'determinant' leaders. In any transition, various trajectories are possible in moving from one state to another, and such trajectories are themselves determined by the transition underway. How then to characterize transitions? How to generate desirable dynamics? What changes to encourage? How to adapt to the inherent uncertainties? How to incite stakeholders to embark in a process of 'innovative design' that is both systemic and systematic (see Figure 8.1 and discussion below), i.e., creating new objects, technologies, organizations, institutions, and not just of 'controlled-design', i.e., changing procedures in the same pathway?

Such 'innovative design' is also a matter of undertaking exploratory approaches, and not only of using what we already know to meet the concerns of the moment (Hubert, Coudel et al., 2012; Hubert, Ison et al., 2012). We must learn to explore the unknown, to imagine, to create! In our own research, we have developed and explored a number of possibilities which are best framed as systemic modalities for knowledge and/or knowing brokering. Prospective scenario methods, for example, can be used so as to explore possible changes (Hubert and Caron, 2009; Paillard et al., 2010; Hubert et al., 2010; Ison, Grant et al., 2014; Delattre-Gasquet et al., 2016).

Systems thinking in practice (STiP) combines Systemic + Systematic praxis

Transformational change requires thinking and practice that is *systemic + systematic*

Together these constitute an holistic response (a DUALITY not a DUALISM)

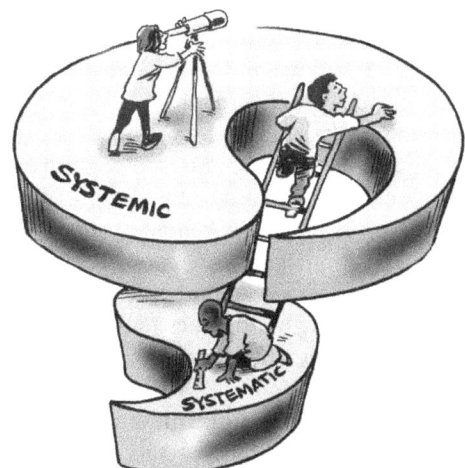

Figure 8.1 Systems thinking in practice involves being both systemic (thinking in terms of relationships and dynamics) and systematic (linear, step-by-step thinking); innovation in complex, uncertain situations is best approached systemically in the first instance

Source: Adapted from Ison, 2010.

By characterizing future states, desired or undesired, foresight can do even more; it can help identify, in advance, the variables that determine the choice of a particular path to take if we prefer one 'future' over another. New modalities of practice (or as we prefer, praxis, theory-informed practical action) can be developed as exemplified in Ison et al. (2011); possible examples include: (i) mapping and diagramming for engaging visually with different stakeholder issues and triggering meaningful interaction, e.g., rich pictures; conversation maps; systems maps; influence; multiple cause, sign, and control model diagrams (Open University, 2006); (ii) media technology: photography and information technology (IT) to provide learning platforms that enable meaningful 'translation' of scientific data, thereby enabling different interests to engage with official 'plans' and policies such as disposable cameras and geographical information systems (GIS) (Steyaert et al., 2007; Toderi et al., 2007); (iii) intermediary objects both living (e.g., a breed of cow; see Steyaert et al., 2007) and non-living can be used as focal points of reference in identifying stakeholders and co-deliberating on stakeholdings, i.e., intermediary objects carry out a facilitative function (Teulier and Hubert, 2008); (iv) development and use of heuristics, conceptual models used to facilitate changes in understandings and practices of stakeholders in complex situations (e.g., SLIM, 2004); and (v) metaphor exploration: actively questioning 'language' used in conveying and developing ideas among different stakeholder groups is an important technique for revealing linguistic and theoretical traps and opening up spaces for innovation (see Ison et al., 2015).

Thus, in agriculture, a key question is that of the social capacity for new technological choices to emerge (given accompanying social, economic, and geographical changes). By technological choice we do not only refer to new artefacts, species, cultivars, and the like, but also changed practices that become socially embedded and which are not restricted to production or health. Past choices may prove difficult to reverse if they have been integrated not only into currently used technical solutions (mechanization, fertilizers, pesticides, genetics, etceteras) but also into cognitive systems (knowledge and know-how, representations of nature, of pollution, of landscapes, and so forth) and the value systems of the main stakeholders involved. Do we not risk being trapped by technical rationalization, a sort of 'lock-in', as experienced by other actors elsewhere, in other economic sectors? To address the question of how we can improve our analysis of these 'lock ins' so that we can get past them, we pose three organizing questions:

- **How to construct a new vision?** How to think differently about innovation? How to deconstruct models that hinder new thinking about innovation? How to explore new directions?
- **How to achieve an agreed purpose?** How to conceive the transition to other models of innovation? How to generate change? What knowledge can be built on?
- **How to organize ourselves?** How to interact better to create knowledge that can engender innovation?

Responses to these challenges ought to be systemic and systematic (Figure 8.1) but our forty years of experience in this field suggest to us that almost universally responses are systematic, rather than systemic. We are on a treadmill of our own making; there is limited intellectual and practical appreciation of how a shift towards the systemic can be achieved, and seemingly little political will to seriously experiment. This chapter is an attempt to redress the balance – and to argue for more investment in situated transformational praxis and conducive institutional innovation built on a genuine appreciation of systems thinking in practice.

The researcher, an agent of change?

Effective engagement with the *problematique* can be undertaken through research that involves denaturalization/regeneration, by the deconstruction of largely accepted evidence, by the uncovering of known but unrecognized facts and unperceived facts, and through the creation of knowledge and new relationships. This endeavour ought to bring actors to question the standards, concepts, and efficiency criteria used so that they can be transformed according to the changes underway. Social learning, which we have been researching for nearly twenty years, provides a conceptual and methodological framework for this form of innovation (Ison et al., 2007; Steyaert and Jiggins, 2007). Social learning from our work is understood as both governance mechanism which can be invested in (just as a symphony orchestra exists as an entity that attracts investment) as well as situational transformation pathway co-constructed by stakeholders whose understandings and practices change as they proceed and who build relational capital as they work together (as happens when a group of musicians comes together for the first time, each with different backgrounds, instruments, and ambitions but who must work together to create an effective performance).

Social learning is not something that necessarily happens naturally or smoothly, nor does it arise out of consensus: it requires social debate, a clash of opinions and visions. How will the concepts and knowledge that may guide the transition be produced and legitimized? Through which networks, within which dominant or marginalized institutions, through which processes will this occur? Concepts that guide thinking, objects that focus attention, indicators and criteria that enable evaluation, chosen thresholds and envisioned processual steps to be taken are all part of a cognitive exercise which invites us to reflect critically and recursively, and which will gradually drive social learning (Ison et al., 2011).

The research process is not separable from decision-making: investigation is central to the design of solutions, requiring explicit choices to be made. By assuming our role of clarifying the implications of different choices, researchers must recognize the political role of all, but particularly social research, potentially empowering actors in innovation. Indeed, faced with prevailing development models, which are taken as being self-evidently valid, it is a challenge for citizens to design alternatives that are not based almost exclusively on these ideological foundations. Science can be expected to provide a range of models in support of social choices to be made in a democratic society. It must be remembered, however, that answers provided by science are shaped by theoretical frameworks and paradigms that are not necessarily definitive.

There have been two waves of engagement by the CGIAR in systems thinking and practice and we have been participants and observers in both; the tensions and, arguably, the underlying reasons for abandoning the first wave can be appreciated from an international conference reported in Remenyi (1985) with further elucidations in Darnhofer et al. (2012). The reasons for the failure of the first wave, as analyzed by Bingen and Gibbon (2012), are systemic in nature, comprising *inter alia* restrictive boundary judgements as to where innovation could be achieved (i.e., too often confined to a species, crop, enterprise, or farm), competing or conflicting epistemologies and thus contrasting appreciations of causality (systematic, linear causality favoured over recursive, circular, systemic causality), and an ongoing failure to recognize the situated nature of innovation and thus the inadequacy of policies and practices designed to 'roll out' or 'scale up' (Hounkonnou et al., 2012; Röling et al., 2014; Ison, in press). This tension is still very evident in the relatively recent focus on 'theories of change', particularly those interpretations which conserve commitments to the linear, systematic model of innovation.

The second wave of engagement in systems-oriented research remains more a possibility than an actuality, since it has not been implemented yet in any meaningful ways. The first new attempts, implemented since 2011 within the CGIAR's new 'System model', that of CRPs, have not been carried forward. There have been encouraging developments in the shifts towards IAR4D (Hawkins et al., 2009), in appreciating that innovation is not restricted to the farm level but along supply and marketing pathways – hence the emergence of interest in innovation platforms and what some have described as 'system innovation' (Hall and Clark, 2010). For a second wave of engagement with systems thinking in practice within the CGIAR to have meaning and the possibility of being effective, there will be a need to address some of the historically derived, but ever present, constraints to transformation.

Constraints to transformation include a widespread lack of epistemic awareness in domains of practice and policy development – there is a crisis of knowing and a lack of awareness of the implications of living in language, i.e., all that is said and done is in language and is governed, biologically, by perception and cognition, areas where there is much conceptual confusion (see Lotto, 2016) which leads many to believe that knowledge, or information, can be transferred from one person or group to another, i.e., the linear model (Ison and Russell, 2007). Inappropriate measures of system performance abound (e.g., measures such as Gross Domestic Product (GDP), or poorly framed key performance indicators); there is a lack of awareness of how objects arise and the implications of reification, the creation of 'things' such as the environment and resources (see Wenger, 1998), and a lack of congruence between what is espoused and what others experience ('talking the talk but not walking the walk') in organizational life. Together these factors give rise to failures to institutionalize systems understandings and practices in manners that create demand pull and sustain institutionalization, and conserve a focus on scientism at the expense of innovative design – particularly the praxis associated with the design of learning systems (Snyder and Wenger, 2010). There is definitely a need for new ways of cooperating to effect purposeful action and to live beyond a 'projectified-world' (Ison, 2010), considering that there is increasing evidence that 'projects' deal poorly with complex, long-term phenomena. In moving forward CGIAR scientists would do well to remember that 'inquiry' is *reflective learning in the literal sense. . . . It is the thinking about thinking, doubting about doubting, learning about learning, and (hopefully) knowing about knowing*' (Churchman, 1971, p. 17).

Change entails a social dynamic in which not only the perspectives and objects of interest are transformed, but also the actors themselves in the collective learning process. The success – and thus the value – of the involvement of researchers rests on their ability to engage in such learning. Not all have this ability; it depends on their scientific culture, their position in the group, and their capacity to engage in a process whose end goal is not known at the outset. In any exploratory research, the elements identified as being important for decision-making are the result of the process and not its starting point. Such an approach requires a full consideration of other possible worlds through joint discussion of construed facts and values. It is by considering together distinct rationalities, knowledge, facts, codes, standards, values, and hybrid actor networks that innovation can develop. And researchers do not usually work in this way, i.e., they follow particular paradigms in which facts are separated from values. How many of us are actually prepared to engage in social learning processes from which we will emerge inevitably transformed, given the inseparability of knowledge and relationships? The challenge is not the individual scientist, but determinants of the overall paradigm in which they operate. Drawing upon research within a science-based organization operating in the research for development (R4D) space, we pose the following question (after Ison, Carberry et al., 2014):

Can a learning system be designed in the CGIAR situation such that reflexive and responsible research for development (R4D) practice is an emergent outcome?

The following points are adapted from an earlier response to this question (see Ison, Carberry et al., 2014) and are equally valid for CGIAR scientists to consider:

- Schön (1983, p. 49) challenged the technical rationality of Herbert Simon.[2] He sought to establish '*an epistemology of practice implicit in the artistic, intuitive processes which [design and other] practitioners bring to situations of uncertainty, instability, uniqueness and value conflict*'.

- Within this tradition 'learning systems' cannot be designed deterministically (i.e., as a blueprint), rather theory-informed contextual design is pursued to create favourable conditions for emergence.
- A 'learning system' can only be said to exist after its enactment, i.e., upon reflection.
- A learning system is an alternative to the common linear model for R&D (transfer of technology model).
- 'Design' of learning systems is a form of systemic action research.

Research to date concludes that it is possible to design such a learning system, but difficult, and that all too often the people in the room, project, program, advisory group, and so forth have the wrong skill sets and inhabit a context where institutional arrangements are less then conducive.

Systems thinking, a way to go forward

In re-engaging with Systems research practice in the CGIAR it is necessary not to fall into the traps that undermined the first cycle of engagement (e.g., Bawden et al., 1985; Hubert et al., 2000; Bawden, 2012; Hubert, Coudel et al., 2012). In some sense, the dramatically changed context that we have outlined makes the challenge both easier but also more urgent. It is also likely to be demanding for those committed to traditional modes of research praxis and who find the traditional institutions comforting, thereby acting perhaps as obstacles to innovation.

Systems thinking can be utilized to facilitate institutional innovation as well as changes in society. But there are different traditions and distinct epistemologies (Ison and Schlindwein, 2015). Those engaged in these diverse traditions conserve, knowingly or not, one of two epistemological positions. What is more, these two positions are already apparent with the equally broad and bifurcating 'complexity' domain. The epistemological positions are:

(i) Objectivist or positivist – for these people 'systems' exist in the world and thus are describable, discoverable, able to be modelled, etc. This is the legacy of General Systems Theory and can be typically spotted whenever the phrase 'systems science' is written. It is Peter Checkland's 'hard' systems tradition (Checkland and Poulter, 2006). It has cultural dimensions – more common in North America and within aspiring or existing 'big science' communities. Within the field of complexity the term 'complexity science' is also a warning sign. This epistemology characterizes the Santa Fe group (see Waldrop, 1992) and its publications as descriptive complexity, i.e., complexity is in the world, describable, able to be modelled, etc. The alternative is:

(ii) Constructivist or interpretivist – for these people the role of the observer is crucial – systems are always brought forth by someone in a context and the 'product' is a system of interest. The act of formulating or seeing *system* is a way of knowing about the world, not an entity in it. A 'system' and the act of constructing a system thus becomes an epistemological device – a way of understanding and learning about situations in the world. It is no longer *sciencia* but *praxis* – but for ongoing innovation new research is needed in both domains. If one acts with epistemological awareness one's commitments do not need to become an either/or choice. Epistemological positioning is a choice one can make; by taking the *as if* position more pathways for action are opened up, e.g., let him/her engage with this situation as if it were complex, or let's learn about this situation through a process of formulating systems of interest, or as if a system existed *a priori*.

Thus, CGIAR researchers need to be aware that many common commitments, or tendencies, of a majority of practitioners within the given approach relate to seeing systems as 'real entities' (ontologies) or heuristic devices (epistemologies). These two epistemological positions now constitute two language communities even though many who participate in them are unaware that they do. They are language communities in the sense that the term is used because members of each community bring forth different traditions of understanding when they do what they do. They admit different claims and thus accept or reject particular explanations. What is more, they often act without awareness; some act in ways that abrogate responsibility by claiming that their explanation is 'the truth' or 'objective' – the path of objectivity to use Maturana's term (Maturana and Varela, 1988). In social relations, this is at the same time a claim to do as they say! However, transformational change requires thinking and practice that is *systemic* + *systematic* as in Figure 8.1: together these constitute a holistic response (a duality not a dualism, where a dualism is an either/or choice in which the choice of one negates the other, e.g., subjective/objective, and a duality is a pair that combines to form a whole, as do the concepts predator/prey from ecology). To summarize:

(i)   both systems and complexity approaches have something to offer when situations are no longer amenable to 'mainstream' practice of analysis based on linear causality or reductionist approaches;
(ii) in the hands of aware practitioners, systems and complexity approaches both offer epistemological devices for shifting mental furniture – for 'managing' complexity;
(iii) both are rich sources of metaphors – and these metaphors have the capacity to trigger new and emergent understandings.

Systems is not a homogeneous field – how it understands itself, just like physics or psychology, is contested (Ison and Schlindwein, 2015). The adaptive whole is one of the key images central to most accounts of Systems. The concept of a whole and the changing nature of the whole in relation to a context, which can be described as a co-evolutionary dynamic, is a key organizing notion. What it conceals, however, is the observer (person or group) dependent formulation – bringing forth – of a system through the distinction of a whole in a context by the act of making a boundary judgment. It is the act, or practice, of distinguishing or formulating a system in a situation as a way of thinking about and acting in that situation that is central to systemic innovation (Ison, in press).

Systems of interest are devices related to purpose, so that the boundary and subsystems will be different in each particular system of interest. Systems of interest even in the same situation are also likely to differ somewhat because each is constructed or formulated by one or more people who have different experiences and backgrounds and possibly purposes. The key systems concepts that are involved in formulating systems of interest are: (i) making boundary judgments; (ii) creating the levels of system, subsystem, suprasystem; (iii) distinguishing a system from an environment – that outside the system boundary – creating a relational dynamic between system and environment mediated by a boundary, rather than 'a thing' called 'system'; (iv) elements and their relationships; (v) attribution of purpose to the system; and (vi) monitoring and evaluating the performance of the system against named measures such as for efficacy, efficiency, effectiveness, ethicality. The adaptive whole is not *the* system but the person(s) + their system(s) of interest learning their way to new understandings and practices which are systemically desirable and culturally feasible.

A system of interest is a chosen way for someone to know about and thus act in a situation. To move to this perspective for some involves making what is known as an epistemological shift – a shift in their way of knowing about systems. This shift as an expansive and ethical shift in that it opens up more choices for pursuing purposeful action to change things for the better. This can be understood when one is aware of the different ways of 'seeing systems', and that how to 'see systems' are purposeful choices that can be made in context-sensitive ways, so that the range of practices at our disposal expands. On the other hand, maintaining a commitment to a 'systems are real' perspective traps those who hold it in a more limited range of practice options. The lack of epistemological awareness and thus flexibility around the concept and practice of Systems in Farming Systems Research has constrained innovation and change, and as a consequence this has limited institutionalization of, and investment in, practices that are informed by systems thinking and practice (Ison, 2012).

In our final section, we return to the question of what constitutes systems practice. Providing a definition is inadequate because in social relations a definition in the social domain is too often interpreted, knowingly or not, as a demand or exhortation to do things 'my' way! Instead, under the aegis of the question, What is it that we do when we do systems practice?, we make connections of different types and quality with a particular history and incorporate concepts into our language games (following Wittgenstein) as explanations and doings which in the social relations we inhabit (including our own reflections) enable claims to be made, or not, that what we do is systems practice.

Praxis innovation – A new field for innovation

Systemic inquiry is a praxis innovation for which there is a need and which has the potential to partly displace 'projects' as an institution and/or at times accompany projects and programs as a meta-framing institution (Ison, 2010). This innovation is needed in order to understand situations in context and especially the history of the situation, addressing questions of purpose, clarifying and distinguishing 'what' from 'how' as well as addressing 'why', facilitating action that is purposeful and which is systemically desirable and culturally feasible, developing a means to orchestrate practices across space and time which continue to address a phenomenon or phenomena of social concern when it is unclear at the start as to what would constitute an improvement. It is being used in our research as an inquiry-based approach that enables managing and/or researching for emergence, a form of practice in which ethics arises in context-related action.

Systemic development, a praxis innovation pioneered by the Hawkesbury group (e.g., Packham, 2011; Bawden, 2012), is another essentially constructivist approach to systemic praxis, developed in order to overcome the mismatch between *what was* and *what could be* in agriculture and rural development. Key systemic features include starting within a situation of concern in which a system of interest is formulated, or bounded in some way, as a means to know about the situation systemically; elements or activities are apparent, connectivity is exhibited, transformation results, and emergent properties are apparent or anticipated. Design/designer features are purposeful to those who participate, are not deterministic, and allow awareness that what is valid knowledge is contested. The Hawkesbury approach to systemic development grew out of an almost twenty-year period of innovation in agricultural education designed to educate a *systems agriculturalist*. Systems students at the UK Open University, who now number in their tens of thousands, have also learnt how to make systems thinking in practice (STiP) practical in their lives, both personally and professionally (Maiteny and Ison, 2000; Ison and Blackmore, 2014).

According to Maturana and Varela (1988), the transformation of our way of 'seeing' is a vital prerequisite for 'doing' things differently, because what we *do* in this world essentially reflects the way we *see* or construe situations and phenomena in it: so, in relation to the concept 'system' and the accompanying practices, to start with a situation or THE system is a choice to be made! Starting out systemically entails, amongst other practices the recognition that we all have agency in deciding how to frame a given situation, amongst which is the choice to see, or bring forth, systems in a situation as means for understanding and transformational change.

Our praxis responses, as systemic designers, to be effective, require embodied understanding and knowing, i.e., the anticipatory (not predicted) possibility of creating effective performances in the face of unfolding surprise. Extant institutional arrangements and praxis continue to present obstacles that constrain systemic learning: globally, there is strong resistance to the appreciation that conservative institutional arrangements and the persistence of framing narratives that are reductionist, deterministic, and highly techno-centric in regard to the innovation processes act as a major bottleneck to development effectiveness. Thus, a key challenge for future studies and practice is to gain a much clearer understanding of the political economy of this conservatism and identify ways of institutionalizing systemic learning as part and parcel of the research and innovation process.

Because of the recognition that humans change the climate of our planet, this realization, as much as the biophysical changes that are wrought, means we are in a period new to human history. Such an appreciation justifies each and every one of us divesting from our current levels of commitment to the historical 'facts of the matter' (facts that for many of us were shaped by where they sat within a normal distribution). Following Bateson (1972), if we wish to commit to making differences that make a difference then we need to frame innovation within the tails of the distribution (historically understood as noise) by asking: can I/we design to create innovations that are systemically desirable and culturally feasible and adaptive with an unfolding context? As Ison, Carberry et al. (2014; p. 10 and 11) note:

> *Without progress in this direction the power of science for the greater good of society will continue to be undermined. Systemic learning is required as a routine element of development investments designed to help multiple actors usefully engage in the process of innovation and change.*

We argue in terms of praxis innovation because researchers who want to enhance their systemic (systematic) praxis and overcome the traps of a too simple view of systems (i.e., limiting themselves to a systematic approach) need to master the path from conceptual and theoretical frames, coming from the literature, to embodied praxis in situations with other researchers and stakeholders. This cannot be done alone; it requires learning by doing in concrete situations with a collective setting to exchange experiences, success, and failures and to build an adapted framework suited to the diversity of working situations. At this historical moment praxis innovation is needed that realizes Bateson's (1972) aphorism of seeking differences that make a difference.

Notes

1 Like IPCC setting up the climate change issue and the contribution of agriculture as well as a cause of GHG emission and an opportunity of C sequestration, or the Millennium Ecosystem Assessment (MA) pointing out the issue of ecosystem services as well as the degradation of many ecosystems, or the IAASTD addressing current agricultural practices.
2 Nobel Prize-winner Herbert Simon coined the term of bounded rationality and satisficing (vs. substantive rationality) and was among the earliest to analyze the architecture of complexity and to propose a preferential attachment mechanism to explain power law distribution. His goal-seeking focus was rejected by later systems theorists – see Ison (2010).

References

Abate, T., Albergel, J., Armbrecht, I., Avato, P., Bajaj, S., Beintema, N., Ben Zid, R., Brown, R., Butler, L.M., Dreyfus, F., Ebi, K.L., Feldman, S., Gana, A., Gonzales, T., Gurib-Fakim, A., Heinemann, J., Herrmann, T., Hilbeck, A., Hurni, H., Huyer, S., Jiggins, J., Kagwanja, J., Kairo, M., Kingamkono, R.R., Kranjac-Berisavljevic, G., Latiri, K., Leakey, R., Lefort, M., Lock, K., Mekonnen, Y., Murray, D., Nathan, D., Ndlovu, L., Osman-Elasha, B., Perfecto, I., Plencovich, C., Raina, R., Robinson, E., Röling, N., Rosegrant, M., Rosenthal, E., Shah, W.P., Stone, J.M.R., Suleri, A. and Yang, H. (2009) 'Executive summary of the synthesis report of the international assessment of agricultural knowledge, science and technology for development (IAASTD)' in McIntyre, B.D., Herren, H.R., Wakhungu, J. and Watson, R.T. (eds) *International assessment of agricultural knowledge, science and technology for development synthesis report*, International Assessment of Agricultural Knowledge, Science and Technology for Development, Washington, DC, pp. 1–11.

Bateson, G. (1972) *Steps to an ecology of mind*, The University of Chicago Press, Chicago.

Bawden, R.B. (2012) 'How should we farm? The ethical dimension of farming systems' in Darnhofer, Ika, Gibbon, David and Dedieu, Benoit (eds), *The farming systems approach into the 21st century: The new dynamic*, Springer, Dordrecht, pp. 119–139.

Bawden, R.J., Ison, R.L., Macadam, R.D., Packham, R.G. and Valentine, I. (1985) 'A research paradigm for systems agriculture' in Remenyi, J.V. (ed) *Farming systems research: Australian expertise for third world agriculture*, ACIAR, Canberra, pp. 31–42.

Beintema, N., Bossio, D., Dreyfus, F., Fernandez, M., Gurib-Fakim, A., Hurni, H., Izac, A.-M., Jiggins, J., Kranjac-Berisavljevic, G., Leakey, R., Ochola, W., Osman-Elasha, B., Plencovich, C., Röling, N., Rosegrant, M., Rosenthal, E. and Smith, L. (2009) *Summary for decision makers of the global report*, International Assessment of Agricultural Knowledge, Science and Technology for Development, Washington, DC, pp. 1–36.

Bingen, J. and Gibbon D. (2012) 'Early Farming System Research and extension experience in Africa and possible relevance for FSR in Europe' in Darnhofer, I., Gibbon, D. and Dedieu, B. (eds) *The farming systems approach into the 21st century: the new dynamic*, Springer, Dordrecht, pp. 47–69.

Calvo, G., Fonte, M., Heinemann, J., Ishii-Eiteman, M., Jiggins, J., Leakey, R. and Flencovich, C. (2009) 'Towards sustainable agriculture', *UNESCO-SCOPE-UNEP Policy Briefs*, 8, pp. 1–6.

Checkland, P.B. and Poulter, J. (2006) *Learning for action: a short definitive account of soft systems methodology and its use for practitioners, teachers and students*, Wiley, Chichester.

Churchman, C.W. (1971) *The design of inquiring systems: basic concepts of systems and organizations*, Basic Books, New York.

Coudel, E., Soulard, C., Devautour, H., Faure, G. and Hubert, B. (eds) (2012) *Renewing innovation systems in agriculture and food: how to go towards more sustainability?* Wageningen Academic Publishers, Wageningen, p. 240.

Darnhofer, Ika, Gibbon, David and Dedieu, Benoit (eds) (2012) *The farming systems approach into the 21st century: the new dynamic*, Springer, Dordrecht.

De Lattre-Gasquet, M., Hubert, B. and Vervoort, J. 'Foresight for institutional innovation and change in agricultural systems: three examples', chapter 9 (this volume).

Hall, A. and Clark, N. (2010) 'What do complex adaptive systems look like and what are the implications for innovation policy?' *Journal of International Development*, 22, pp. 308–324.

Hawkins, R., Heemskerk, W., Booth, R., Daane, J., Maatman, A. and Adekunle, A.A. (2009) 'Integrated Agricultural Research for Development (IAR4D), a concept paper for the Forum for Agricultural Research in Africa (FARA) Sub-Saharan Africa Challenge Programme (SSA CP)' FARA, Accra, Ghana.

Hounkonnou, D., Kossou, D., Kuyper, T.W., Leeuwis, C., Nederlof, E.S., Röling, N., Sakyi-Dawson, O., Traoré, M. and van Huis, A. (2012) 'An innovation systems approach to institutional change: smallholder development in West Africa', *Agricultural Systems*, 108, pp. 74–83.

Hubert, B. (2004) *Pour une écologie de l'action. Savoir agir, apprendre, connaître*. Ed. Arguments, Paris, 430 p.

Hubert, B., Brossier, J., Caron, P., Fabre, P., de Haen, H., Labbouz, B., Petit, M. and Treyer, S. (2010) 'Forward thinking in agriculture and food: a platform for a dialogue to be continued', *Perspective- Research*, Cirad, no. 6, sept., 4 pp.

Hubert, B. and Caron, C. (2009) 'Imaginer l'avenir pour agir aujourd'hui, en alliant prospective et recherche: l'exemple de la prospective Agrimonde', *Natures Sciences Sociétés*, 17 (4), pp. 417–423.

Hubert, B., Coudel, E., Coomes, O., Soulard, C., Faure, G. and Devautour, H. (2012) 'Conclusion: en route . . . but which way?' in Coudel, E., Soulard, C., Devautour, H., Faure, G. and Hubert, B. (eds) *Renewing innovation systems in agriculture and food: how to go towards more sustainability?* Wageningen Academic Publishers, Wageningen, pp. 221–230.

Hubert, B. and Ison, R.L. (2011) 'Institutionalising understandings: from resource sufficiency to functional integrity' in Kammili, T., Hubert, B. and Tourrand, J.F. (eds) *A paradigm shift in livestock management: from resource sufficiency to functional integrity*, Cardère éditeur, Lirac, France, pp. 11–16.

Hubert, B., Ison, R.L. and Röling, N. (2000) 'The 'problematique' with respect to industrialised country agricultures' in LEARN (eds) *Cow up a tree. Knowing and learning for change in agriculture. Case studies from industrialised countries*, INRA (Institut National de la Recherche Agronomique) Editions, Paris, pp. 13–30.

Hubert, B., Ison, R.L., Sriskandarajah, N., Blackmore, C., Cerf, M., Avelange, I., Barbier, M. and Steyaert, P. (2012) 'Learning in European agricultural and rural networks: building a systemic research agenda' in Darnhofer, I., Gibbon, D. and Dedieu, B. (eds) *The farming systems approach into the 21st century: the new dynamic*, Springer, Dordrecht pp. 179–200.

Ison, R.L. (2010) *Systems practice: how to act in a climate-change world*, Springer, London and The Open University.

Ison, R.L. (2012) 'Systems practice: making the systems in farming systems research effective' in Darnhofer, Ika, Gibbon, David, and Dedieu, Benoit (eds) *The farming systems approach into the 21st century: The new dynamic*, Springer, Dordrecht, pp. 141–158.

Ison, R.L. (in press) 'What is systemic about innovation systems? The implications for policies, governance and institutionalisation', in Francis, Judith and van Huis, Arnold (eds) *Innovation systems: towards effective strategies in support of smallholder farmers*, CTA/WUR, Wageningen.

Ison, R.L., Allan, C. and Collins, K.B. (2015) 'Reframing water governance praxis: does reflection on metaphors have a role?' *Environment & Planning C: Government & Policy*, 33, pp. 1697–1713.

Ison, R.L. and Blackmore, C. (2014) 'Designing and developing a reflexive learning system for managing systemic change', Systems Education for a Sustainable Planet, Special Issue, *Systems*, 2 (2), pp. 119–136.

Ison, R.L., Carberry, P., Davies, J., Hall, A., McMillan, L., Maru, Y., Pengelly, B., Reichelt, N., Stirzaker, R., Wallis, P., Watson, I. and Webb, S. (2014) 'Programs, projects and learning inquiries: institutional mediation of innovation in research for development', *Outlook on Agriculture*, 43 (3), pp. 165–172.

Ison, R.L., Collins, K.B., Colvin, J.C., Jiggins, J., Roggero, P.P., Seddaiu, G., Steyaert, P., Toderi, M. and Zanolla, C. (2011) 'Sustainable catchment managing in a climate changing world: new integrative modalities for connecting policy makers, scientists and other stakeholders', *Water Resources Management*, 25 (15), pp. 3977–3992.

Ison, R.L., Grant, A. and Bawden, R.B. (2014) 'Scenario praxis for systemic and adaptive governance: A critical framework', *Environment & Planning C: Government & Policy*, 32 (4), pp. 623–640.

Ison, R.L., Röling, N. and Watson, D. (2007) 'Challenges to science and society in the sustainable management and use of water: investigating the role of social learning', *Environmental Science & Policy*, 10 (6), pp. 499–511.

Ison, R.L. and Russell, D.B. (eds) (2007) *Agricultural extension and rural development: breaking out of knowledge transfer traditions*, Cambridge University Press, Cambridge, UK, p. 239.

Ison, R.L. and Schlindwein, S. (2015) 'Navigating through an "ecological desert and a sociological hell": a cyber-systemic governance approach for the Anthropocene', *Kybernetes*, 44 (6/7), pp. 891–902.

Kammili, T., Brossier, J., Hubert, B. and Tourrand, J.F. (eds) (2009) *INRA-CIRAD open science network meeting: partnerships, innovation, agriculture*, FI4IAR, Paris.

Lotto, B. (2016) 'Optical illusions show how we see', http://www.froebeldecade.com/perception/ (accessed 2nd February 2016).

Maiteny, P.T. and Ison, R.L. (2000) 'Appreciating systems: critical reflections on the changing nature of systems as a discipline in a systems learning society', *Systems Practice & Action Research*, 16 (4), pp. 559–586.

Maturana, H.R. and Varela, F. (1988) *The tree of knowledge*, Shambala Press, Boston and London.

Open University (2006) *Techniques for environmental decision making*, The Open University (UK), Milton Keynes.

Packham, R. (2011) 'The farming systems approach' in Jennings, J., Packham, R. and Woodside, D. (eds) *Shaping change: natural resource management, agriculture and the role of extension*, APEN, Australia, pp. 32–51.

Paillard, S., Treyer, S. and Dorin, B. (2010). *Agrimonde: scenarios and challenges for feeding the world in 2050*, Editions QUAE, Versailles (France).

Remenyi, J.V. (ed) (1985) *Farming systems research, Australian expertise for third world agriculture*, ACIAR, Canberra.

Schön, D. (1983) *The reflective practitioner: how professionals think in action*, Temple Smith, London.

Röling, N., Jiggins, J., Houkonnou D. and van Huis A. (2014) 'Agricultural Research from recommendation domains to arenas for interaction. Experiences from West Africa', *Outlook on Agriculture*, 43 (3), pp. 179–185.

Schön, D. (1995) 'The new scholarship requires a new epistemology', *Change*, Nov/Dec., pp. 27–34.

Scoones, I. (2009) *'The politics of global assessments: the case of the International Assessment of Agricultural Knowledge, Science and Technology for Development (IAASTD)'*, The Journal of Peasant Studies, 36 (3), pp. 547–571.

SLIM (2004) 'SLIM framework: social learning as a policy approach for sustainable use of water', https://sites.google.com/site/slimsociallearningforiwm/ (accessed 2nd February 2016).

Snyder, W.M. and Wenger, E. (2010) 'Our world as a learning system: a communities of practice approach' in Blackmore, C.P. (ed) *Social learning systems and communities of practice, Springer and the Open University, London*, pp. 107–143.

Steyaert, P., Barzman, M.S., Brives, H., Ollivier, G., Billaud, J.P. and Hubert, B. (2007) 'The role of knowledge and research in facilitating social learning among stakeholders in natural resources management in the French Atlantic coastal wetlands', *Environmental Science and Policy*, 10, pp. 537–550.

Steyaert, P. and Jiggins, J. (2007) 'Governance of complex environmental situations through social learning: a synthesis of SLIM's lessons for research, policy and practice', *Environmental Science & Policy*, 10 (6), pp. 575–586.

Teulier, R. and Hubert, B. (2008) 'Des concepts intermédiaires pour la conception collective. Les situations d'action collective avec acteurs hétérogènes', in Mélard, F. (ed), *Ecologisation. Objets et concepts intermediaries*, Peter Lang, Bruxelles, pp. 163–186.

Toderi, M., Powell, N., Seddaiu, G., Roggero, P.P. and Gibbon, D. (2007) 'Combining social learning with agroecological research practice for more effective management of nitrate pollution', *Environmental Science and Policy*, 10, pp. 551–563.

Waldrop, M. (1992) *Complexity: the emerging science at the edge of order and chaos*, Viking, London.

Wenger, E. (1998) *Communities of practice: learning, meaning and identity*, Cambridge University Press, New York.

9 Foresight for institutional innovation and change in agricultural systems

Three examples

Marie de Lattre-Gasquet, Bernard Hubert and Joost Vervoort

Introduction

In the 1960s and '70s, research and extension promoted Green Revolution, which resulted in many technical innovations. However, this did not solve all the problems and today, new challenges that threaten food and nutritional security are emerging. Some of them are climate change, tensions related to resources and evolutions of the energy mix, environmental degradation, demographic developments, urbanization, changes in diets and livelihoods. The complexity of these challenges is provoking debates on the way the future is to be approached and modelled and calling for renewed research approaches. It calls for holistic thinking among various stakeholders in their quest to find solutions to the challenges. It requires technical solutions as well as social and organizational solutions, thus a suite of changes in research questions and methodologies, sources of funding, partnerships and policies.

In this chapter, after defining what foresight, system thinking (Hubert and Ison, 2016) and institutional change are, we will examine three examples of foresight exercises and see the kinds of institutional changes they led to, and the key elements of their success. We will conclude by making recommendations about the conditions for going from foresight to institutional innovation or policy-making.

Foresight and innovation: Two systemic and complementary activities

The characteristics of foresight

Foresight is 'a systematic, participatory and multi-disciplinary approach to explore mid- to long-term futures and drivers of change' (Forward Thinking Platform, 2014, p. 11). Systems thinking is a fundamental perspective for futures studies. It means not looking at one element, but looking at a set of interconnected elements that is coherently organized. It also means considering not only the set of factors but also the actors. Although the future is unpredictable, different potential directions for developments might be anticipated and alternatives explored, i.e.

plausible and/or desirable futures (Van Vuuren et al., 2012). Therefore, there is the possibility of preparing for the future or trying to shape it directly (Hatem, 1993; Cornish, 2004). Foresight is neither prophecy nor prediction but invites us to consider the future as something we can engage meaningfully with, to create or build in the context of uncertainty, rather than something already decided (de Jouvenel, 2000). Actually, few examples exist that show how foresight can be used to guide planning processes effectively (Bourgeois, 2012).

Foresight methods

Foresight is a process, as it involves a logical step-wise progression, and an attitude, as it means being 'pre-active' and 'pro-active', where pre-active infers to preparing for an anticipated change, and pro-active implies acting to provoke a desired change (Godet et al., 2008). Future studies are a mosaic of approaches, objectives and methods as the three cases that will be presented demonstrate.

Key elements to the success of foresight as postulated by Martin and Irvin (1989) in their 5Cs are: communication, concentration on the long-term goals, coordination, consensus and commitment. For Godet et al. (2008), going from Anticipation to Action via Appropriation is essential as 'it is through the process of emotional investment (appropriation) that projects ultimately succeed'. This is illustrated in Figure 9.1. In addition, other key elements identifiable

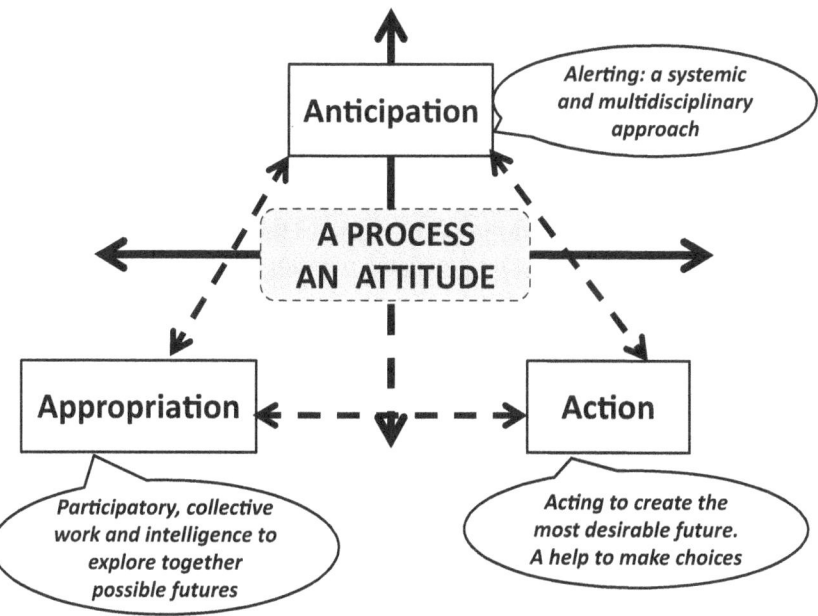

Figure 9.1 The characteristics of foresight
Source: Adapted from Godet et al., 2008.

are: focus on specific policy guidance, consideration of the systemic nature of reality, explicit expression of divergences, integration of scientific uncertainties by revealing the assumptions underlying alternative scenarios, collective learning and development of relationships and trust with policy-makers and other stakeholders, timing of the foresight exercise in relation to surrounding events and policy cycles, flexibility and multidimensionality of scenarios as well as their credibility in terms of narratives and model results and legitimacy (Paillard et al., 2011).

Innovation: A systemic activity

Innovation is 'the implementation of a new or significantly improved product (good or service), or process, a new marketing method, or a new organizational method in business practices, workplace organization, or external relations' (OECD and EUROSTAT, 2005, p. 46). Innovation is a systemic activity involving a variety of actions within the system, of which the innovating organization forms part. Therefore, foresight is very useful to actors involved in innovation processes (Smits, 2002; Salo and Cuhls, 2003; Smits and Kuhlman, 2004).

Innovation processes differ greatly from sector to sector in terms of development and rates of change. Some innovations lead to rapid and radical change, whereas others create more subtle, gradual and incremental change. Institutional innovation can be radical like the adoption of integrated research programs, or incremental like, for instance, the reallocation of funds in an organization. Organizations redesign themselves and generate new products, clients, management systems or business models (Hagel and Brown, 2013). Institutional innovation is very important for private enterprise that must pay attention to consumers, competitors, shareholders, etc. Public sector organizations must also look for new ways to fulfil their public mission (Daglio et al., 2014), but the pressure is not the same as in the private sector. However, as they often face very complex and cross-boundary problems, innovations are difficult to identify and implement.

In the following sections, we will look at three foresight processes and examine how they led to institutional innovations or new policies.

Defining research questions and building partnerships: Cocoa and *hevea* foresights

In January 1997, following a period of turbulence in both the cocoa commodity chain and national research institutions, Cirad (Centre de coopération internationale en recherche agronomique pour le développement) launched a cocoa foresight exercise to help its researchers get a better grasp of possible future scenarios, identify new key research questions, review changes in their research priorities and to help build new partnerships (de Lattre-Gasquet et al., 2003; de Lattre-Gasquet, 2006). Three years later, a similar foresight exercise was also launched on the *hevea* (i.e. rubber tree) by researchers from the rubber program.

Scenario development

The time horizon set for the two foresight exercises was 2015. The process involved Cirad's local and overseas research teams working on cocoa and rubber. There were three phases: organization of the exercise, scenario building (setting up knowledge base, identifying key variables and operators strategy, consulting experts and preparing scenarios) and then strategy thinking and dissemination. There was no separate operating budget since the future studies were included in the normal activities of researchers.

In both cases, a knowledge base with information on producing and consuming countries, actors in the commodity chain, cross-cutting themes such as product quality, prices and legislation, and science was set up. From its analysis, two types of potential variables were identified: uncontrollable variables (for instance climate change, demographic trends, geopolitical shifts, economic growth, exchange rates), and controllable variables which were as independent of each other as possible, so that changes in one variable could not influence changes in other variables. The key variables identified were: the balance between the three factors of production (land, labour and capital), parasites and diseases, quality of production, consumption and operators of the commodity chain. For the identification of the variables, outside experts were consulted on an ad hoc basis. Individual contacts were often more constructive than group meetings because representatives of firms and policy-makers would not give clear-cut opinions in open meetings since they did not wish to reveal their strategies to competitors.

In both cases, three scenarios were constructed by linking the micro-scenarios of each driver. There was (1) a 'business as usual' scenario, (2) an 'optimistic' scenario based on sustainable development and improvements along the value chain, and (3) a 'pessimistic' scenario that detailed a production crisis (Figure 9.2).

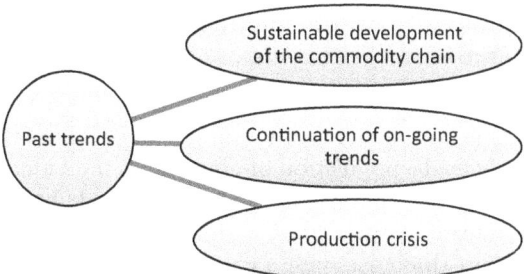

Figure 9.2 The three scenarios of the cocoa and rubber foresight exercises

Scenario use and consequent innovations

The two foresight exercises helped in crystalizing the on-going thinking on the need to do research that favours sustainable development of the commodity chain. No impact study has been done to assess the impact of the foresight exercises. Nevertheless, we can say that within Cirad, the cocoa and rubber foresight contributed to modifying the research approach of the programs. There was more collaboration between researchers from different disciplines and new research projects focused on sustainable cocoa production systems for smallholders. At that period, collaborations with other cocoa initiatives were reinforced and sustainability became an important issue for the industry. The institutional changes that took place at the time of rubber foresight exercise have been less visible and did not lead to visible institutional changes. The Sustainable Natural Rubber Initiative was set up as a voluntary and collaborative industry project by IRSG (International Rubber Study Group) only in 2014.

In the late 1990s, the foresight approach was not common in research programs, and looking back we can say it contributed to three types of innovations: new research projects, new research questions and methods, and new partnerships. Discussions among researchers played an important role in the learning cycle, and slowly a consensus emerged about a possible optimistic scenario for the commodity system, about changes that should be implemented and new research activities. Then, foresight analysis helped the researchers to build a common view about the past situation and the possible future scenarios for a commodity. Increased awareness of challenges led to identifying adjustment of priorities. The programs changed their focus from improvement of productivity to sustainable development of production. They adopted an interdisciplinary approach to make up for the insufficiencies that arose from the disciplinary approaches. Researchers enlarged their network and got to know better the industry and the end users of their research. They started to go from a closed system into the political arena (Mintzberg, 1983). Discussions with the private sector contributed to changing opinions and to the construction of a common language.

A multi-stakeholder forward thinking exercise to guiding research orientations: Agrimonde

The Agrimonde foresight exercise was launched in 2006 by Cirad and Inra (Institut national de la recherche agronomique) in order to build scenarios on how to adequately feed a population of nine billion individuals in 2050 while at the same time preserving the ecosystems. Agrimonde had three objectives: to guide future research at Inra and Cirad in the field of agronomy and food security; to structure discourse within Cirad and Inra on global food security; and to build capacity so that French experts could participate more effectively in international debates on the futures of agriculture (Paillard et al., 2011;

Treyer, 2011). It was influenced by the Millennium Ecosystem Assessment (MA, 2005) and carried out at the same time as the International Assessment of Agricultural Knowledge, Science and Technology for Development (IAASTD, 2009).

A two-tier package involving a panel of experts and combining quantitative and qualitative approaches to develop scenarios

Agrimonde was designed as a platform for the preparation, analysis and discussion of scenarios to facilitate the collective analysis of the challenges facing the world's food and agricultural systems. The platform consisted of a project team and an expert panel of researchers and decision-makers bringing different points of view and visions of the future.

It consisted of three main steps: (1) choosing the scenarios and the principles underlying their construction; (2) building quantitative scenarios and checking the consistency of major assumptions; and (3) building complete scenarios by integrating quantitative scenarios with qualitative assumptions (Paillard et al., 2011). Agrimonde decided to have a baseline scenario which was inspired by the Global Orchestration scenario of the Millennium Assessment and called it Agrimonde GO. Then it developed a contrasted scenario and called it Agrimonde 1. The variables considered were multifarious; they are: global context, international regulations, dynamics of agricultural production, actors' strategies, knowledge and technologies, and sustainable development. During the course of the foresight, the panel analyzed and discussed academic work as well as other foresight studies, which enabled it to explore systematically possibilities of discontinuity and gradually end up with contrasting futures.

The scenario-building method was based on the complementarities of quantitative and qualitative analyses. A quantitative tool called Agribiom was devoted to the analysis of the world's production, trade and use of biomass (Dorin and Le Cotty, 2011). When the experts formulated assumptions on diet, land use, yields or inter-regional trade, they had to analyze all their implications and ramifications. Through this process, they enhanced the basic quantitative assumptions.

Three outputs

Beyond the figures and storylines, the comparison of the two scenarios raised novel questions that had not found their way into the mainstream scientific discourse and the facts reported or the assumptions put forward called for new approaches by the policy-makers in implementing policies in regards to the type of future wished for. Three main challenges that the panel of experts had not really anticipated were identified (Hubert et al., 2011).

First, agronomists must take into account issues of nutrition and quality of diet. They are at the heart of the public health problems concerning all societies and particularly those undergoing transition. Nutritional issues raise questions on supply in all its dimensions. The ways of addressing the problem of food security on a global scale and the difficulties encountered in the process will differ depending on the food and nutritional options taken.

Second, the challenge is to develop a range of technical options which are applicable in diverse situations and which may be complementary to one another in terms of the types of resources, knowledge and know-how. These options must be oriented towards a more sustainable use of resources by taking into account the way in which they are produced on various spatial and temporal scales. Performance criteria should be defined to evaluate the effectiveness and efficiency of renewed production systems, from the point of view of their intrinsic productivity, their effects on environment and their economic viability and social sustainability with regards to the organization and difficulty of the tasks to accomplish.

Third, international trade, which is needed to make up for the deficit of domestic production in countries that are net importers of food calories and allow for increases in food production in stable countries, should be secure. Agrimonde does not offer a solution but it emphasizes the role of global trade in agricultural and agri-food systems, its security, its stability and relations with international environmental and social regulations.

In conclusion, this foresight exercise highlights the need for a broader perspective on the agricultural activities in changing rural areas, in all regions of the world and all economic dynamics, including inter-sectoral approaches. Unfortunately, we missed the target of influencing policy decision-makers (out of research policy) because of the kind of experts we chose, the length of the process and the commitment it required from participants for a long lasting period, including some homework between the collective workshops. Inra and Cirad have the ambition, if not the duty, to continue and to intensify the effort in this direction.

Supporting policy design for agriculture in the face of climate change: The example of CCAFS foresight

As unprecedented climate change interacts with demographic, economic, political and socio-cultural changes to create an uncertain future for the global food system, governments and non-state actors are looking for ways to act strategically in the face of such uncertainty (Vermeulen et al., 2013). The CGIAR program on Climate Change, Agriculture and Food Security (CCAFS) established a Scenarios Project, focusing on scenario development in East and West Africa, South and Southeast Asia, the Andes, Central America and the Pacific. The goal of the Scenarios Project has been to use scenarios for the guidance of significant plans and policies around food security, livelihoods and environments in the face of climate change (Vervoort et al., 2014). The theory of change of how to successfully use scenarios for the guidance of planning has changed over the years – ultimately resulting in an approach that has yielded significant policy impact across its seven global regions (Schubert and Vervoort, 2015).

The project started with a model of scenarios as a product of stakeholder participation, and an emphasis on scenario development only – following the logic that if relevant actors are involved in the development of future scenarios, the products (for instance, reports containing scenario narratives, driver analyses and quantified versions of stakeholder scenarios) would be more likely to be used. However, involved stakeholders quickly made it clear that while they were interested in using the results of the scenarios, this use of the scenarios should be incorporated in the participatory process, and facilitated by the scenario researchers.

Having recognized this need to incorporate the use of scenarios into foresight processes, the approach which was then used was to organize scenario workshops at the regional level (the level at which scenarios were developed) – to consider approaches to tackle regional challenges in the face of the different scenarios. This approach, mainly applied in East Africa, yielded a number of strategies for improved food security, livelihoods and environments under climate change that were tested for robustness and flexibility across different scenarios, and generated enthusiasm among participants for the implementation of such plans. However, since many of the proposed strategies involved new regional connections and the development of new institutional arrangements on top of on-going activities, achieving such goals proved challenging. A new, more focused approach was used and scenarios were developed for food security, environments and livelihoods under climate change and other drivers at the regional level, and then used to guide multiple specific policy processes at the national level in these regions. These policy guidance processes involved close collaboration between researchers, governments and other actors, and ultimately led to a range of policies and plans which were tested and developed through scenario exercises. Some of the countries where such processes have been carried out are Cambodia, Bangladesh, Honduras, Burkina Faso, Tanzania, Uganda and Costa Rica (Schubert and Vervoort, 2015).

In the above examples, we applied the following approach to scenario-guided policy formulation:

Scenario development

In preparation for a regional scenario exercise, stakeholder/power-mapping is conducted to identify key participants. Then, an emergent identification of drivers is conducted (clustering around 500 driver suggestions into 20–30 drivers), which are then structured to create a 'skeleton' of the most diverse scenarios based on the 4 to 6 most important/uncertain drivers of change, and 2 to 4 future states for each driver. Thereafter, most diverse scenarios are identified using a custom-designed tool, OLDFAR (Lord et al., 2016), and any impossible combination of driver states according to participants are knocked out. Participants then break up into diverse groups to develop scenario narratives to flesh out the scenarios; they then proceed to provide content on their scenario for all the drivers that were not made part of the basic skeleton, plus drivers needed as model inputs. In a final session, participants identify potential policy or (non-state) strategy processes that are projected to happen in the next year or so, and where they believe the use of these regional scenarios can be of great value. The scenarios are also quantified using multiple simulation models including GLOBIOM (Valin et al., 2013) and IMPACT (Rosegrant et al., 2012).

Scenario use

Immediately after the scenario development workshop, a regional coordinator engages governments or other partners in conversations to plan how the regional scenarios could be used as a way to test and develop policy. Multiple opportunities have emerged across and in different countries – with 81 policy/strategy pathways having been proposed.

In the most feasible processes, where partners are most committed to use the scenarios for policy guidance, the case is taken to harmonize scenario guidance with the goals of the policy-makers and the timing of policy cycles – so that expectations are set correctly in the science–policy partnership. A meeting of policy-makers, stakeholders representing diverse interests and vulnerable groups, and the research team is then convened to test and develop the draft of the policy or plan in question. The structure of this meeting depends on the development stage of a given plan or policy. If a draft plan exists, the group first reviews and provides a first revision of the plan to integrate their key concerns. A list of key issues related to the plan's scope also emerges. Then, the regional scenarios are adapted and re-interpreted based on the national context and the plan's specific focus. These scenarios are used to review critically the plan in terms of strengths and weaknesses from the perspectives of different future worlds, and to suggest changes. Key strengths, weaknesses and recommended changes that have been made across multiple scenarios are then incorporated into a new draft of the plan. If no plan exists, scenarios are first adapted to the national and policy context. Then, priority strategies are developed in each of the different scenarios, and thereafter tested for robustness and flexibility across the other scenarios, resulting in a single integrated plan draft. In both cases, the regional process coordinators/researchers keep working closely with governments after this meeting, to ensure uptake of the proposed changes. If the process has been set up correctly and expectations are clear, this process is not challenging, because policy-makers will have been involved in and aware of the benefits of the scenario guidance. In some cases, follow-up meetings to further test and develop plans and policies are needed when they reach more mature stages of development, or when, for instance, sub-national implementation plans are being developed, involving different stakeholders.

In all seven regions, variations of the above process have been followed and have led to the co-design and finalization of the mentioned plans and policies, as well as further strategic planning around implementation plans and sub-national plans. For instance, the example of the Cambodian Climate Change Priorities Action Plan included further work on the role of non-state actors in the plan's finalization; in the example of the National Strategy for Climate Change Adaptation in Agriculture developed by the Honduras Secretariat for Agriculture and Livestock, sub-national implementation plans, led by farmers and other stakeholders, were developed.

The CCAF Scenarios Project case study offers a strong example, based on a range of processes, of how scenarios can be used for policy impact – and this can be useful for governments, non-state actors (NGOs/CSOs and private sector) as well as for researchers seeking greater interaction between research and policy. However, challenges remain:

- Moving beyond tailor-made processes that fit perfectly with policy cycles and interests to creating new institutional structures and international links in a more transformative fashion remains more challenging.
- Developing in-house capacity for foresight in government.
- Overcoming resistance in governments and the practice of forecasting approach that assumes that the world is wholly controllable and that only one future is most likely.

Conditions for going from foresight to institutional innovation or policy-making

The three foresight exercises presented in this chapter have taken place at different periods of time (late 1990s for cocoa and rubber, mid 2000s for Agrimonde and early 2010s for CCAFS) and the objectives and approaches reflect this difference. However, these cases demonstrate that foresight improves the collective and strategic intelligence and the relational capital. It does not replace decision-making, since foresight starts with context rather than plans, but it can improve and support it (Kok et al., 2011). To improve decision-making for institutional innovation or policy-making, three recommendations are proposed.

First, it is important to identify the actors (organizations and individuals) that have an impact on institutional change and policy-making. It can be useful to prepare a 'topography of actors', i.e. to identify institutions and industries that will be concerned, and even final decision-makers and their advisors, intermediary decision-makers, opinion leaders, experts, journalists, members of think-tanks, business men that are concerned with the topic of the foresight. Some of them can be part of the foresight group, but others can be involved at the end of the process. Having been introduced by common relations and experiences of having worked together in the past provide good pre-conditions to facilitate appropriation of the foresight outcomes and future action with diverse actors.

Second, it is useful to understand policy processes and planning cycles. It is rare to make radical decisions or adopt completely new innovations at once. Most of the time, it is a succession of incremental changes that leads to a radical decision or innovation. The decision-making cycle is not always rational and well-planned. Anticipating the cycle is very important to be able to have an impact. Therefore, the ground must be prepared far in advance and maintained.

Getting the attention of institutional decision-makers or policy-makers is not achieved by a one-off action. The relationship must be built for the long-term. Because a foresight process can be long, frequent, widely distributed updates on the progress of the project are important. Each phase of the foresight process should have an official document that categorizes the output of that stage and relates it to the institution or the policy (Farrington et al., 2012)

Third, the presentation of the foresight work must lead the decision-makers to understand what is at stake. It can be easier to present the foresight work in existing instances than creating one to communicate the results (CGARM, 2009). Bringing about institutional change can require a tailor-made communication. Four ingredients can be highlighted (adapted from CGARM, 2009):

- Good content: appropriate analysis, coherent and plausible scenarios.
- Knowledge of and integration with the targeted decision-makers' needs.
- Scenario presentation that is short, memorable, understandable and targeted towards decision-makers. Diverse modes of communication are recommended such as websites, videos, reports, oral presentations, but there can also be interactive format. Beyond mere presentation, highly co-creative processes that create co-ownership of foresight, deep understanding of the material and strong integration with policy cycles is key.
- Good conditions to be heard: good timing, confidence, credibility and trust, human relations, good advocates and champions of the work.

In conclusion, foresight helps researchers and decision-makers not to fall into the trap of inaction, on the pretext that we do not know everything about such complex issues. This is because it is today's decisions that will determine the trajectories which will enable tomorrow's world to meet adequately all its citizens' needs in food, energy and biomaterials, to reduce poverty and inequalities and to curb the deterioration of environmental goods and services. In a world of rare resources, the rarest of all may be time. Clearly, it will be useful, if not necessary, to learn to bring together effectively two worlds that have different underlying values and paradigms: the world of foresight and that of research. By helping research to identify the right questions on the future, but which need to be addressed today, foresight can help us avoid intellectual dead-ends focused on the production of evidence-based arguments which are supposed to contribute to debates at the highest level.

Indeed, there is currently virtually no international organization, country or NGO that does not intervene in one way or another in research and debates on the future of global agricultural and food production and their interactions with the objective of sustainable development and by extension its implications on international relations and public policies. It is true that the profusion of information, data and results are not an encouragement to examine the underlying hypotheses – whether scientific or ideological – or to give the space they warrant to analyses that do not correspond to what decision-makers would prefer to hear, and that bring up uncomfortable insights. Yet, we must aim to enhance and expand these studies so that we can illuminate the future options towards which we should be tending.

References

Bourgeois, R. (2012) 'Foresight Report to CGARD II: The State of Foresight in Food and Agriculture and the Roads toward Improvement', Global Forum on Agricultural Research (GFAR), Rome, Italy.

CGARM (Conseil général de l'armement) (2009) *Prospective et décision publique*, Ministère de la Défense, Paris, France.

Cornish, E. (2004) *Futuring: The Exploration of the Future*, World Future Society, Bethesda, MD.

Daglio, M., Gerson, P. and Kitchen, H. (2014) 'Building Organisational Capacity for Public Sector Innovation', *Background Paper prepared for the OECD Conference, 'Innovating the Public Sector: From Ideas to Impact'*, 12–13 November 2014, Paris, France.

de Jouvenel, H. (2000) 'A brief methodological guide to scenario building', *Technological Forecasting and Social Change*, vol 65, no 1, pp. 37–48.

de Lattre-Gasquet, M. (2006) 'The use of foresight in setting agricultural research priorities', in Box, L. and Engelhard, R. (eds) *Science and Technology Policy for Development, Dialogues at the Interface*, Anthem Press, London, UK, pp. 191–214.

de Lattre-Gasquet, M., Petithuguenin, P. and Sainte-Beuve, J. (2003) 'Foresight in a research institution: a critical review of two exercises', *Journal of Forecasting*, vol 22, no 2–3, pp. 203–217.

Dorin, B. and Le Cotty, T. (2011) 'Agribiom: a tool for scenario-building and hybrid modelling', in Paillard, S., Treyer, S. and Dorin, B. (eds) *Agrimonde, Scenarios and Challenges for Feeding the World in 2050*, QUAE, Versailles, France, pp. 25–54.

Farrington, T., Henson, K. and Crews, C. (2012) 'The use of strategic foresight methods for ideation and portfolio management', *Research-Technology Management*, March–April, pp. 26-33.

Forward Thinking Platform (2014) *A Glossary of Terms Commonly Used in Future Studies*, GFAR, Rome.

Godet, M., Durance, P. and Gerber, A. (2008) 'Strategic Foresight, La Prospective, Use and Misuse of Scenario Building', *LIPSOR research working paper*, no 10 http://en.laprospective.fr/dyn/anglais/ouvrages/sr10veng.pdf.

Hagell, III, J. and Brown, J.S. (2013) *Institutional Innovation: Creating Smarter Organizations to Scale Learning*, Deloitte University Press, Westlake.

Hatem, F. (1993) *La prospective: Pratiques et méthodes*, Economica, Paris.

Hubert, B., Caron, P. and Guyomard, H. (2011) 'Conclusion', in Paillard, S., Treyer, S. and Dorin, B. (eds) *Agrimonde. Scenarios and Challenges for Feeding the World in 2050*, QUAE, Versailles, France, pp. 233–239.

Hubert, B. and Ison, R. (2016) 'Systems thinking: towards transformation in praxis and situations', chapter 8 (this volume).

IAASTD (2009) 'Agriculture at a crossroads, global report', in McIntyre, B.D., Herren, H., Wakhunan, J. and Watson, R.T. (eds) *International Assessment of Agricultural Knowledge, Science and Technology for Development*, Island Press, Washington, DC, 590 p.

Kok, K., van Vliet, M., Baerlund, I., Dubel, A. and Sendizimir, J. (2011) 'Combining participative backcasting and exploratory scenario development: experiences from the SCENES project', *Technological Forecasting and Social Change*, vol 78, pp. 835–851.

Lord, S., Helgott, A. and Vervoort, J. (2016) 'Choosing diverse sets of plausible scenarios in multidimensional exploratory futures techniques', *Futures*, vol 77, March, pp. 11–27.

Martin, B. and Irvine, J. (1989) *Research Foresight: Priority-Setting in Science*, Pinter, London.

Millennium Ecosystem Assessment (2005) *Ecosystems and Human Well-Being: Synthesis*, Island Press, Washington, DC.

Mintzberg, H. (1983) *Power In and Around Organizations*, Prentice Hall, Englewood Cliffs, NJ.

OECD and EUROSTAT (2005) *Guidelines for Collecting and Interpreting Innovation Data. Oslo Manual*, 3rd edition. The measurement of scientific and Technological activities, OECD Publishing, Paris. DOI: http://dx.doi.org/10.1787/9789264013100-en.

Paillard, S., Treyer, S. and Dorin, B. (eds) *Agrimonde. Scenarios and Challenges for Feeding the World in 2050*, QUAE, Versailles, France.

Rosegrant, M.W., Ringler, C., Msangi, S., Sulser, T.B., Zhu, T. and Cline, S.A. (2012) *International Model for Policy Analysis of Agricultural Commodities and Trade (IMPACT): Model Description*, IFPRI, Washington, DC.

Salo, A. and Cuhls, K. (2003) 'Technology foresight – past and future', *Journal of Forecasting*, vol 22, no 2/3, pp. 79–82.

Schubert, C. and Vervoort, J. (2015) 'Future scenarios work informs climate and agriculture policies in seven countries', CCAFS. https://ccafs.cgiar.org/blog/future-scenarios-work-informs-climate-and-agriculture-policies-seven-countries#.V13NZlROLIU [website consulted on June 10, 2016].

Smits, R. (2002) 'The new role of strategic intelligence', in Tübke, A., Ducatel, K., Gavigan, J.P. and Moncada-Paternò-Castello, M. (eds) *Strategic Policy Intelligence: Current Trends, the State of Play and Perspectives*, IPTS Technical Report Series, JRCIPTS, Sevilla, Spain, pp. 1–29.

Smits, R. and Kuhlmann, S. (2004) 'The rise of systemic instruments in innovation policy', *The International Journal of Foresight and Innovation Policy*, vol 1, no 1/2, pp. 4–32.

Treyer, S. (2011) 'Comment se nourrira la planète en 2050 ? La place de l'exercice Agrimonde dans la multiplication récente des prospectives agricoles et alimentaires mondiales', *Agronomie, Environnement et Sociétés*, vol 1, no 2, pp. 27–35.

Valin, H., Havlik, P., Mosnier, A., Herrero, M., Schmid, E. and Oberstein, M. (2013) 'Agricultural productivity and greenhouse gas emissions: trade-offs or synergies between mitigation and food security?' *Environmental Research Letters*, vol 8, p. 035019.

van Vuuren, D.P., Riahi, K., Moss, R., Edmonds, J., Thomson, A., Makicenovic, N., Kram, T., Berkhout, F., Swart, R., Janetos, A., Rose, S.K. and Arenell, N. (2012) 'Scenarios in global environmental assessments: key characteristics and lessons for future use', *Global Environmental Change*, vol 22, no 4, pp. 884–895.

Vermeulen, S.J., Challinor, A.J., Thornton, P.K., Campbell, B.M., Eryagama, N., Vervoort, J.M., Kinyangi, J., Jarvis, A., Läderach, P., Ramirez-VBillegas, J., Nicklin, K.J., Hawkins, E. and Smith, D.R. (2013) 'Addressing uncertainty in adaptation planning for agriculture', *Proceedings of the National Academy of Sciences of the United States of America*, vol 110, pp. 8357–8362.

Vervoort, J.M., Thornton, P.K., Kristjanson, P., Förch, W., Ericksen, P.J., Kok, K., Ingram, J.S., Herrero, M., Palazzo, A., Helfgott, A.E.S., Wilkinson, A., Havlik, P., Mason-D'Croz, D. and Jost, C. (2014) 'Challenges to scenario-guided adaptive action on food security under climate change', *Global Environmental Change*, vol 28, pp. 383–394.

10 Exploring futures of aquatic agricultural systems in Southern Africa

From drivers to future-smart research and policy options

Ranjitha Puskur, Sarah Park, Robin Bourgeois, Emma Hollows, Sharon Suri and Michael Phillips

Introduction

More than 800 million people currently live on less than US$1.25 per day.[1] Despite the availability of adequate food supplies, they are acutely or chronically undernourished and around two billion people suffer from micronutrient deficiency or 'hidden hunger' (Sundaram, 2014). Increasing the quantity, quality, stability of production and access to agricultural production is championed as a key to addressing these challenges. Yet the natural resource base upon which food production is predicated is increasingly undermined by overexploitation of land, unsustainable water use and climate change (CGIAR, 2015).

The progress made towards halving the population of undernourished by 2015, whilst being significant, has been regionally uneven. In 2015, Africa had the highest share of food insecure population at 28% in the world and this is projected to rise to over 30% by 2025 (Rosen, 2015). Food insecurity does not primarily stem from a lack of food production, but from a lack of access to food caused by the disempowerment of the world's poor (Loos et al., 2014). While food may be available, access may be restricted by low incomes, high unemployment rates particularly among youth, weak markets for staples and export commodities, poor infrastructure, weak regulation enforcement in intra-country and cross-border trade, and instability in markets resulting in food price spikes (Asuming-Brempong, 2015).

The Sustainable Development Goals (SDGs) aim to ensure access to safe, nutritious and sufficient food for all people by 2030.[2] The debates on sustainable intensification have highlighted that issues of food and nutrition security cannot be addressed effectively unless equitable distribution and empowerment are addressed (Loos et al., 2014). This underscores the need for adopting a trans-disciplinary and systems perspective and catalyzing broader social action to ensure that the people who are food insecure today are not food insecure in 2030 and others do not become food insecure.

Investing in agri-food systems has significant potential to contribute simultaneously to multiple SDGs, particularly those that relate to reducing poverty, improving food and nutrition security for health, and improving natural resources systems and ecosystem services (CGIAR, 2015). Reganold et al. (2011) argue that two types of changes are especially relevant to agriculture: incremental and transformative changes. We argue that multi-dimensional food insecurity cannot be solved through incremental changes which narrowly focus on techniques and technological fixes, but require transformative changes.

Using the case of aquatic agricultural food systems in Southern Africa and drawing on a multi-stakeholder participatory scenario-building exercise, this chapter throws light on the drivers influencing the evolution of these complex, multi-functional systems, and illustrates that multi-dimensional research, development and policy options are required to realize transformations in agri-food systems.

Aquatic agricultural food systems in Southern Africa

Aquatic agricultural systems (AAS) are diverse production and livelihood systems where families cultivate a range of crops, raise livestock, farm or catch fish, gather fruits and other tree crops, and harness natural resources such as timber, reeds and wildlife. They occur along freshwater floodplains, coastal deltas and inshore fresh and marine water bodies (AAS, 2012). They are important components of agri-food systems for rural and urban consumers and provide a source of dietary diversity and quality through the supply of fish from wild fisheries, as well as supporting aquaculture production. The role of fish in addressing nutrition through the provision of micronutrients and protein is well established (Thilsted and Wahab, 2014).

Aquatic agricultural systems are multi-functional in terms of their users and uses. They provide food and other ecosystems services, and generate employment and income. Multiple global trends shape these dynamic production and food systems. These trends include globalization, urbanization, increased climate variability, enhanced connectivity, changes in consumption patterns, demographic changes, technology developments and rising inequalities (Bourgeois, 2015). These changes have substantial implications for the future ability of aquatic agricultural food systems to support food and nutrition security and enhance livelihoods.

Approximately 83 million people in Africa depend on AAS for their livelihoods. Of these, about 47 million live in poverty (Béné and Teoh, 2015). Large numbers of poor dependent on AAS in Southern Africa are concentrated in Madagascar, Mozambique, Malawi and Zambia (Table 10.1). Out of the 6.8 million people that are AAS-dependent in these four countries, 75% are poor. It is estimated that in Zambia and Malawi, 100% of the nation's total fish production comes from AAS while for Mozambique it is 69% and Madagascar 65% (Romijn, 2015). These countries exhibit levels of hunger that are alarmingly high and the proportion of children under five who are stunted ranges from 30% to more than 40% (Harvest Choice, 2014).

Table 10.1 People dependent on aquatic agricultural systems (AAS) in Southern Africa

Country	Estimated AAS area (km²)	AAS-dependent population	Estimated number of AAS-dependent poor[1]	Estimated number of AAS-dependent poor[2]
Madagascar	40,956	2,948,329	2,101,146 (71%)	1,710,619 (58%)
Malawi	24,055	1,053,046	822,343 (78%)	662,822 (63%)
Mozambique	39,662	1,900,533	1,495,329 (79%)	1,036,717 (55%)
Zambia	25,900	911,229	691,518 (76%)	628,121 (69%)

[1] Multidimensional Poverty Index.
[2] Harvest Choice, International Food Policy Research Institute estimates.
Source: Original data from Béné and Teoh, 2015.

Foresight as a systemic approach to aquatic agricultural system transformations

Foresight refers here to a systematic, participatory and multi-disciplinary approach to explore drivers of change and mid- to long-term futures of complex socio-ecological systems.[3] Foresight exercises encourage stakeholders and experts to explore future changes by qualitatively and quantitatively analyzing plausible future developments and challenges. Through unveiling uncertainties, foresight can support stakeholders to shape actively their future by influencing the development and implementation of strategies and actions taken today (GFAR, 2014).

The field of foresight includes a diversity of qualitative and quantitative methods and tools tailored to address relevant questions and meet different objectives including generating knowledge and interactions and/or catalyzing joint action to address anticipated challenges (Reilly and Willerbockel, 2010; Vervoort et al., 2014). Multi-stakeholder scenario building, as part of a foresight approach, can be particularly useful in identifying alternative futures, unveiling uncertainties in the interactions between human and natural systems. Scenarios are an effective tool to capture the complexity of connected socio-economic, environmental and institutional processes and to support decision-making in the face of uncertainty. The scenarios for alternative futures provide a basis for identifying robust research and policy responses that can shape the pathways towards the desired futures. They help to take the political, economic, technological and other 'known unknown' aspects into account while apprehending local, national and global influences on the systems. The outcome of the scenario development can be informative or instrumental (Könnölä et al., 2011) and can be used to inform planning for desired future pathways (Vervoort et al., 2013).

Such approaches, however, have only recently been applied to the complex problems around food systems and agriculture in the developing world (Chaudhury et al., 2013). Recent analysis of the foresight approach highlighted that it is significantly absent from agriculture planning in the world's least developed countries, particularly in sub-Saharan Africa (Bourgeois, 2012). Holistic systems approaches that involve exploration of entire food systems – linking agriculture, environmental and human systems together – are needed to understand and meet the future food challenges of developing regions (Vervoort and Ericksen, 2012).

Exploring futures of aquatic agricultural systems in Southern Africa

Consultative Group on International Agricultural Research (CGIAR) Research Program (CRP) AAS, together with the Forum for Agricultural Research in Africa (FARA) and Africa Union – New Partnership for African Development (AU-NEPAD), organized a multi-stakeholder participatory scenario-building workshop during 14–18 July 2015 in Lusaka, Zambia. The aim was to identify and explore drivers of change and plausible futures of aquatic agricultural systems. The geographic focus was on Madagascar, Mozambique, Malawi and Zambia due to their high levels of poverty and food insecurity, as well as the significant land area under AAS (Puskur et al., 2016).

Participants were selected to ensure a diversity of perspectives. They were asked to step out of the role of representing their organizations and focus on contributing their knowledge of the multiple facets of AAS in their countries and region. Diversity in age, gender, ethnicity and power were important considerations for selection, so was their willingness to discuss and tolerate other opinions. Participants were invited from the four countries of focus, as were stakeholders who offered regional and continental perspectives. Amongst the 22 participants, the rich mix of expertise and experience consisted of African environmental history, fisheries and wildlife, agricultural economics, gender, public health, organizational management and change, innovation systems, modelling, climate change, water resources management, private sector aquaculture management, private sector technologies for agricultural information provision, markets and trade, women farmers, civil society organizations and policies.

The Participatory Prospective Analysis (PPA) was used in the workshop to build scenarios. It consists of an adaptation of various methods combined into a comprehensive and rapidly operational framework. Its cognitive nature can be characterized as a 'focus on interactions and consensus building' (Bourgeois and Jésus, 2004). Participants worked through the seven steps of the method over five days (Figure 10.1).

The participants focused on inland aquatic agricultural systems in the four countries while recognizing and appreciating that there was heterogeneity between and within the countries. A time frame of 15 years (up to 2030) was used to build scenarios.

Through brainstorming, participants identified 49 external and internal forces[4] that had a past, present and future influence on the evolution of AAS in Southern Africa (Table 10.2). These forces captured the social, technical, economic, environmental, policy and political (STEEP) dimensions which reflect the complexity and dynamics of these systems. Temperature changes and

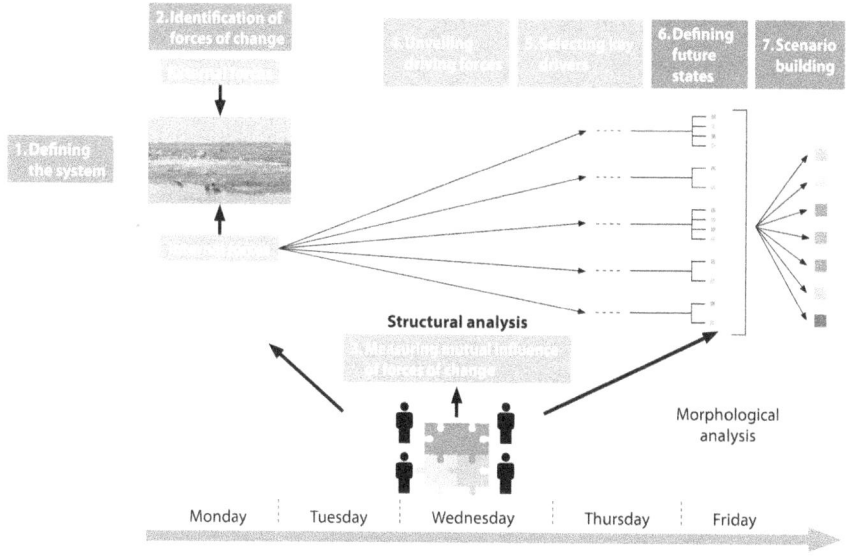

Figure 10.1 Steps in Participatory Prospective Analysis (PPA)

Table 10.2 Forces of change influencing evolution of aquatic agricultural systems (AAS) in Southern Africa

Social
- Learning opportunities
- Population dynamics and role of youth in AAS
- Gender relations and economic and social inequalities
- Relative attractiveness of AAS areas; local perceptions towards potential development; interest in investing in AAS development
- Community participation
- Relations between government and traditional leaders
- Interactions between socio-economic interest groups
- Health status of the communities

Technical
- Capture fisheries management practices
- **Agricultural and fishing practices**
- Production, processing and loss management technologies
- Quality of inputs
- Connectivity and information/data availability and flows
- Leadership capacities
- Local capabilities and opportunities for capability development
- Research for Development

Economic
- **Competition for land and land tenure systems**
- **Access to land and water**
- Access to financial resources, alternative livelihood opportunities, entrepreneurial opportunities for women and youth
- Demand for fish and AAS products, access to input and output markets, transportation infrastructure, market openness

(Continued)

Table 10.2 (Continued)

• **Profitability of AAS activities**
• Socio-economic groups engaged in AAS activities
Environmental
• Stakeholder management of natural resources
• Land and water availability and quality
• Soil quality and biodiversity
• Multi-functionality of AAS
Policy and political
• Political will reflected in policy priorities and enabling environment to implement
• **Policy implementation**
• Stakeholder participation in policy making
• **Presence and orientation of trade policies**

Source: Adapted from Puskur et al., 2016.

climate variability were identified as external forces and excluded from further analysis. A definition for each force was then crafted and agreed by the participants.[5] The participants recognized that these forces they identified for AAS in Southern Africa were generally applicable to other similar systems in Africa, with some contextual nuances.

Then participants discussed the direct influence[6] of each of the internal forces on each other. Understanding the relationships between forces is an important component of the method because it shows what 'drives' the system and how it 'moves'. The existence of a direct influence between two forces was recorded in an influence/dependence matrix using a binary score. A structural analysis software using Excel was used to calculate the mutual influences and provide a quantitative basis to identify key driving forces, which are most influential in determining the future orientation of the system. Over 2,000 binary interactions between the 49 forces were analyzed. The results indicate that AAS in the four countries in Southern Africa are complex systems affected by multiple direct and indirect forces, each exerting a different level of influence. The key driving forces of AAS from this analysis are presented in Table 10.3. Drivers are the strongest, highly influential variables and are unlikely to be influenced by other variables; they determine the direction of change in the system. Leverages, on the other hand, are both influential and dependent. They can drive the system but are also driven by the way the system evolves. Leverages amplify the direction given to the system by the changes in the drivers.

Using more variables will make the construction of scenarios more problematic and complicated; fewer variables will lead to an oversimplification and a very narrow capacity for exploration of the futures (Bourgeois and Jésus, 2004). Six driving forces were selected to develop future scenarios of AAS in 2030 (highlighted in bold in Table 10.3). The participants worked in two groups to define a number of possible contrasted and mutually exclusive future states in 2030 for each selected driving force. Thirty plausible future states for

Table 10.3 Key drivers and leverages in aquatic agricultural systems (AAS) in Southern Africa

Drivers	Leverages
• **Presence and orientation of trade policies**	• **Access to water**
• **Policy implementation**	• **Agricultural and fishing practices**
• **Land tenure systems and competition for land**	• **Profitability of AAS activities**
• Water availability	• Access to land
• Access to financial resources	• Entrepreneurial opportunities for women and youth
• Economic and social inequalities	• Stakeholder management of natural resources
• Access to output markets	• Water quality
• Demand for fish and AAS products	• Connectivity and information flows
	• Capture fisheries management
	• Soil quality
	• Production technologies

the Southern Africa AAS were identified across these driving forces. The participants assessed the compatibility of the states of each force with the states of each of the other forces. Two states are incompatible if the described elements of each future cannot logically and plausibly co-exist.

Participants created scenarios combining the future states of the driving forces for the purpose of exploring a broad range of contrasted transformations. They used software based on Excel for checking compatible states and building coherent scenarios. They expanded six plausible scenarios[7] for the futures of AAS in Southern Africa and four of those which are most contrasted and which are briefly described in Box 10.1 as illustrations to highlight the key challenges and priorities.

Box 10.1   Four plausible scenarios for aquatic agricultural systems in Southern Africa

1 Towards sustainable development goals

In 2030, there is a strong political will to invest in the development of AAS. An equitable land distribution policy with designated land use for AAS is in place. Land titling systems guarantee availability, access and tenure security for AAS actors and minimize competition from non-AAS actors. Access to water is universal and equitable, including for maintenance of important ecosystem services. Communities and both formal and local traditional institutions are effectively managing the natural resource base. Research actors are interacting with local stakeholders about their needs and opportunities, and setting priorities and developing socio-economically relevant technologies. Good quality inputs are

---

accessible and affordable to the poor. Good connectivity, market information availability and strong transportation infrastructure facilitate access to output markets. This is further strengthened by high demand for fish and AAS products and unrestricted trade movement in the region. The poorest people have surplus income. The highly profitable nature of AAS activities attracts private investment in AAS. Positive gender relations are prevalent, with women actively participating in community activities and decision-making, and having equal access to productive resources. Access to suitable financial instruments, tools and financial resources allow women and youth to engage in entrepreneurial opportunities.

2 From the grassroots

In 2030, AAS are characterized by local control with community-driven decision-making and participation in policy-making processes. Customary rules or co-management practices govern land tenure and competition. Traditional leaders and the government build a harmonious relationship with clear delineation of roles, responsibilities and authority. There is secure access to land and water. Due to total protectionism, there is restricted movement of fish and other AAS products within the region. However, the high and increasing demand for AAS products gives communities strong bargaining and financial power. There is access to financial resources, as well as entrepreneurial opportunities to diversify beyond fish. Low external input-based agricultural and fishing practices are used, capitalizing on skills, knowledge, culture and values present within the community. There is sustainable use of natural resources. Research organizations are engaged in a dynamic dialogue with communities to assess their needs and develop technologies and management practices that are profitable and sustainable. Poor people, women and youth are capitalizing on entrepreneurial opportunities. The poorest people in AAS have surplus income to meet their needs, including for maintaining their production activities, providing education for their children and attaining productive health status. Local individuals and institutions are positive about the future development of AAS. There is a lower level of out-migration. Due to the limited connectivity and restricted trade flows, the longer term opportunities for diversification and growth are questionable.

3 Save yourself if you can

In 2030, there is no established system for land management and control. Intense competition for land, including from urbanization, prevails. Relations between the traditional leaders and government are dysfunctional. Access to and the amount of land and water negatively impacts the production activities of the poor. Community participation is overridden by the scramble for land, water and profitability. There is full regional trade liberalization with unrestricted movement of fish and AAS products across the region. Individual profit motives influence the relations and interactions between socio-economic interest groups. Local stakeholders are increasingly using unsustainable approaches to managing natural resources. High-tech mechanized, automated practices are widely used to fuel agricultural production, but they exclude the poor. Socio-economic inequalities are high and increasing. Access to input and output markets is limited, despite access to communication infrastructure. An increase in the frequency of extreme climate events and natural disasters, such as drought and flooding, has resulted in low incomes. The diversification options are far and few. The food prices are high. There is widespread migration to urban areas by local youths, leaving women, children and the old to manage the households.

4 Highway to poverty

In 2030, customary or co-management based local decision-making processes govern land tenure and land competition. There is universal and equitable water access. The relationship between government and traditional leaders is tenuous, but communities have a strong policy influence. No specific trade policies about fish and aquatic agricultural products exist. With high demand for fish and AAS products but no effective governance, local stakeholders are using destructive agricultural and fishing practices to maximize short-term gains. Increasing extreme climate events and natural disasters, such as drought and flooding, perpetuate household poverty. There is limited access to financial resources for diversification possibilities. The entrepreneurial opportunities for the poor, particularly women and youths, are almost non-existent. Local individuals and institutions are negative about the potential development of AAS in the future. There is widespread out-migration, particularly by the youth.

Research, development, policy directions

There is a need for research to take a systems approach to understand the challenge of food insecurity, to recognize that human technical and societal innovations and the environment influence one another at multiple spatial and temporal scales (Garnett and Godfrey, 2012). While accelerating food production to meet the increasing demand is envisaged in several scenarios with the use of technologies and external investments, consideration of consequences on sustainability, inequalities and impacts on the poor who are food insecure is important. For example, in Scenarios 3 and 4 increased production is envisaged;

however, who benefits from this and for how long is an issue. Rising inequalities will be a cause for concern as they provide grounds for civil strife. Inequalities appear to be both a cause and consequence and the poor are locked in a vicious cycle. The drivers and scenarios described in the previous sections give insights into the challenges associated with a range of plausible futures and indicate priority areas for research, policy and action that can work across a range of scenarios – not just the 'what' but the equally important 'how'.

The strongest and least dependent drivers were around policies and institutional arrangements related to trade, land access and tenure, and water access. The drivers and scenarios highlight the need for functional structures for implementation, including accountability mechanisms and adequate resource allocation. Security of access to land and water resources will have a significant influence on the participation of marginalized socio-economic groups in profitable economic activities and the consequent impacts on socio-economic inequalities and food security. The example from Zambia highlighted the case where despite the seemingly adequate availability of land, customary laws and institutions prevent women from accessing it (Kwashimbisa and Puskur, 2014; Puskur et al., 2016). Policies can be double-edged swords. Liberal or protective trade policies could have positive or negative outcomes in different contexts, places and times for different socio-economic groups. For example, while liberalization policies make local enterprises viable in Scenario 1, they place the local economy at a disadvantage in Scenario 3. While protectionist policies benefit the local enterprises in Scenario 2, their competitiveness is questionable and might influence future growth opportunities.

In addition, it is important to note that food security is influenced by not just agricultural and natural resource policies, but also environmental, economic and health policies, amongst others. Policy coherence is of utmost importance to achieve the food security goals. The importance of stakeholder voices and participation in policy processes is critical, to ensure they are relevant and address the needs of the poor, vulnerable and food insecure and avoid elite capture. Ex-ante economic, social and environmental impact assessments of policies should be a regular practice within research.

Access to resources including water, land, technologies, connectivity and information evolved to be important leverages. These have significant implications for entrepreneurial opportunities for women and youth. Production, processing and loss management technologies have been highlighted as areas requiring attention in all future scenarios. It is important for research to understand the impacts of changes in agricultural technologies and practices on the poor and food insecure. Input-intensive technologies are not neutral. They often exclude the poor and women and can increase inequalities. Affordable and relevant technologies that can support women and youth to capitalize on entrepreneurial possibilities would go a long way in reducing gender disparities and catalyze them to play a significant role in development. Access to financial resources will be important to support diversification and, consequently, resilience. The research agenda should, therefore, span political economy and social science. Farmers' own expertise needs to feed into processes of research and innovation, and systems for extending and translating knowledge into changed practices need to be improved (The Royal Society, 2009).

Community-driven resource management plays a particularly crucial role in AAS. The aquatic food system is highly complex and potentially vulnerable, consisting of a mix of large-scale operations with significant political influence and small-scale or artisanal fishing, which provides an important source of direct food security and an income safety net for poor people (Godfray et al., 2010). While the marine fisheries resources have been harvested to full capacity or over-exploited with troubling implications for ecosystem health, stock resilience and long-term output and value (Garcia and Rosenberg, 2010), inland fishery systems are intensely exploited in Africa and there is no room for further expansion and resources are at high risk (Welcomme et al., 2010). Therefore, community empowerment and use of strengths-based approaches will be crucial to ensure equity, sustainability and resilience. Capabilities and leadership in the communities will influence their engagement in local level decision-making and also participation in policy processes at other levels. While Scenario 2 highlights the positive consequences of community empowerment and leadership, Scenario 4 illustrates how individually motivated short-term profit-maximizing motives can lead to damaging economic and environmental consequences in the long-run.

Conclusion

The results of the analysis illustrate that the drivers of the futures of AAS are numerous, multi-dimensional and interconnected. The resulting scenarios call for a systems approach linking scientists, local actors and policy makers to design the type of investment (human, capital, knowledge) required for achieving desirable pathways towards the preferred future. The pathways towards desirable scenarios and prevention of undesirable scenarios require combining the different domains in a holistic perspective involving recursivity.

For example, achieving Scenario 1 would require systemic intervention through three interconnected components:

- Research directed to interact with communities supporting them in managing the natural resource base and developing socio-economically relevant technologies with women actively participating in community activities and decision-making,
- Tenure security and equitable land distribution synergized with universal and equitable access to water for maintenance of important ecosystem services,
- Public investment in connectivity, market information, trade and transportation infrastructure synergized with high demand for fish and AAS products attracting private investment in AAS, also allowing women and youth to engage in entrepreneurial opportunities.

This advocates for conducting research on AAS in a way that will not isolate the technical components from the actors' reality and the policy environment. Similarly, it means that research needs to include investigating institutional and social dimensions along with inclusion of more technological advances. This also requires changing the research process and engaging with stakeholders at the local level (both policy makers and local communities, and other actors).

The scenarios developed in this participatory foresight analysis offer a rich resource for informing the development of a range of social, technical, economic, environmental, policy and political strategies needed to influence the food and agricultural systems associated with AAS towards more robust and equitable trajectories. Importantly, this information has evolved from a trans-disciplinary process with a wide range of stakeholders and through a systems lens, and as such, reflects the need for strategies and action plans to similarly reflect integrated and multi-sector responses. While scientific knowledge and advances are critical, values shape stakeholders' attitudes to the food system and their views on what the way forward should be (Garnett and Godfray, 2012). Recognizing that there is heterogeneity amongst and within the four countries of focus, there is a need to scale down the scenarios at national, sub-national and community levels to develop tailored and context-specific priorities and policy directions. Such an approach offers the linking of global transformation and challenges, to site-specific and local agriculture and rural development problems and decision-making processes. These scenarios can be used as tools for facilitating dialogue and building coalitions between rural communities, scientists, policy makers and civil society to inform future-focused research and development policies, practices and investments that shape pathways towards food and nutrition-secure futures.

Notes

1 http://www.undp.org/content/undp/en/home/mdgoverview/post-2015-development-agenda/goal-1.html. Accessed 5 October 2015.
2 http://www.un.org/sustainabledevelopment/hunger/. Accessed 5 October 2015.
3 http://bit.ly/FTPglossary [Accessed 10 April 2015].
4 A force of change is something that has the capacity to transform a system through its influence on outcomes. Internal forces of change are those where some or all system actors have the capacity to modify the future state of. External forces are those that cannot be influenced.
5 See Puskur et al. (2016) for detailed definitions of the forces of change.
6 Direct influence is said to exist if a change in one force would cause an immediate change in another force without needing any other force to act, and this change can be clearly and logically explained.
7 For further details of the analysis and detailed scenario narratives, see Puskur et al. (2016).

References

AAS (2012) CGIAR Research Program on Aquatic Agricultural Systems, Program Proposal, AAS-2012–07, Penang, Malaysia, p. 1.

Asuming-Brempong (2015) 'Food security challenge: Africa', Pre-Conference Presentation, European Commission/ICAE, Milan, Italy.

Béné, C. and Teoh, S.J. (2015) *Estimating the Numbers of Poor Living in Aquatic Agricultural Systems*, Unpublished Program report. CGIAR Research Program on Aquatic Agricultural Systems. Penang, Malaysia.

Bourgeois, R. (2012) 'Foresight Report to CGARD II: The State of Foresight in Food and Agriculture and the Roads toward Improvement', Global Forum on Agricultural Research (GFAR), Rome, Italy.

Bourgeois, R. (2015) 'Seven plausible rural transformations', *Development*, 58, 2/3 expected availability second semester 2016.

Bourgeois, R. and Jésus, F. (eds) (2004) *Participatory prospective analysis: exploring and anticipating challenges with stakeholders*, Bogor: UNESCAP-CAPSA, 114 p. (CAPSA Monograph, 46). http://www.uncapsa.org/Publication/cg46.pdf.

CGIAR (2015) *Strategy and Results Framework 2016–2030*. Version 18 May 2015. Consulted in June 2015.

Chaudhury, M., Vervoort, J., Kristjanson, P., Ericksen, P. and Ainslie, A. (2013) 'Participatory scenarios as a tool to link science and policy on food security under climate change in East Africa', *Regional Environmental Change*, 13, 2, pp. 389–398.

Garcia, S.M., and Rosenberg, A.A. (2010) 'Food security and marine capture fisheries: characteristics, trends, drivers and future perspectives', *Philosophical Transactions of the Royal Society B*, 365, pp. 2869–2880.

Garnett, T. and Godfray, C. (2012) *Sustainable intensification in agriculture. Navigating a course through competing food system priorities*, Food Climate Research Network and the Oxford Martin Programme on the Future of Food, University of Oxford, UK.

GFAR (2014) 'Forward Thinking Platform (2014) A Glossary of Terms Commonly Used in Futures Studies', Global Forum on Agricultural Research, Rome, Italy.

Godfrey, H.C.J., Crute, I.R., Haddad, L., Lawrence, D., Muir, J.F., Nisbett, N., Pretty, J., Rosbinson, S., Toulim, C. and Whiteley, R. (2010) 'The future of the global food system', *Philosophical Transactions of the Royal Society B*, 365, pp. 2769–2777.

Harvest Choice (2014) *Atlas of African agriculture research & development*, International Food Policy Research Institute, Washington, DC and University of Minnesota, St. Paul, MN. http://harvestchoice.org/node/9704.

Könnölä, T., Scapolo, F., Desruell, P. and Mu, R. (2011) 'Foresight tackling societal challenges: Impacts and implications on policy-making', *Futures*, 43, 3, pp. 252–264.

Kwashimbisa, M. and Puskur, R. (2014) 'Gender Situational Analysis of the Barotse Floodplain', *Program Report: AAS-2014–43*, Penang, Malaysia: CGIAR Research Program on Aquatic Agricultural Systems.

Loos, J., Abson, D.J., Chappell, M.J., Hanspach, J., Mikulcak, F., Tichit, M. and Fischer, J. (2014) 'Putting meaning back into sustainable intensification', *Frontiers in the Ecology and Environment*, 12, 6, pp. 356–361. doi:10.1890/130157.

Puskur, R., Park, S., Hollows, E. and Bourgeois, R. (2016) 'Futures of Inland Aquatic Agricultural Systems and Implications for Fish Agri-Food Systems in Southern Africa', *Program Report: AAS-2016–01*, Penang, Malaysia: CGIAR Research Program on Aquatic Agricultural Systems.

Reganold, J.P., Jackson-Smith, D., Batie, S.S., Harwood, R.R., Kornegay, J.C., Bucks, D., Flora, C.B., Hanson, J.C., Jury, W.A., Meyer, D., Schumacher Jr., A., Sehmsdorf, H., Shennan, C., Thrupp, L.A. and Willis, P. (2011) 'Transforming US agriculture', Policy Forum, *Science*, 6 May 2011, 332, 6030, pp. 670–671.

Reilly, M. and Willenbockel, D. (2010) 'Managing uncertainty: a review of food system scenario analysis and modelling', *Frontiers in the Ecology and Environment B*, 365, pp. 3049–3063.

Romijn, L. (2015) *Aquatic Agricultural Food Systems: Towards an Estimation of Impacts beyond the Geographical Boundaries of Aquatic Agricultural Systems.* Unpublished draft report. CRP AAS, Penang, Malaysia.

Rosen, S. (2015) 'International Food Security Assessment: Past Progress and Projections Global Food Security Challenges', Paper presented at the ICAE Pre-Conference Event. Milan, Italy.

The Royal Society (2009) *Reaping the benefits: science and the sustainable intensification of agriculture*, Royal Society, London, UK. https://royalsociety.org/~/media/Royal_Society_Content/policy/publications/2009/4294967719.pdf.

Sundaram, J.K. (2014) 'Tackling the nutrition challenge: a food systems approach', *Development*, 57, 2, pp. 141–146.

Thilsted, S.H. and Wahab, M.A. (2014) 'Increased production of small fish in wetlands combats micronutrient deficiencies in Bangladesh', *Policy Brief: AAS-2014–10*, Penang, Malaysia: CGIAR Research Program on Aquatic Agricultural Systems.

Vervoort, J.M., Bourgeois, R., Ericksen, P., Kok, K., Thornton, P., Förch, W., Chaudhury, M. and Kristjanson, P. (2013) 'Linking multi-actor futures for food systems and environmental governance', University of Oxford Environmental Change Institute. Paper presented at the Earth System Governance Tokyo Conference.

Vervoort, J.M. and Ericksen, P. (2012) 'No foresight, no food? Regional scenarios for Africa and South Asia', *The Futures of Agriculture*, Brief No. 03 – English, Global Forum on Agricultural Research (GFAR), Rome, Italy.

Vervoort, J.M., Thornton, P.K., Kristjanson, P., Förch, W., Ericksen, P.J., Kok, K., Ingram, J.S., Herrero, M., Palazzo, A., Helfgott, A.E.S., Wilkinson, A., Havlik, P., Mason-D'Croz, D. and Jost, C. (2014) 'Challenges to scenario-guided adaptive action on food security under climate change', *Global Environmental Change*, 28, pp. 383–394.

Welcomme, R.M., Cowx, I.G., Coates, D., Béné, C., Funge-Smith, S. and Lorenzen, K. (2010) 'Inland capture fisheries', *Philosophical Transactions of the Royal Society*, 365, pp. 2881–2896.

Part II

Sustainable intensification in practice

11 System productivity and natural resource integrity in smallholder farming

Friends or foes?

Bernard Vanlauwe, Edmundo Barrios, Timothy Robinson, Piet Van Asten, Shamie Zingore and Bruno Gérard

Introduction

In sub-Saharan Africa (SSA), increases in yields of the major crops in small-holder farming systems have failed to match population growth, with increased production resulting rather from agricultural area expansion (Worldbank, 2007), very often at the expense of the natural resource base, such as carbon-rich and biodiverse forest land (e.g. Gockowski and Sonwa, 2011). Intensification of smallholder agriculture is a must under high population densities but also desirable in less populated areas in order to protect natural ecosystems. Smallholder farming communities and systems in SSA are heterogeneous, both at the community and farm level, driven by varying and often limited access for production resources (land, labour, capital) (Tittonell et al., 2010). At the community level, variable resource endowments and production objectives are often conceptualized through the construction of farm typologies. At farm level, preferential management of specific plots within a farm has resulted in within-farm soil fertility gradients, often with soils of higher fertility near the homestead, and more degraded soils towards the outer limits of the farm. For many households and regions, agriculture alone will not be able to provide rural populations with adequate livelihoods due to limited farm size and access to land (Harris and Orr, 2014; Jayne et al., 2014). Besides heterogeneity at farm and community level, enabling conditions for intensification, often expressed as access to agro-inputs, markets, and credit, quality of rural infrastructure, or con-ducive policies, also vary considerably. Intensification of smallholder farming systems will thus require co-learning among research, development, and private sector actors for the tailored integration of both technical and institutional innovations (Giller et al., 2011; Coe et al., 2014).

Sustainable Intensification (SI), though ill-defined, encompasses the need to enhance productivity, whilst maintaining or improving ecosystem services and system resilience. Although the discourse on what constitutes SI and how it relates to other intensification paradigms is very active nowadays (e.g., van Noordwijk and Brussaard, 2014; Petersen and Snapp, 2015; Wezel et al., 2015), advancing this discourse is beyond the scope of the current paper. In this chapter, the productivity dimension of SI refers to total farm productivity, using the unit of land as the denominator for areas with high population densities and limited smallholder access to land. Total farm productivity can be expressed in various ways, partly related to the objectives of why such data are collected. Summing up dry matter yields over various crops does not really make sense since crops such as cassava (with up to 25 ton dry matter ha^{-1} year^{-1}) have a much higher yield potential than others such as cowpea (with up to 2 seasons of 3 ton ha^{-1} or 6 ton ha^{-1} year^{-1}). Farm-level productivity can be aggregated based on total value generated (in USD farm^{-1}), total energy (kJ farm^{-1}), protein (kg farm^{-1}), or other nutritional components generated (energy (kJ farm^{-1}), protein (kg farm^{-1})), or total biomass produced (in kg dry matter farm^{-1}). While the first indicator is important in the context of income generation, the second set of indicators is relevant in relation to the food and nutrition security discourse while the latter indicator is relevant when focussing on carbon and nutrient stocks and flows that affect natural resource integrity.

Ecosystem services that require maintenance under SI are many and are regulated at different scales. Following TEEB (2010), ecosystem services are classified as provisioning, regulating, habitat, and cultural services, whereby (i) provisioning services refer mainly to goods that can be directly consumed, and include food, water, raw materials, such as fibre and biofuel, and genetic, medicinal, and ornamental resources; (ii) regulating services comprise regulation of climate, air quality, nutrient cycles, and water flows; moderation of extreme events; treatment of waste; preventing erosion; maintaining soil fertility; pollination; and biological controls of pests and diseases; (iii) habitat services are those that maintain the life cycles of species or maintain genetic diversity; and (iv) cultural services refer to the aesthetic, recreational and tourism, inspirational, spiritual, cognitive development, and mental health services provided by ecosystems.

In the context of this chapter, we focus on plot/farm level intensification and provisioning services are covered by the enhanced productivity dimension of SI. Only regulating services operating at plot/farm level are considered and these include regulation of climate, regulation of nutrient cycles and water flows, preventing erosion, and maintaining soil fertility. All of these can be positively affected by increased soil organic carbon (SOC) contents. For instance, (i) increased SOC contents enhance climate change mitigation; (ii) SOC interacts positively with the biological (e.g., provision of energy for biological activity), chemical (e.g., exchange capacity for nutrient retention), and physical dimensions (e.g., enhanced aggregate stability) of soil fertility; and (iii) application of mulch and increased aggregate stability reduce soil erosion and increase water infiltration. For the remainder of the chapter, SOC or soil fertility status is used as an indicator of ecosystem service maintenance, thereby recognizing that

SOC contains several distinct pools or fractions, each with their own functions and consequent contributions to specific ecosystem services (Lehmann and Kleber, 2015). We also recognize that beyond biophysical dimensions, social, economic, and human dimensions are critical for SI (e.g., Loos et al., 2014), but these are outside the scope of this chapter.

The objectives of this paper are (i) to conceptualize the yield reduction and soil fertility degradation processes and how these interact; (ii) to conceptualize and provide evidence from long-term soil management trails for potential rehabilitation trajectories as proposed by various intensification paradigms; and (iii) to evaluate the potential impact on yield and SOC of those paradigms in response to the question posed in the title of this chapter: can SI interventions simultaneously address the need for more produce and the delivery of other soil-based ecosystem services, or are trade-offs in space and time inevitable?

Soil degradation and yield decline

After conversion of natural fallows to agricultural land (Figure 11.1a), it has been frequently observed that in the absence of the use of external nutrients, crop yields decline over time as do soil fertility conditions, often expressed as SOC content. In the short-term, the first degradation process that is commonly initiated is nutrient mining, resulting in deficiencies of those nutrients of which removal by a crop quickly exceeds the nutrient replenishment potential of the soil (Stoorvogel et al., 1993). In most cases in cereal-based systems, these nutrients are N and P (e.g., Sanchez et al., 2001). Note that under specific circumstances, other degradation processes can also be immediately initiated, e.g., soil erosion on fields with steep slopes and lack of surface cover. As a result, not only crop yields decline but also the amount of crop residues produced that can either be retained in the plot or recycled through livestock feeding systems. Consequently, declining crop yields are accompanied by declining soil fertility conditions, with both processes reinforcing each other. In the initial stages of land conversion (e.g., the first 5–10 years, depending on the soil type), solutions to these trends can be found in the application of those nutrients that are limiting crop growth, with N, P (and K) fertilizer being the most commonly available. Application of these nutrients can help rehabilitate crop yields, and the provision of crop-residue-related biomass, thereby contributing to reduce the rate of SOC loss.

Where the soil degradation process is not addressed and thus allowed to proceed, several other degradation processes gradually take effect reducing the effectiveness of nutrient applications, ultimately resulting in non-responsive soils (Vanlauwe et al., 2010) (Figure 11.1b). Such degradation processes include acidification caused by the removal of crop residues (Van Breemen et al., 1983), soil erosion due to a reduced surface cover (Valentin et al., 2004), or the generation of nutrient imbalances causing secondary and micronutrient deficiencies (Turmel et al., 2015). Other degradation processes that can trigger non-responsiveness include soil crusting reducing infiltration and germination

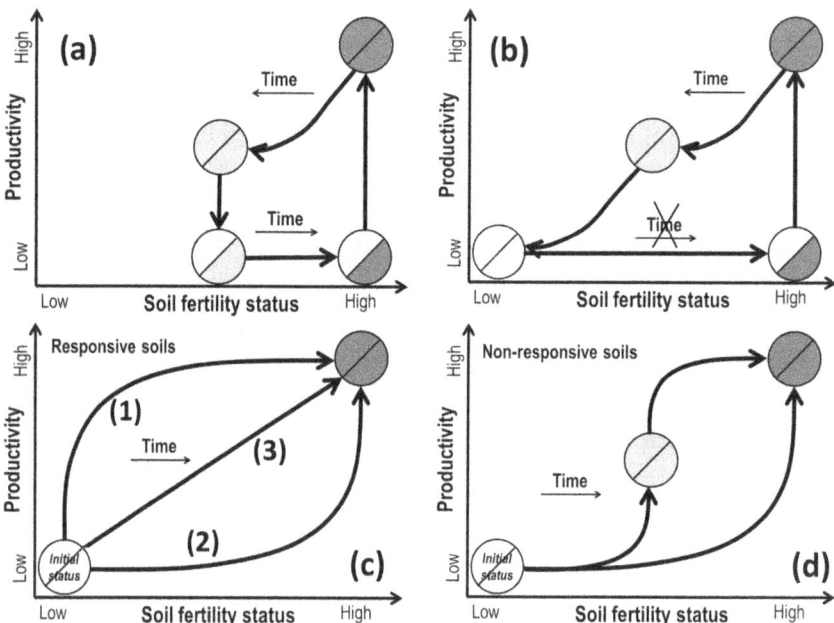

Figure 11.1 Conceptual depiction of the slash-and-burn cycle whereby (a) land is left to return to fallow after a certain period with continuous yield and soil fertility decline, and (b) how an extension of the cropping cycle with increasing population density can generate fields with low yields and severely degraded soils. The conceptual pathways from a current situation (degraded soil with low productivity; lower left circle) to a SI situation (healthy soil and high productivity; upper right circle) are described by two possible situations: (c) where limitations to crop growth can be readily addressed (e.g., application of N fertilizer on N-deficient soils, application of lime on acid soils), and (d) where soil rehabilitation will require more complex and longer-term investments (e.g., soils with multiple chemical, biological, and/or physical deficiencies). The upper left half of each circle refers to the status of productivity and the lower right half to soil fertility conditions with darker (half)circles indicating better conditions

(Rouw and Rajot, 2004) or hardpan formation reducing rooting depth (Lahmar et al., 2012). In such conditions, soil fertility rehabilitation is a pre-condition for inducing increased crop growth and the nature of the degradation status should determine whether it still makes economic sense to rehabilitate certain soils, especially in the absence of specific incentive schemes such as food-for-work programmes aiming at establishing physical and/or biological erosion control measures.

Sustainable Intensification (SI) could be placed in the top right circle in Figure 11.1b. A substantial acreage of cropland in densely populated areas of SSA, characterized by low crop productivity and poor soil health due to long-term nutrient mining and SOC decline, can be situated within or near the conceptual 'initial status' circle of Figure 11.1c. That said, as described above, within each farm, various fields can be positioned at different locations within the two-dimensional space. While homestead plots commonly have good yields and better soil fertility conditions, and are often reserved for high-value crops, degraded outfields can often be mapped at the lower left of this space and reserved for other uses, e.g., woodlots.

Pathways towards SI and potential entry points

Sustainable Intensification requires increases in productivity and maintenance/ restoration of ecosystem services, and in this chapter we focus on field-based ecosystem services that are regulated by soil conditions. Various pathways can be followed to turn the 'lower left' situation of Figure 11.1b back to a 'top right' situation. Path (1) of Figure 11.1c depicts a pathway focussing primarily on increases in crop productivity, thus assuming that this will not only result in higher yields but also in higher amounts of biomass in the form of crop residues which can then be re-invested in rehabilitating soil fertility conditions. Integrated Soil Fertility Management (ISFM) (Vanlauwe et al., 2010) follows this logic, using fertilizer as entry point towards the SI of smallholder agriculture. ISFM also recognizes that non-responsive conditions require other entry points (Vanlauwe et al., 2010, Vanlauwe et al., 2015), including the application of lime on acid soils, or the use of high rates of farmyard manure (Zingore et al., 2008). Zingore et al. (2005), for instance, demonstrated that on clayey soils, commercial farmers were able to retain substantially higher SOC contents when using high yielding maize varieties and fertilizer in comparison with communal farmers who were practicing low input agriculture on the same soils.

On the other hand, path (2) follows a logic whereby investments in improved soil fertility conditions will gradually improve crop yields and thus move towards SI. Some agroforestry practices fit this logic since tree establishment can take some years before these deliver their full benefits to productivity and soil fertility (e.g., Garrity et al., 2010). Nevertheless, other agroforestry practices like improved fallows (Barrios et al., 1998; Chirwa et al., 2003), as well as those which rely on existing tree cover, like farmer-managed natural regeneration (Dossa et al., 2012) and the Quesungual agroforestry system (Fonte et al., 2010), have a more rapid impact and can contribute more quickly to enhancing crop productivity, as do other paradigms, such as conservation agriculture (e.g., Kassam et al., 2009), 'push–pull' intercropping (e.g., Khan et al., 2000), or crop–livestock integration (e.g., Achard and Banoin, 2003), which likely follow intermediate paths (3). Note that for severely degraded soils with multiple deficiencies, path (1) is not an option and rehabilitation of soil fertility, e.g., through the application of large amounts of manure for several years (Zingore et al., 2008), may be required to restore the responsiveness of soils to standard fertilizer and other amendments. For instance, in southern Benin Republic, deep-rooting *Senna siamea* trees were able to access the relatively fertile subsoil of a site with severe topsoil degradation and thus restore crop productivity (Aihou et al., 1998). More details on the above paradigms are given in relation to their potential impact on yield and soil fertility conditions in Table 11.1. Recognizing the strong demand for crop residue as livestock feed in West-African smallholder systems, Lahmar et al. (2012), based on CA (Conservation Agriculture) principles, explored the option to use prunings of native evergreen multipurpose woody shrubs to provide field permanent soil cover and rehabilitate degraded land through an aggradation than a conservation phase. The work of Fatondji et al. (2009) demonstrated the potential of significantly increasing cereal yields on degraded land using the Zai technology, whereby crops are grown in planting pits for harvesting water and spot-placing organic inputs and fertilizer. However, to our knowledge, there is no published research on the long-term management after the initial labour-demanding rehabilitation of degraded land through Zai.

Yield and SOC data from long-term trials in support of potential entry points

Besides restoring crop productivity and rehabilitating degraded soils, it is equally important to ensure that yields and soil fertility conditions are maintained on fields with favourable conditions ('top right' situations in Figure 11.1c). Data from two long-term trials (Figure 11.2), established on sites that had been cleared from natural fallows with supposedly favourable soil fertility conditions at their establishment, provide some insight into how this could be achieved. Data from a long-term agroforestry trial in Ibadan (Figure 11.2a) demonstrate that only the treatment with *Senna siamea* alley cropping and fertilizer application succeeds in retaining yield and relative SOC within the vicinity of the data at trial establishment, although some decline in yields is obvious. Without fertilizer, both yields and SOC decline in the Senna alley cropping treatment. Yields in the no-input control treatment decline very rapidly to values near zero while SOC decline takes more time. With fertilizer application, the decline in yield and SOC was reduced but did reach unacceptably low levels after 20 years (Figure 11.2a). This decline in SOC with fertilizer application contradicts what was earlier observed in Zimbabwe (Zingore et al., 2005), which is very likely related to the fact that maize yields in Nigeria were less than half those of Zimbabwe, with consequent lower inputs of maize crop residues and the lighter texture of the soil in Nigeria, thus providing less physical protection for applied organic C (Six et al., 2002). In a conservation agriculture trial in Zambia (Figure 11.2b), standard practices tend to result in decreasing yields and SOC, especially since maize residues were removed in this treatment (Thierfelder et al., 2013). Only the treatment with inclusion of cotton and sum hemp appears to retain yields at original levels (though SOC contents did not appear to decrease under all treatments with direct seeding and residue retention).

   While both trials in Figure 11.2 started from relatively good soil fertility conditions, experiments in Figure 11.3 started on degraded and non-degraded

Table 11.1 Selected content of currently promoted intensification paradigms

Paradigm	Major principles	Potential impact on yield	Potential impact on soil fertility status
Integrated Soil Fertility Management	1. Integrate improved varieties, fertilizer, organic resources, and other soil amendments 2. Target resources in relation to soil fertility gradients, resource endowments, and status of enabling conditions	1. Immediate increases in yield if the right inputs are applied to the right field type (e.g., the use of improved germplasm and improved varieties on responsive soils) 2. Rehabilitation of non-responsive soils will be required before increased yields can be expected on such soils	1. Enhanced crop yields also generate a larger amount of crop residues that can be recycled either directly or as manure after feeding to livestock thus potentially increasing SOC[1] stocks 2. Integration of organic resource production systems (e.g., dual purpose legumes) can further enhance the availability of (higher quality) organic inputs
Conservation Agriculture	1. Reduce tillage 2. Keep soil covered 3. Diversify cropping systems, e.g., with rotations 4. Apply fertilizer (or other sources of nutrients)[2]	1. The impact of CA on crop yields is usually expressed after 2–3 years 2. Sufficient nutrient inputs are required to produce the required amount of crop residues for mulch otherwise yields can decrease; reduced/no tillage without mulch application reduces crop yields 3. Mulch commonly improves soil moisture conditions and thus resilience of crops to drought stress	1. CA can enhance soil fertility conditions through reduced soil disturbance although care needs to be taken to manage degradation processes that are alleviated by tillage (e.g., soil crusting) 2. The impact of CA on SOC stocks varies between negative and positive, depending on many factors affecting C mineralization and stabilization
Crop–livestock integration	1. Integrate fodder options (e.g., fodder legumes) in cropping systems 2. Integrate appropriate feed and manure management systems 3. Store, compost, and recycle farmyard manure	1. Enhanced availability of livestock feed of appropriate quality improves livestock weight gains and manure production 2. Manure recycled to crops after proper composting/storage can create immediate increases in crop productivity 3. Competition for organic resources of high quality (e.g., legumes) can negatively affect crop yields especially if manure is not recycled	1. Manure is shown to retain more of its C in the SOC pool, probably related to C stabilization processes during manure production 2. Removal of crop residues for feed can negatively impact on SOC status, especially if the manure produced is not recycled in the same plot 3. Much of the potential benefits of manure depends on the way feeding regime of the livestock and the way in which the manure is managed and recycled

(Continued)

Table 11.1 (Continued)

Paradigm	Major principles	Potential impact on yield	Potential impact on soil fertility status
Agroforestry	1. Integrate perennials in cropping systems	1. Crop yield increases in 2–3 years in N-limited soils as a result of organic inputs from fast-growing N-fixing trees (e.g., *Gliricidia sepium*) 2. Selective slash and mulch management of existing tree cover increases yields in 1–3 years 3. Planting of trees and crops with compatible attributes to generate multi-strata agroforestry systems generates yield increases and a variety of products in the longer term	1. Above ground and below ground contributions of N-rich biomass increases soil organic N stocks and N availability to crops 2. Significant biomass additions, commonly greater than amounts added as crop residues, contribute to permanent soil cover, erosion control, increased SOC, greater soil water-holding capacity and nutrient availability, creation of habitats for beneficial soil organisms
'Push–pull' intercropping systems	1. Plant maize in the live intercropped Desmodium mulch 2. Harvest Desmodium fodder at least 3 times per year	1. Desmodium takes one year to establish so yield effects will take at least one year to appear 2. Once Desmodium is fully established and managed well, maize yields are substantially higher in comparison with current practices	1. In comparison with current practices, the larger amounts of maize stover and Desmodium above and below ground biomass can increase SOC stocks 2. N stocks can be enhanced due to the N added to the soil through biological N fixation 3. The Desmodium live mulch protects the soil from erosion and improves soil moisture conditions

[1] SOC means 'Soil organic carbon'.

[2] In recent publications, the need for external nutrient inputs to ensure a sufficient quantity of crop residues to keep at least 30% of the soil covered was proposed as an additional principle, first by Vanlauwe et al. (2014), and later confirmed by Lal (2015).

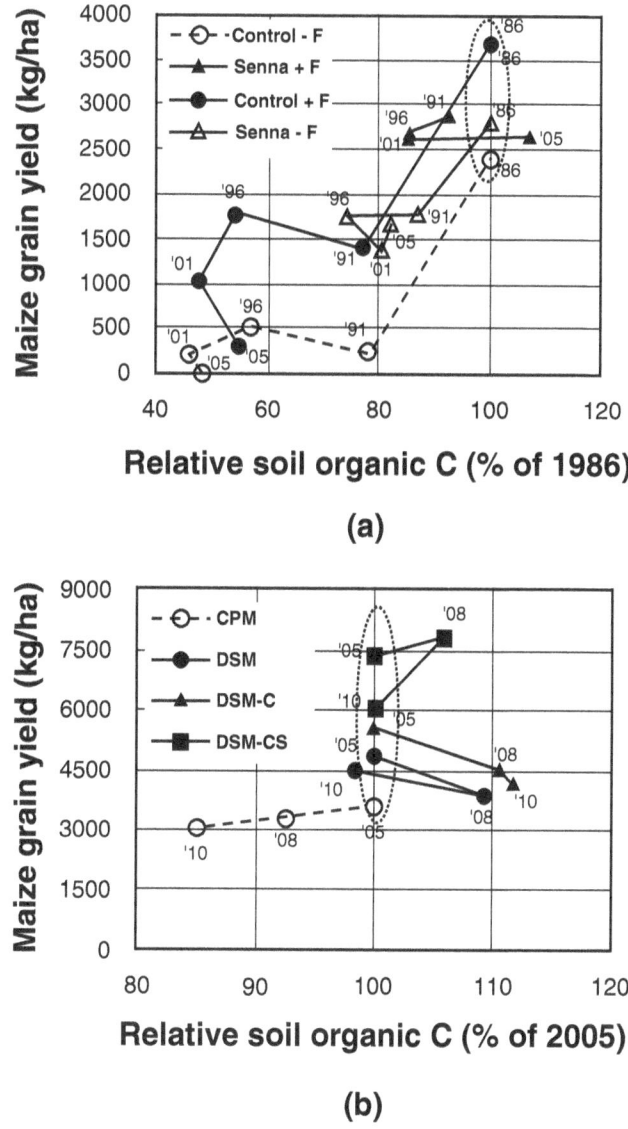

Figure 11.2 Trends in yields and relative soil organic carbon (SOC) contents for (a) a long-term (1986–2005) agroforestry trial in Nigeria, and (b) a long-term (2005–2010) conservation agriculture trial in Zambia. In (a) SOC data from 1991 were interpolated between those of 1986 and 1996. In (b) 'CPM' means 'conventional ploughing, residue removal, sole maize', 'DSM' means 'animal traction direct seeding, residue retention, sole maize', 'DSM-C' means 'animal traction direct seeding, residue retention, maize-cotton rotation', and 'DSM-CS' means 'animal traction direct seeding, residue retention, maize-cotton-sun hemp rotation'. Fertilizer was applied in all treatments. Dashed oval shapes indicate yields and SOC data at the start of the trials

Source: (a) Diels et al., 2004; Vanlauwe et al., 2005; Vanlauwe et al., 2012; (b) Thierfelder et al., 2013.

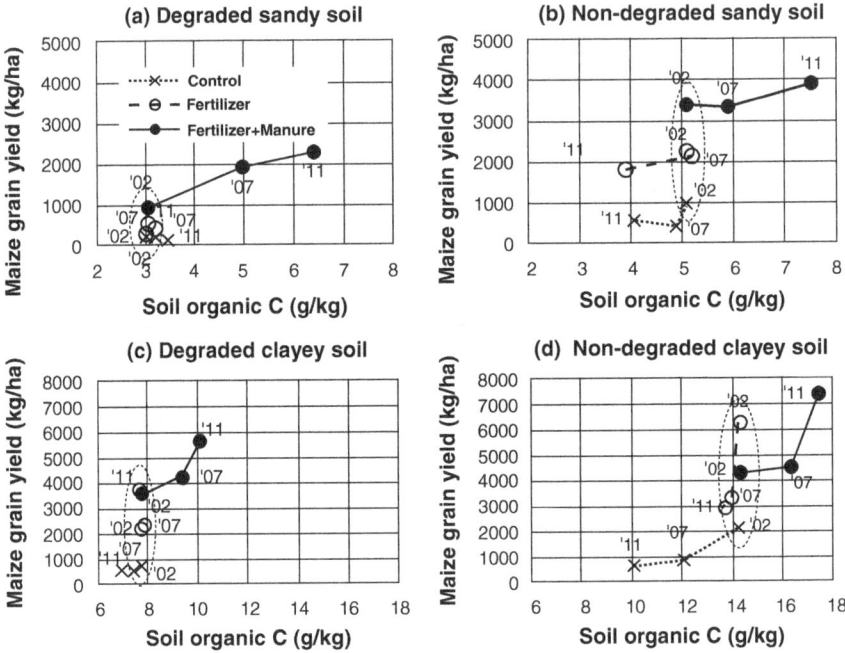

Figure 11.3 Trends in yields and relative soil organic carbon (SOC) contents for a set of long-term (2002–2011) trials established in Zimbabwe on (a) a degraded or (b) non-degraded sandy soil, and (c) on a degraded or (d) non-degraded clayey soil. Fertilizer application in the sole fertilizer treatment consisted of 100 kg N ha[-1] and 30 kg P ha[-1], applied annually, while in the mixed treatment, 100 kg N ha[-1] was applied as fertilizer in combination with 15 ton farmyard manure ha[-1], applied annually. Dashed oval shapes indicate yields and SOC data at the start of the trials

Source: Rusinamhodzi et al., 2013.

sites with sandy and clayey soils in Zimbabwe (Rusinamhodzi et al., 2013). In situations where crop residues are removed from the field, only the treatments with high application of manure managed to increase yields and SOC contents; however, fertilizer application doubled yields in 2011 on the clayey soil (Figure 11.3). As expected, yields and SOC declined on both non-degraded soils without the application of fertilizer and/or manure. Again, only in the treatment with application of manure do SOC contents increase substantially while yields in the fertilizer-only treatment are marginally lower than those in the combined treatment on the clayey soil (Figure 11.3). As for the data presented in Figure 11.2a, recycling maize crop residues does not appear sufficient to increase SOC contents.

Similar long-term data assessing the status of yields and SOC (and other ecosystem services) are required to make objective inferences about the SI nature of various soil management paradigms and are unfortunately in short supply, especially for sub-Saharan Africa.

Productivity and natural resource integrity: Friends or foes?

System productivity and natural resource integrity are inherently foes since opening up natural ecosystems for agriculture consistently reduces their C stocks, above as well as below ground, and agriculture results in a net removal of nutrients from available soil stocks, thus initiating nutrient mining and a consequent suite of degradation processes. In addition, conversion from natural to agricultural land can strongly reduce ecosystem diversity. Traditional systems under low intensification levels succeeded in managing trade-offs between agriculture and nature by limiting the agricultural phase to a relatively short period allowing nature to regenerate during relatively long fallow periods. Although in many situations such a model is no longer feasible and/or desirable, continuous agriculture without inputs of nutrients and organic matter either through fertilizer, biomass transfer, or integration of trees extracting nutrients below the crop root zone consistently leads to yield declines and soil degradation (Figures 11.2 and 11.3). In the short-term, more crop residues can be removed, e.g., to feed livestock, or crops having a higher yield and nutrient extraction rate can be chosen, but in the long-term, these practices cannot be sustained, unless organic resources are imported from outside the plot/farm, at the expense of other plots or natural lands. Some researchers suggest that an increase in livestock should be part of the solution, but Bekunda and Woomer (1996) and Sseguya et al. (1999) have shown that the use of cattle manure is closely related to farm size and that the latter is continuously shrinking under increasing land pressure. Unless cattle feed is imported from outside the farm, the use of fodder and crop residues for feeding zero-grazing cattle generally decreases nutrient replenishment at the plot level. The collapse of traditional 'nutrient transfer systems' under current population growth has also been demonstrated by Baijukya et al. (2005).

The main question then is how farmers can move from current, degraded, and low productivity conditions to SI and thereby ensure that improvement in either productivity or natural resource integrity does not occur at the expense of further degradation of the other. Considering the plot level, based on the data from the long-term trials, to maintain productivity and SOC conditions at the initial, relatively high levels (Figures 11.2a, 11.2b, 11.3b, 11.3d), under most conditions, simultaneous interventions are needed that address both crop productivity and SOC status. While fertilizer alone resulted in yield declines over time, except when maize yields were really high (Figure 11.3d), applying fertilizer in combination with tree prunings (Figure 11.2a), high biomass intercrops (Figure 11.2b), or farmyard manure (Figure 11.3b, 11.3d) allowed yields and SOC conditions to stabilize (Figures 11.2a, 11.2b), or further increase (Figures 11.3b, 11.3d). For degraded conditions (Figures 11.3a and 11.3c), while application of fertilizer results in gradual increases in crop yield for the clayey soil, the co-application of fertilizer and manure increases both yields and SOC contents.

Notwithstanding the continuing reference in literature to thresholds for SOC, it will remain hard or impossible to derive these for various soil and climatic conditions since SOC regulates various functions that will probably require different levels of SOC. For instance, Diels et al. (2002) noted that to increase the amount of plant-extractable water in the topsoil, an increase of 8 to 13 g kg^{-1} SOC would store an extra 1 mm of water in the top 15 cm of soil while the cation exchange capacity (CEC) function of SOC is only relevant for soils of which the CEC of the mineral fraction is less than 2 cmol$_c$ kg^{-1} (e.g., Arenosols or coarse-textured Ferralsols). On the other hand, plant-available N supply from the soil organic matter pool is known to commonly increase with higher SOC content. Studies on sandy soils in Zimbabwe showed that non-responsiveness was associated with SOC contents less than 4 g kg^{-1} (Mtambanengwe and Mapfumo, 2005; Zingore et al., 2008), although the SOC functions that influence crop response in these cases were complex and not clearly understood (Zingore et al., 2008). Interventions addressing crop productivity are 'friends' of natural resource integrity mainly when crop yields are high and crop residues recycled, while interventions addressing SOC can have a positive effect on crop yield only if substantial amounts of organic inputs with the right quality characteristics (e.g., high N content, low lignin, and soluble polyphenol contents) are applied (Palm et al., 2001).

Pathways towards SI can be considered as consisting of consecutive phases. An initial phase focussed on increasing productivity and thus in-situ biomass accumulation (ISFM paradigm) followed by a stabilization phase in which other paradigms take over. For instance, agroforestry, after some time, can facilitate SI by addressing the challenge of optimizing crop productivity while maintaining the provision of other ecosystem services (Barrios et al., 2012). Vanlauwe et al. (2014) argued that fertilizer is needed to kick-start CA since at low crop yields insufficient crop residue biomass is produced to keep the soil covered.

At farm level, farmers can make decisions on where to apply inputs and organic resources within their heterogeneous farms. Such decisions affect the productivity and natural integrity status of individual fields and the total farm whereby it is common for farmers to degrade certain plots (e.g., outfields furthest from the homestead) in favour of others (e.g., homestead plots), very often through the transport of crop residues for livestock feed and the consequent recycling of farmyard manure produced. Rowe et al. (2006) observed that regular applications of manure to only part of a farm, common on farms with limited manure availability, rapidly led to large gradients in crop yield while spreading manure evenly at a lower rate would give greater whole-farm yields. Fertilizer should be applied only on fields where grain yield is responsive to higher nutrient inputs, and not on infields which are nutrient-saturated or on degraded outfields. Of course, to improve the relevance of fertilizer and manure recommendations, it is necessary to consider resource limitations and production at the farm scale, and the effects of applying nutrient resources not only on current crop yield but also on the development of soil fertility of different fields. In highland conditions, with farms covering steep gradients, larger applications of nutrients in the uphill end of the farm, combined with live barriers following contour lines, may favour greater whole-farm yields given the natural redistribution of nutrients in steep terrain as a result of leaching and soil erosion which favours higher fertility soils downslope.

In reality, trade-offs in time and/or space between productivity and soil fertility rehabilitation will be the rule rather than the exception since not all the required inputs, amendments, and implements will be available to most smallholder farmers at the required time for the required space. Most smallholder farmers are resource-constrained and the earlier-mentioned soil fertility gradients are a manifestation of spatial trade-offs between productivity on homestead fields at the cost of degradation of outfields, mostly via biomass transfer to livestock feed and manure recycling strategies. Indeed, one can expect that crop-livestock farmers favour feeding their livestock, which contribute to multiple livelihood functions, to the detriment of long-term maintenance of their SOC status. Moreover, since decisions made by farmers on resource allocations in time and space will depend on their production objectives, resource endowments, and/or attitudes towards farming, assisting farmers with decision support tools that can facilitate decision-making is likely to have more impact in the route towards SI than providing 'best' recommendations for all.

The current chapter focussed on two important dimensions of SI of smallholder agriculture, thereby recognizing that achievement of SI will require institutional, economic, and social dimensions to be aligned. While agro-input and output market forces can provide the necessary incentives to invest in enhanced productivity, investing in natural resource rehabilitation that is independent of immediate benefits generated through improved productivity will require other incentives such as subsidy or payment for ecosystem service schemes and changes in land tenure systems with land ownership being a major driver for long-term investments in improving soil fertility and land quality. Moving towards SI requires investments from farmers and farming communities in terms of capital and labour and where many households are trapped in poverty and lack the necessary resources to invest (Tittonell and Giller, 2013), the move towards SI at scale will require substantial support and facilitation. Without this, the issue of 'friends or foes' is irrelevant.

Conclusions

In many cases, after clearing natural fallows, nutrient mining is the first degradation process kick-starting a number of other degradation processes, if not contained in time. Declining soil fertility drives crop yields down and triggers a mutually reinforcing vicious cycle of resource degradation which can often be reverted at early stages with the application of nutrient inputs. After years of soil degradation, soils can become non-responsive to fertilizer applications and must be rehabilitated before becoming productive again. Different SI trajectories, and land management paradigms associated with such trajectories, are discussed, and their potential impact on productivity increases and soil fertility conditions are evaluated. This is supported by yield and SOC data from ISFM, CA, and agroforestry trials, established on sandy to clayey soils.

The question of whether system productivity and natural resource integrity in smallholder farming are friends or foes does not have a simple answer. When population pressure over land is low, the potential for 'friendship' is high because there is often room to manage negative interactions and trade-offs through changes in the temporal and spatial arrangements across fields. As population pressure on land increases, and the flexibility for land use arrangements is limited or not possible, soil degradation is invariably initiated in the absence of nutrient inputs. External nutrient inputs are thus needed to prime farming systems, thus breaking the downward spiral of soil degradation. The biophysical context (e.g., non-responsive soils), however, can determine which nutrient input type would be effective (e.g., manure) under such circumstances. To make ends meet, poor smallholder families often curtail their investment horizons, resulting in a bias towards short-term returns which might jeopardize long-term land productivity.

Lastly, more long-term trials related to various intensification options are needed to guide meaningful inferences on the SI nature of those options, including aspects of resilience to biophysical stresses.

Acknowledgements

The authors gratefully acknowledge the many donors that have supported natural resources management R4D for the past decades and, more specifically, the Bill and Melinda Gates Foundation (BMGF), the Directorate General for Development (Belgium), the International Fund for Agricultural Development (IFAD), and the United States Agency for International Development (USAID). Partial support to produce this chapter was provided by the Humidtropics, the Forest, Trees, and Agroforestry (FTA), and the MAIZE CGIAR Research Programs.

References

Achard, F. and Banoin, M. (2003) 'Fallows, forage production and nutrient transfers by livestock in niger', *Nutrient Cycling in Agroecosystems*, vol 65, pp. 183–189.

Aihou, K., Sanginga, N., Vanlauwe, B., Lyasse, O., Diels, J. and Merckx, R. (1998) 'Alley cropping in the moist savanna of West-Africa: I. Restoration and maintenance of soil fertility on 'terre de barre 'soils in Bénin Republic', *Agroforestry Systems*, vol 42, pp. 213–227.

Baijukya, F.P., de Ridder, N., Masuki, K.F. and Giller, K.E. (2005), 'Dynamics of banana-based farming systems in Bukoba district, Tanzania: changes in land use, cropping and cattle keeping', *Agriculture, Ecosystems and Environment*, vol 106, pp. 395–406.

Barrios, E., Kwesiga, F., Buresh, R.J., Coe, R. and Sprent, J.I. (1998) 'Relating preseason soil nitrogen to maize yield in tree legume-maize rotations', *Soil Science Society of America Journal*, vol 62, pp. 1604–1609.

Barrios, E., Sileshi, G.W., Shepherd, K. and Sinclair, F. (2012) 'Agroforestry and soil health: Linking trees, soil biota and ecosystem services' in D.H. Wall, R.D. Bardgett, V. Behan-Pelletier, J.E. Herrick, T.H. Jones, K. Ritz, J. Six, D.R. Strong and W. van der Putten (eds) *Soil Ecology and Ecosystem Services*, Oxford University Press, Oxford, pp. 315–330.

Bekunda, M.A. and Woomer, P.L. (1996), 'Organic resource management in banana-based cropping systems of the Lake Victoria Basin, Uganda', *Agriculture, Ecosystems and Environment*, vol 59, pp. 171–180.

Chirwa, T.S., Mafongoya, P.L. and Chintu, R. (2003) 'Mixed planted-fallows using coppicing and non-coppicing tree species for degraded acrisols in eastern zambia', *Agroforestry systems*, vol 59, pp. 243–251.

Coe, R., Sinclair, F. and Barrios, E. (2014) 'Scaling up agroforestry requires research 'in' rather than 'for' development', *Current Opinion in Environmental Sustainability*, vol 6, pp. 73–77.

Diels, J., Aihou, K., Iwuafor, E.N.O., Merckx, R., Lyasse, O., Sanginga, N, Vanlauwe, B. and Deckers, J. (2002) 'Options for soil organic carbon maintenance under intensive cropping in the West-African Savanna' in B. Vanlauwe, J. Diels, N. Sanginga and R. Merckx (eds) *Integrated Plant Nutrient Management in sub-Saharan Africa: From Concept to Practice*, CABI, Wallingford, UK, pp. 299–312.

Diels, J., Vanlauwe, B., Van der Meersch, M.K., Sanginga, N. and Merckx, R. (2004) 'Long-term soil organic carbon dynamics in a subhumid tropical climate: 13 C data in mixed C 3/C 4 cropping and modeling with ROTHC', *Soil Biology and Biochemistry*, vol 36, pp. 1739–1750.

Dossa, E.L., Diedhiou, I., Khouma, M., Sene, M., Lufafa, A., Kizito, F., Samba, S.A.N., Badiane, A.N., Diedhiou, S., Dick, R.P. (2012) 'Crop productivity and nutrient dynamics in a shrub (*Guiera senegalensis*)–based Farming System of the Sahel', *Agronomy Journal*, vol 104, pp. 1255–1264.

Fatondji, D., Martius, C., Zougmore, R., Vlek, P.L.G., Bielders, C.L. and Koala, S. (2009) 'Decomposition of organic amendment and nutrient release under the zai technique in the Sahel,' *Nutrient Cycling in Agroecosystems*, Springer, vol 85, no 3, pp. 225–239.

Fonte, S.J., Barrios, E. and Six, J. (2010) 'Earthworms, soil fertility and aggregate-associated soil organic matter dynamics in the Quesungual agroforestry system', *Geoderma*, vol 155, pp. 320–328.

Garrity, D.P., Akinnifesi, F.K., Ajayi, O.C., Weldesemayat, S.G., Mowo, J.G., Kalinganire, A., Larwanou, M. and Bayala, J. (2010) 'Evergreen agriculture: a robust approach to sustainable food security in Africa', *Food Security*, vol 2, pp. 197–214.

Giller, K.E., Tittonell, P., Rufino, M.C., van Wijk, M.T., Zingore, S., Mapfumo, P., Adjei-Nsiah, S., Herrero, M., Chikowo, R., Corbeels, M., Rowe, E.C., Baijukya, F., Mwijage, A., Smith, J., Yeboah, E., van der Burg, W.J., Sanogo, O.M., Misiko, M., de Ridder, N., Karanja, S., Kaizzi, L., K'ungu, J., Mwale, M., Nwaga, D., Pacini, C. and Vanlauwe, B. (2011) 'Communicating complexity: integrated assessment of trade-offs concerning soil fertility management within African farming systems to support innovation and development', *Agricultural Systems*, vol 104, pp. 191–203.

Gockowski, J. and Sonwa, D. (2011) 'Cocoa intensification scenarios and their predicted impact on CO_2 emissions, biodiversity conservation, and rural livelihoods in the Guinea rain forest of West Africa', *Environmental Management*, vol 48, pp. 307–321.

Harris, D. and Orr, A. (2014) 'Is rainfed agriculture really a pathway from poverty?', *Agricultural Systems*, vol 123, pp. 84–96.

Jayne, T.S., Chamberlin, J. and Headey, D.D. (2014) 'Land pressures, the evolution of farming systems, and development strategies in Africa: a synthesis', *Food Policy*, vol 48, pp. 1–17.

Kassam, A., Friedrich, T., Shaxson, F. and Pretty, J. (2009) 'The spread of conservation agriculture: justification, sustainability and uptake', *International Journal of Agricultural Sustainability*, vol 7, pp. 292–320.

Khan, Z.R., Pickett, J.A., Berg, J. van den, Wadhams, L.J. and Woodcock, C.M. (2000) 'Exploiting chemical ecology and species diversity: stem borer and striga control for maize and sorghum in Africa', *Pest Management Science*, vol 56, pp. 957–962.

Lahmar, R., Bationo, B.A., Lamso, N.D., Guéro, Y. and Tittonell, P. (2012) 'Tailoring conservation agriculture technologies to West Africa semi-arid zones: building on traditional local practices for soil restoration', *Field Crops Research*, vol 132, pp. 158–167.

Lal, R. (2015) 'Sequestering carbon and increasing productivity by conservation agriculture', *Journal of Soil and Water Conservation*, vol 70, pp. 55A–62A.

Lehmann, J. and Kleber, M. (2015) 'The contentious nature of soil organic matter', *Nature*, vol 528, pp. 60–68.

Loos, J., Abson, D.J., Chappell, M.J., Hanspach, J., Mikulcak, F., Tichit, M. and Fischer, J. (2014) 'Putting meaning back into sustainable intensification'. *Frontiers in Ecology and the Environment*, vol 12, pp. 356–361.

Mtambanengwe, F. and Mapfumo, P. (2005), 'Organic matter management as an underlying cause for soil fertility gradients on smallholder farms in Zimbabwe', *Nutrient Cycling in Agroecosystems*, vol 73, pp. 227–243.

Palm, C.A., Gachengo, C.N., Delve, R.J., Cadisch, G. and Giller, K.E. (2001), 'Organic inputs for soil fertility management in tropical agroecosystems: application of an organic resource database', *Agriculture, Ecosystems and Environment*, vol 83, pp. 27–42.

Petersen, B. and Snapp, S. (2015), 'What is sustainable intensification? Views from experts', *Land Use Policy*, vol 46, pp. 1–10.

Rouw, A. De and Rajot, J. (2004) 'Soil organic matter, surface crusting and erosion in Sahelian farming systems based on manuring or fallowing', *Environment*, vol 104, pp. 263–276.

Rowe, E.C., van Wijk, M.T., de Ridder, N. and Giller, K.E. (2006), 'Nutrient allocation strategies across a simplified heterogeneous African smallholder farm', *Agriculture, Ecosystems and Environment*, vol 116, pp. 60–71.

Rusinamhodzi, L., Corbeels, M., Zingore, S., Nyamangara, J. and Giller, K.E. (2013) 'Pushing the envelope? Maize production intensification and the role of cattle manure in recovery of degraded soils in smallholder farming areas of Zimbabwe', *Field Crops Research*, vol 147, pp. 40–53.

Sanchez, P.A., Jama, B.A., Vanlauwe, B., Diels, J., Sanginga, N. and Merckx, R. (2001) 'Soil fertility replenishment takes off in East and Southern Africa' in B. Vanlauwe, J. Diels, N. Sanginga and R. Merckx (eds), *Integrated Plant Nutrient Management in Sub-Saharan Africa: From Concept to Practice*, CABI, Wallingford, UK, pp. 23–45.

Six, J., Conant, R.T., Paul, E.A. and Paustian, K. (2002) 'Stabilization mechanisms of soil organic matter: implications for C-saturation of soils', *Plant and Soil*, vol 241, pp. 155–176.

Sseguya, H., Semana, A.R. and Bekunda, M.A. (1999), 'Soil fertility management in the banana-based agriculture of central Uganda: farmers constraints and opinions', *African Crop Science Journal*, vol 7, pp. 559–567.

Stoorvogel, J.J., Smaling, E.A. and Janssen, B.H. (1993) 'Calculating soil nutrient balances in Africa at different scales', *Fertilizer Research*, vol 35, pp. 227–235.

TEEB (The Economics of Ecosystems and Biodiversity) (2010) *Ecological and Economic Foundations*, Earthscan, London.

Thierfelder, C., Mwila, M. and Rusinamhodzi, L. (2013) 'Conservation agriculture in eastern and southern provinces of Zambia: long-term effects on soil quality and maize productivity', *Soil and Tillage Research*, vol 126, pp. 246–258.

Tittonell, P. and Giller, K.E. (2013), 'When yield gaps are poverty traps: The paradigm of ecological intensification in African smallholder agriculture', *Field Crops Research*, vol 143, pp. 76–90.

Tittonell, P., Muriuki, A., Shepherd, K.D., Mugendi, D., Kaizzi, K.C., Okeyo, J., Verchot, L., Coe, R. and Vanlauwe, B. (2010), 'The diversity of rural livelihoods and their influence on soil fertility in agricultural systems of East Africa – a typology of smallholder farms', *Agricultural Systems*, vol 103, pp. 83–97.

Turmel, M.-S., Speratti, A., Baudron, F., Verhulst, N. and Govaerts, B. (2015) 'Crop residue management and soil health: a systems analysis', *Agricultural Systems*, vol 134, pp. 6–16.

Valentin, C., Rajot, J.-L. and Mitja, D. (2004) 'Responses of soil crusting, runoff and erosion to fallowing in the sub-humid and semi-arid regions of West Africa', *Agriculture, Ecosystems & Environment*, vol 104, pp. 287–302.

van Breemen, N., Mulder, J. and Driscoll, C.T. (1983) 'Acidification and alkalinization of soils', *Plant and Soil*, vol 75, pp. 283–308.

Vanlauwe, B., Bationo, A., Chianu, J., Giller, K.E., Merckx, R., Mokwunye, U., Ohiokpehai, O., Pypers, P., Tabo, R., Shepherd, K.D., Smaling, E.M.A., Woomer, P.L. and Sanginga, N. (2010) 'Integrated soil fertility management operational definition and consequences for implementation and dissemination', *Outlook on Agriculture*, vol 39, pp. 17–24.

Vanlauwe, B., Descheemaeker, K., Giller, K.E., Huising, J., Merckx, R., Nziguheba, G., Wendt, J. and Zingore, S. (2015) 'Integrated soil fertility management in Sub-Saharan Africa: unravelling local adaptation', *Soil*, vol 1, pp. 491–508.

Vanlauwe, B., Diels, J., Sanginga, N. and Merckx, R. (2005) 'Long-term integrated soil fertility management in south-western Nigeria: crop performance and impact on the soil fertility status', *Plant and Soil*, vol 273, pp. 337–354.

Vanlauwe, B., Nziguheba, G., Nwoke, O.C., Diels, J., Sanginga, N., Merckx, R. (2012) 'Long-term integrated soil fertility management in south-western Nigeria: crop performance and impact on the soil fertility status' in A. Bationo, B. Waswa, J. Kihara, I. Adolwa, B. Vanlauwe and K. Saidou (eds), *Lessons Learned from Long-Term Soil Fertility Management Experiments in Africa*, Springer, Dordrecht, the Netherlands, pp. 175–200.

Vanlauwe, B., Wendt, J., Giller, K.E., Corbeels, M., Gerard, B. and Nolte, C. (2014) 'A fourth principle is required to define conservation agriculture in sub-Saharan Africa: the appropriate use of fertilizer to enhance crop productivity', *Field Crops Research*, vol 155, pp. 10–13.

van Noordwijk, M. and Brussaard, L. (2014), 'Minimizing the ecological footprint of food: closing yield and efficiency gaps simultaneously?' *Current Opinion in Environmental Sustainability*, vol 8, pp. 62–70.

Wezel, A., Soboksa, G., McClelland, S., Delespesse, F. and Boissau, A. (2015), 'The blurred boundaries of ecological, sustainable, and agroecological intensification: a review', *Agronomy for Sustainable Development*, vol 35, pp. 1283–1295.

Worldbank (2007) *World Development Report 2008: Agriculture for Development*, The International Bank for Reconstruction and Development, Washington, DC.

Zingore, S., Delve, R.J., Nyamangara, J. and Giller, K.E. (2008) 'Multiple benefits of manure: the key to maintenance of soil fertility and restoration of depleted sandy soils on African smallholder farms', *Nutrient Cycling in Agroecosystems*, vol 80, pp. 267–282.

Zingore, S., Manyame, C., Nyamugafata, P. and Giller, K.E. (2005) 'Long-term changes in organic matter of woodland soils cleared for arable cropping in Zimbabwe', *European Journal of Soil Science*, vol 56, pp. 727–736.

12 Using local knowledge to understand challenges and opportunities for enhancing agricultural productivity in Western Kenya

Mary Mutemi, Maureen Njenga, Genevieve Lamond, Anne Kuria, Ingrid Öborn, Jonathan Muriuki and Fergus L. Sinclair

Introduction

Global agricultural systems are relied on to produce enough food to support a rapidly growing population (Mann et al., 2009; Godfray et al., 2010, Bremner, 2012), but many of these systems are currently under threat from land degradation (Deininger et al., 2003; Holden and Shiferaw, 2004), climate change (Schmidhuber and Tubiello, 2007; Lasco et al., 2014; Luedeling et al., 2014), and socio-economic and political forces. Insecurity of tenure, inequalities in access and control of land, poor farming practices, and weak policies and institutions have all been shown to undermine agricultural productivity (Gebremedhin and Swinton, 2003; Musemwa et al., 2015). Furthermore, it has been shown that small-scale agricultural production in many developing countries is dropping, even in 'high potential' areas. This has been attributed to decreasing farm sizes as a direct result of increasing populations and subdivision of land through customary inheritance (Lambin and Meyfroidt, 2011). Agricultural expansion and intensification have resulted in land degradation where sustainable practices have not been implemented (Waithaka et al., 2006).

Despite agriculture being the backbone of the economy in Sub-Saharan Africa (SSA), it is expected that farm sizes will continue to decrease due to customary inheritance and a threshold will be reached, if not already, in terms of farms being able to meet livelihood needs (Conelly and Chaiken, 2000; Waithaka et al., 2006; Masters et al., 2013; Öborn et al., 2015). Studies have shown that food security (in terms of quantity and quality of food) can be seriously jeopardized when farm sizes are too small to meet household needs (Conelly and Chaiken, 2000). According to Waithaka et al. (2006), there appears to be a minimum threshold of 0.4 ha (land area needed being dependent on household size), below which it becomes impossible for households to satisfy their dietary needs from subsistence agriculture alone.

With smallholder agriculture acting as the foundation of food security and an important part of the socio-ecological landscape in SSA (HLPE, 2013), sustainable intensification of these smallholder systems has been suggested as one way of enhancing livelihoods of smallholder farmers. Sustainable intensification has been defined as increasing productivity while maintaining the natural resource base (e.g. soil health) and delivery of ecosystem services, as well as enhancing social and ecological resilience to shocks and stresses including climate change (Vanlauwe et al., 2014). Other definitions emphasize the social dimensions of sustainable production systems, for example, Pretty et al. (2004) add to the above definition by stating that a sustainable production system is also one that makes productive use of human capital in the form of knowledge and capacity to adapt and innovate, and uses social capital to resolve common landscape level problems.

Despite different livelihood strategies being employed for economic improvement and increased agricultural productivity in areas that are facing population pressures and reduced farm sizes, sustainable intensification has largely not been achieved due to the myriad challenges faced by smallholder farmers. There is therefore a need to understand these challenges, identify opportunities, and develop sound scientific interventions that incorporate farmers' knowledge so as to improve productivity in these areas, shifting from a purely technical to a more inclusive approach (Barrios et al., 2012; Ginger, 2014).

In this chapter we explore farmers' local knowledge of the challenges they face in intensifying their farming systems, the livelihood strategies they employ to sustain their households, and the opportunities within these systems for enhancing agricultural productivity. Our aim was to combine local and scientific knowledge in order to design innovative interventions that are customized to local context and circumstances (Coe et al., 2014). The research was designed to inform activities being undertaken and planned by the CGIAR research program Humidtropics and its partners in Western Kenya.

Methodology

Site selection

Research was conducted across four villages in Western Kenya: Urudi and Bar Ohinga villages in Kisumu County and Uradi A and Ojalo villages in Siaya County. The four sites were all dominated by agricultural activities with similar crops being grown and livestock kept and were selected based on variation in vegetation cover, soil types, and market access. Mixed farming systems dominated in all villages and livestock was mostly local, and sometimes improved, breeds of cows, goats, and chickens. In Kisumu County, Bar Ohinga had visibly more forested areas than Urudi where there were fewer trees on farms and more intensive crop cultivation. Urudi had better access to market centres than Bar Ohinga. In Siaya County, Uradi A was a good representation of villages in the area in terms of soil type (red clay) while Ojalo had different soil types. Uradi A had better access to market centres than Ojalo. As a whole, the sites appeared to be fairly representative of the humid tropics where farming activities tend to be integrated (combining trees, crops, and livestock) and land use is intensive.

Local knowledge acquisition

Knowledge about agro-ecological interactions at the farm and landscape level was elicited from smallholder farmers, whose livelihoods are largely dependent on mixed farming systems, using the Agro-ecological Knowledge Toolkit (AKT), a knowledge-based systems approach (Sinclair and Walker, 1998; Walker and Sinclair, 1998; Dixon et al., 2001). Three stages of the AKT methodology were applied and complemented with participatory rural appraisal methods. The initial 'scoping' stage included a transect walk across each of the villages with the aid of a village leader and/or community worker. Single sex focus group discussions (FGDs) were held in each of the villages, with youths actively involved in each group. Participatory methods used during these sessions included resource mapping, historical timelines, and seasonal calendar exercises. The second 'definition' stage involved setting the boundaries to the study and deciding the sampling strategy which was purposive, with informants stratified according to topography (lower, mid, and upper slope), farm size (small 0.1–1 ha and medium 1–6 ha), and gender. This led into the third 'compilation' stage involving an iterative cycle of semi-structured interviews with a purposive sample of 60 willing and knowledgeable people (15 in each of the four sites). Interviews were processed and knowledge represented in two knowledge bases using the Agro-ecological Knowledge Toolkit software (AKT5), and then analyzed descriptively using the software's inbuilt tools (Sinclair and Walker, 1998; Walker and Sinclair, 1998).

Results

The knowledge bases contain a combined total of 635 unitary statements representing the knowledge of 60 farmers. The majority of statements (74%) show farmers' explanatory knowledge about agro-ecological interactions within their direct environment, while other statements serve to describe attributes of trees, livestock, and crops that they had experience of. In this section we start by characterizing the study sites and then move into looking more deeply at the shared and site-specific challenges and opportunities for enhancing livelihoods in the study area.

Shared and unique features within the sites

The four study sites shared some common socio-ecological features but also had some significant differences (Table 12.1). They all experienced a bimodal rainfall pattern; long rains from March to June and short rains from September to November. Although the rains tended to be within the same range of months of the year, they were of different durations across the sites. Soil types

Table 12.1 Farm characteristics and challenges and opportunities for sustainable intensification in four villages in Western Kenya

Bar Ohinga village (1440–1474 masl – gentle to steep slopes – red loam and murram soils)

Component	Description	Challenges	Opportunities
Crops	Maize, beans, sorghum, cowpeas, sweet potatoes, groundnuts, slender leaf rattlebox	Wildlife raids by monkeys, baboons, and weaverbirds; low profit from maize and beans sales due to exploitation by brokers; low prices due to market flooding (lack of crop diversification or specialization); insect pests such as stem borers and aphids; diseases (East African cassava mosaic virus); parasitic weeds (*Striga hermonthica*); water scarcity	*Identified by local people*: Alternative livelihoods like bee-keeping to reduce reliance on crops. *Identified by researchers*: Improve integration of farmers in crop value chains and formation of cooperatives for collective marketing; integrated pest management; introduction of disease resistant crop varieties; integrated weed management

Urudi village (1456–1503 masl – gentle slopes – red loam soils)

Component	Description	Challenges	Opportunities
Crops	Maize, beans, cowpeas, cassava, sweet potatoes, soya beans, *sukuma wiki*, nightshade, slender leaf rattlebox	Scarcity of cropland; low yields due to declining soil fertility; pests and diseases affecting crops and livestock; parasitic weeds (*Striga hermonthica*); water scarcity	*Identified by researchers*: Integrated pest management; introduce disease resistant crop varieties; integrated weed management; enhance existing kitchen gardens

Livestock	Local and improved goats, local and improved cows, local chickens, rabbits	Wildlife raids on free-range chickens; inadequate water resources	*Identified by local people:* Intensify chicken production through cage system *Identified by researchers:* Train and build capacity on rainwater harvesting techniques	Livestock	Local and improved goats, local and improved cows, local chickens, Kenbro chicken	Limited land for grazing livestock and keeping free-range chickens; low milk production	*Identified by local people:* Intensify chicken production through cage system *Identified by researchers:* Intensify dairy farming in zero grazing units
Trees	Woodlots (mostly *Eucalyptus* spp.); natural forest; boundary planting (*Markhamia lutea*, *Grevillea robusta*); home compound (*Mangifera indica*, *Persea americana*)	Indigenous forest species threatened by charcoal production; low profit from mango fruit sales due to exploitation by brokers; poor access to tree germplasm; pest and disease in *Mangifera indica*	*Identified by Researchers:* Invest in fuel efficient stoves and/or use local waste to make briquettes instead of using charcoal; improve integration of farmers in fruit value chains and market access; establish tree nurseries within the village; improve advisory services about tree pests and diseases	Trees	Woodlots (*Eucalyptus* spp., *Grevillea robusta*, *Casuarina equisetifolia*), boundary planting (*Grevillea robusta*, *Euphorbia tirucalli*); home compound (*Persea americana*, *Senna siamea*, *Casimiroa edulis*)	Poor access to tree germplasm; *Eucalyptus* spp. competes with crops; scarcity of land for planting trees; low resources for firewood; disease affecting *Persea americana*	*Identified by researchers:* Establish tree nurseries within the village; improve advisory services about tree pests and diseases; outsource alternative sources of fuel energy (e.g. LPG gas); invest in fuel efficient stoves

(*Continued*)

Table 12.1 (Continued)

Uradi A village (1299–1358 masl – flat to gentle to steep slopes – red clay soil)				Ojalo village (1281–1341 masl – gentle slopes – red clay, black cotton, mixed brown soils)			
Component	Description	Challenges	Opportunities	Component	Description	Challenges	Opportunities
Crops	Maize, beans, groundnuts, sorghum, cassava, sweet potatoes, cowpeas, traditional vegetables (e.g. slender leaf rattlebox)	Scarcity of land; low profit from maize and beans sales due to exploitation by brokers; low prices due to market flooding (lack of crop diversification or specialization); pests and diseases affecting crops and livestock; parasitic weeds (*Striga hermonthica*); limited water sources	*Identified by researchers:* Improve integration of farmers in crop value chains and formation of cooperatives for collective marketing; integrated pest management; introduce disease resistant crop varieties; integrated weed management; train and build capacity on rainwater harvesting techniques	**Crops**	Maize, beans, cassava, sweet potatoes sorghum, kales, traditional vegetables	Competition from gold mining for labour and land; low yields due to declining soil fertility (top soils are mixed with mined soils); pests and diseases affecting crops and livestock; parasitic weeds (*Striga hermonthica*)	*Identified by local people:* Promote alternative livelihoods like fish farming and bee-keeping. *Identified by researchers:* Plant trees and bananas for rehabilitation after gold mining; crop diversification; integrated pest management; introduce disease resistant crop varieties; integrated weed management

	Current situation	Constraints	Interventions	Current situation	Constraints	Interventions
Livestock	Local breeds of cows, goats, sheep, and chickens, free-range grazing mainly practiced	Lack of water resources; pests and diseases	*Identified by researchers:* Train and build capacity on rainwater harvesting techniques; improve efficiency of extension services	Local breeds of cows, goats, sheep, chickens, and pigs; some improved breeds of cows and pigs	Pests and diseases affecting livestock	*Identified by local people:* Improve efficiency of extension services *Identified by researchers:* Intensify dairy farming in zero grazing units
Trees	Boundary planting of *Euphorbia tirucalii*, *Markhamia lutea*, home compound (*Mangifera indica*, *Persea americana*); woodlots of *Eucalyptus* spp. on some farms	Poor access to tree germplasm; inadequate water for tree nurseries; limited land to plant trees	*Identified by local people:* Establish tree nurseries within the village *Identified by researchers:* Rainwater harvesting to provide water for tree seedlings; establish a tree nursery of multipurpose trees suitable for integrating on small farms	Boundary planting of *Markhamia lutea*; home compound (*Mangifera indica*, *Persea americana*); woodlots dominated by *Eucalyptus* spp.	Poor access to tree germplasm; destruction of young trees by free ranging livestock	*Identified by local people:* Establish tree nurseries within the village *Identified by researchers:* Increase protection of trees from livestock; establish a tree nursery of multipurpose trees suitable for integrating on small farms

ranged from red clay, red loam, murram (gravelly lateritic material) to mixed brown clay soils (Table 12.1).

The terrain varied from gentle sloping to steep within different parts of the villages. The average farm size of informants was 0.9 ha (ranging from 0.05 ha to 6 ha). The dominant farming system was mixed cropping with farmers utilizing their small lands for both subsistence (e.g. maize, beans, groundnuts, cassava, bananas, sorghum, sweet potatoes, and green vegetables) and cash crops such as fruit trees (e.g. mango and avocado), and for keeping livestock (e.g. cows, goats, sheep, poultry, and, in one village, pigs). The main agricultural land use practices were annual cropping, woodlots and boundary tree planting, and livestock keeping which was done through a mixture of zero grazing, tethering systems, and free grazing. Tree density was generally low with trees planted in homesteads, along farm boundaries, scattered on crop fields, and in woodlots, but it did vary widely between the four villages. Bar Ohinga village, for example, in contrast to the other three sites, had more tree cover with indigenous trees in forested areas (some species enrichment had also taken place) and exotic species in woodlots. Small-scale rock mining was commonly practiced alongside farming activities. Water was a scarce resource in Bar Ohinga and Uradi A villages. Zero grazing of dairy cows was mainly practiced in Urudi village while free grazing was practiced in the other three villages. Gold mining was unique to Ojalo village.

Constraints and opportunities for increasing agricultural productivity

It was found that farmers faced site-specific as well as shared challenges that acted as constraints to increasing agricultural yields to meet household needs and generate cash income. There was a mixture of natural resource based issues as well as labour constraints and market influences. Besides being knowledgeable of the challenges they were facing, farmers were also willing to discuss ideas for resolving some of the challenges and improving agricultural production/ livelihoods in their local areas. Where farmers were not able to offer potential solutions, the researchers identified opportunities based on the challenges posed by them (Table 12.1).

Common challenges and potential entry points

The common challenges identified by people across the four sites were: high population pressures and land fragmentation; decreased soil fertility; and pests and diseases affecting crops (Figure 12.1). Increasing populations had led to land fragmentation through subdivision of land based on male lineage, subsequently leading to agricultural intensification efforts in order to provide enough food for the households. The continuous cultivation of land rather than allowing fallow periods was having negative impacts on soil fertility, in turn leading to decreased crop yields. External inputs were considered expensive and out of reach for most farmers. Low crop yields meant that harvests were mainly used

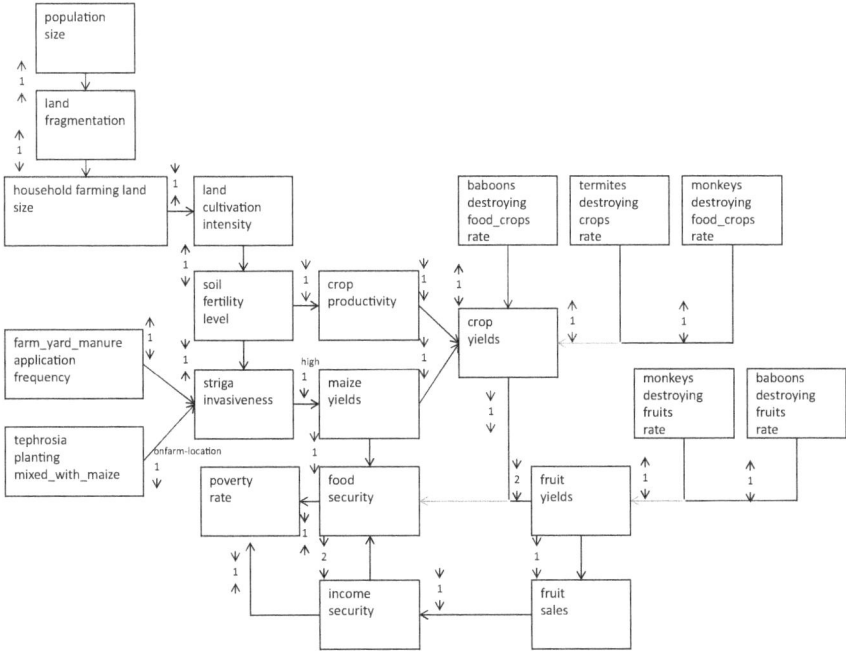

Figure 12.1 Causal diagram showing the agro-ecological interactions expressed by farm-ers and the challenges they faced when cultivating crops in Nyahera sub-loca-tion, Kisumu, Kenya. Nodes (boxes with straight edges) represent attributes of objects, processes, or actions; arrows connecting nodes show the direction of causal influence. The first small arrow on a link indicates either an increase (↑) or decrease (↓) in the causal node, and the second refers to the effect node. An asterisk (★) indicates attributes of objects processes or actions that do not have an increase or decrease value. Numbers between small arrows indicate whether the relationship is two-way (2), in which case an increase in A causing a decrease in B also implies that a decrease in A would cause an increase B, or one-way (1), where this reversibility does not apply

Source: Nyahera knowledge base.

for subsistence purposes rather than having surplus to sell at markets to earn cash income. Small landholdings also negatively affected the incorporation of trees on farms as trees were said to take up a lot of space on land that could otherwise be used for food crops.

Land shortages and decreasing farm sizes due to population pressures were an issue across sites but there appeared to be different entry points for potential sustainable intensification measures. In Urudi and Bar Ohinga, kitchen gardens were a common practice for household vegetable production unlike the other two villages. There was observed to be opportunities for enhancing the existing practices using raised beds and knowledge sharing between sites. Small farm sizes also posed a challenge to farmers keeping livestock and a move towards zero grazing and more intensive chicken farming were seen as potential opportunities for increasing production (Table 12.1).

Building ecological resilience into the system, such as soil fertility improvement practices using integrated nutrient cycling approaches, and looking into proper soil/land management techniques, has been said to be a pre-requisite for sustainable intensification (Folke, 2006). Combining proven and effective methods to make agriculture more productive, attractive, and sustainable, and also to prevent the natural environment from further degradation, was identified as a pressing issue by the researchers in discussion with farmers.

Pests and diseases were major challenges affecting both crop and livestock production. Maize was highly affected by stem borers while aphids were a nuisance pest especially to beans, cowpeas, kale, and nightshade. Weaverbirds affected sorghum and millet. Ticks, mites, and tsetse flies affected cows. Although farmers did not know the names of all the diseases affecting their crops, trees, and livestock, symptoms included: wilting in maize and beans, rolling of bean leaves, yellowing of bean leaves and some leaves appearing burnt, bean rot, stunted growth in maize, swelling of the crop, and diarrhoea in chickens. The diseases with known names included: cassava mosaic disease; bacterial wilt affecting tomatoes and Irish potatoes; blight in tomatoes; halo blight in beans; yellow sigatoka in bananas; mastitis and foot and mouth disease in cattle; and typhoid in chickens. Pests and diseases were also affecting mango productivity and profitability in Bar Ohinga and the same for avocado in Urudi (Table 12.1). Most farmers were not able to afford pesticides and disease treatments so their crops, trees, and livestock would suffer as a result, inevitably impacting their livelihoods. Based on these challenges, improving advisory services regarding the identification of pests and diseases and effective and affordable control methods would be of great benefit to these communities.

Farmers reported that maize yields had decreased significantly due to the presence of Striga, a parasitic weed that causes stunting in maize and millet crops. The cause of this weed was unknown though some farmers attributed its increase to the use of inorganic fertilizers (which may itself be related to low soil fertility). Although most farmers interviewed did not know how to control Striga, some had observed a decrease in its occurrence when using organic manure (Figure 12.1). The need for advisory services and knowledge sharing on control methods was also very apparent in this case.

To address the issue of Striga, integrated weed management by combining push–pull technologies and livestock manure could be a viable intervention (Hassanali et al., 2008). A similarly integrated approach would also be feasible for pests and diseases, whereby repellent plants are intercropped and others are used to 'pull' and trap pests or disease vectors around the perimeter of the crop (Glover et al., 2012; Pickett et al., 2014). Research organizations based in the region were carrying out demonstration trials of push–pull technologies in the study area, led by the International Centre of Insect Physiology and Ecology (icipe), so this may lead to wide adoption of the technology if accepted by farmers.

Site-specific challenges and potential entry points

The main challenges discussed by farmers specific to some of the villages included: crop raids by wildlife; water scarcity; overexploitation of natural forests; firewood scarcity; small-scale gold mining taking labour away from farming activities; and exploitation by 'middlemen' when marketing their products (Table 12.1). Besides mentioning the challenges there was also discussion around potential opportunities for positive change.

WILDLIFE HUMAN CONFLICTS

Although Bar Ohinga village experienced bimodal rains, meaning it was possible to have two cropping seasons per year, crop yields were limited due to raids by monkeys and baboons (Table 12.1). These animal pests caused huge damage and farmers were abandoning crop production in order to limit losses. It was not only agricultural crops that were affected by baboons; they also attacked chickens and made free-range chicken production unviable. Farmers were not compensated for their losses by the government by any means and this had resulted in feelings of resentment and anger.

The researchers identified a need to come up with practical solutions for reducing human–wildlife conflicts without damaging wild animal populations (Hoffman and O'Riain, 2012). Farmers could potentially consider adopting non-food crop production and/or venture into alternative means of livelihood like bee-keeping; this was something people were interested in being trained on (Table 12.1). Some farmers were already opting to concentrate on petty trade of fish, fruits, and vegetables sourced from neighbouring villages while others sought off-farm employment in the urban centres.

WATER MANAGEMENT

Water scarcity was a major challenge in Bar Ohinga and Uradi A. In Bar Ohinga, this was attributed to erratic rains and the absence of any local water sources. Water was sourced from boreholes in a neighbouring village at a fee and it took on average an hour to and from the nearest water source. Out of the 15 farmers interviewed in Bar Ohinga, two had modern rainwater harvesting tanks while 11 had old water harvesting tanks and two had none. There was evidently some knowledge about rainwater harvesting but the water harvested could not last these households for more than a week and those interviewed were asking for support to buy newer tanks. In contrast, Uradi A village had one spring and one borehole where residents drew water for domestic use, but out of the 15 farmers interviewed in this village, only three farmers practiced rainwater harvesting using tanks or gutters along the roof for collecting rainwater. This lack of water storage led to water shortages particularly during the dry season when the water volumes of the spring and borehole would run low. To curb the challenges of water scarcity, farmers expressed the need for training on water harvesting and to be supported with water harvesting equipment.

FOREST AND ON-FARM TREE RESOURCES

Charcoal production was a major economic activity in Bar Ohinga, particularly since crop production was not feasible due to wildlife raids; however, the activity posed a threat to natural forests and the indigenous treespecies they harbour. Lack of firewood for domestic use was a major challenge in Urudi, although trees such as *Grevillea robusta* and *Senna siamea* have been planted on the farms, and farmers opted to buy firewood or to trade their products (e.g. avocados) for firewood in neighbouring Bar Ohinga. It was explained that the small farm sizes constrained planting of trees for firewood in Urudi. Although this was often given as the main reason, there did appear to be interest in establishing local tree nurseries to improve access to tree germplasm and this could be an opportunity to identify suitable agroforestry species for integrating on small farms (this was largely applicable across sites). Due to limited land availability for planting more trees, and the need to reduce pressure on existing forest resources, an option could also be the adoption of alternative fuel technologies such as fuel efficient stoves (Abdelnour and Branzei, 2010) and/or using local waste to produce briquette (Njenga et al., 2009), to solve the firewood and charcoal problem.

LABOUR SHORTAGES AND LAND USE CONFLICTS

Competing land uses and labour shortages were major challenges in Ojalo village where small-scale gold mining was a major economic activity. Gold mining had started in the late 1990s and was gaining popularity due to the discovery of economically viable deposits. Men were mainly involved in the mining, which was a very labour intensive activity, while some women would help in the digging and carrying of the soil. This resulted in on-farm labour shortages. The burden would fall on women to maintain the farm but when farm work was too much, and there were no resources to hire labour, essential tasks such as ploughing or weeding would not be done. This had an effect on the overall productivity of the farms. In addition, those involved in mining were not sure where gold deposits were located so holes had been dug haphazardly on farms. This was resulting in wastage of limited land that would have otherwise been used for crop production. While some famers were planting bananas in the holes, others were leaving them open. They explained that gold mining had a negative effect on the composition and fertility of the soil. Unfertile soil dug from deep layers of the soil was mixed with relatively fertile topsoil, leading to reduced soil fertility. Based on what was observed and discussed with farmers, planting trees and bananas to rehabilitate areas after gold mining appeared feasible and farmers were interested in alternative income sources, for example bee-keeping and fish farming (Table 12.1).

MARKETING FARM PRODUCE

In Uradi A village the main cash crops were maize and beans, with little diversification from these staples. Due to flooding of the market with the same products at harvesting time, low prices were common and discouraged farmers from selling their produce. Some farmers were able to store their maize and beans for up to three months after harvesting which meant they could sell later at a higher price. Urudi village had a similar issue with avocados and Bar Ohinga with mangoes; in these cases, middlemen took advantage of opportunities to exploit farmers by buying from them at very low prices and reselling at high profits. Avocados being highly perishable goods, it was said to be difficult to store them to sell later on and there was no local fruit processing plant. Producers therefore felt exploited. One way of tackling exploitation and low prices could be through forming cooperatives for marketing farm produce; coupled with value addition this could increase market returns for products such as mangoes and avocados. Further, diversification of crops could also be a viable option for reducing both losses incurred due to flooding of markets with similar crop products and those from wildlife raids.

Discussion

Using integrated approaches to resolve complex agricultural challenges

The knowledge elicited gives insights into the challenges people were facing in intensifying their farming practices and brought to light potential entry points for improvement of food security and incomes of smallholder farmers in Western Kenya. With rapidly growing populations, pressure on the natural resource base has been increasing in the region for many years (Conelly and Chaiken, 2000). The impacts of this are evident in the low agricultural productivity, not to mention poverty and malnutrition levels of these rural communities, because of poor natural resource management practices (Bloss et al., 2004). As demonstrated by Lambin and Meyfroidt (2011), there are ways of economically developing while at the same time protecting the natural resources (e.g. forests, agricultural land, water sources) that people are reliant on, but there needs to be effective policies in place for this to happen.

As mentioned earlier, previous studies have shown a clear threshold in terms of farm size for households to satisfy their income and food security objectives from agriculture alone (Waithaka et al., 2006). Of the 60 interviewed households in the present study, 33% had farms of less than 0.4 ha in size and almost half of these were in Ojalo village. This serves to demonstrate the very real challenges farmers face in sustaining their livelihoods through farming and the need to seek alternative sources of income if landholdings are small. As presented in the results, options could include venturing into activities that do not require much land such as chicken, bee-keeping, and fish farming, or engaging in off-farm activities to supplement what people get from agriculture.

Similar to smallholders in other parts of SSA, mixed farming systems in Western Kenya have been widely adopted with little to no specialization (Conelly, 1994). A lack of specialization and marketing prowess explains why a majority of the farmers interviewed in Uradi A and Urudi villages produced similar products leading to flooded markets and low prices. Similar to earlier studies in Western Kenya (Kongstad and Mönsted, 1980; Francis and Hoddinott, 1993; Conelly, 1994; Crowley and Carter, 2000), this study revealed that with decreasing land productivity, coupled with unviable highly fragmented small landholdings, farmers are gradually abandoning agriculture for other on-farm and off-farm activities, such as rock and gold mining. However, if alternative and viable ways of making a living from these small farms were presented and advisory services were improved in the area, perhaps land productivity could be improved while also meeting livelihood needs. There is a need for integrated approaches to ensure sustainable agricultural production.

Best practices in terms of soil fertility management using improved crop varieties, fertilizers, and organic inputs adapted to local conditions need to be shared (Vanlauwe et al., 2014). An essential component of building ecological resilience of soils is through promoting the use of organic material for building soil organic matter and promoting nutrient cycling to complement inorganic fertilizer use (Pretty and Hine, 2001; Folke et al., 2010; Pretty et al., 2011). Farmers lacked access to inputs such as chemical fertilizers due to high cost so this is an area that would need addressing. Pretty et al. (2004) argue that smallholder farming systems can increase and sustain production when farmers are provided with inputs. Research organizations and the government should therefore invest more in providing the farmers with the necessary support and subsidized inputs for increased production (The Montpellier Panel, 2013).

There is also the need to invest in integrated management regimes to control pests, weeds, and diseases using locally available, easily accessible, and affordable technologies (Pretty et al., 2011; The Montpellier Panel, 2013). The results of the present study concur with several authors (Berner et al., 1995; Khan et al., 2006) who found that Striga causes major damage to maize, which is a major staple food crop for households in Western Kenya. Intercropping two or more crops at the same time, e.g. maize and beans or maize and *Desmodium* spp., has been shown to reduce the risk of total harvest losses due to Striga (Khan et al., 2006; Khan et al., 2009). According to Waithaka et al. (2007), high soil organic matter content tends to reduce Striga infestation, which is in agreement with those farmers using organic fertilizers who reported a decrease in the occurrence of the weed. In addition, manure and crop residues release nutrients to the soil slowly and help soils to build organic matter with long-term benefits (Palm et al., 1997; Place et al., 2003). Exposing farmers to information on improved farming methods could help in pest, weed, and disease control efforts (Chitere and Omolo, 2008). Sharing such knowledge is vital if smallholder farmers are to address the challenges they are facing. Not only is it important to recognize the role that local knowledge can play in informing scientific research, scientific research also needs to be communicated through appropriate channels to those people that would benefit from the results.

As shown, crop raids by wildlife have caused huge economic losses to farmers in parts of Western Kenya. Although governments have policies on enhancing wildlife's societal values, there is also a need to understand the underlying drivers of human–wildlife conflicts and how this can be mitigated (Terry, 2000).

This would ideally lead to the design of conflict resolution policies that are integrated in nature, including compensating farmers for loss of crops to wild-life (Okello, 2005).

Charcoal production in many rural areas of SSA has been opted as an income-generating activity especially in poverty stricken areas. The activity has not been sustainable since the survival rates of charcoal producing tree species have been low. Also, not many people in these areas give priority to planting suitable trees since they are slow growing (Iiyama et al., 2015). These challenges coupled with poor policy environment have led to overexploitation of naturally occurring tree species. To resolve these paradoxes, there has to be an understanding of the causes of engaging in charcoal production, and incentives that can be used to reduce poverty-driven charcoal production (Iiyama et al., 2015).

Water scarcity is a major problem not only in Western Kenya but also across other parts of SSA where smallholder farmers rely on rain-fed agriculture (Helmreich and Horn, 2009). Rainwater harvesting technologies have been shown to play a key role in addressing this challenge, especially in the wake of a changing climate and unreliable rainfall (Malesu et al., 2007; Thorlakson et al., 2012). Simple techniques such as roof catchments using corrugated iron sheets and ground surface collection are very feasible in many rural areas of developing countries since they are suited to local conditions (Sturm et al., 2009).

Wood products have continued to be the most universal fuel for rural areas in developing countries (May-Tobin, 2011). With the decreasing lands for retaining only trees, trees on farms are increasingly becoming popular worldwide and agro-forestry practices have been shown to help in meeting firewood needs and reducing pressures on natural forests (May-Tobin, 2011; Zomer et al., 2014). However, social and economic demands such as firewood, fodder, soil nutrients, and other needs need to be considered before steps are taken to promote a particular practice and invest heavily in its adoption (The Montpellier Panel, 2013). Whenever integrating trees on farms is not feasible due to extremely small pieces of land owned by individuals, adopting alternative sources of fuel energy for cooking would be a good solution. These can be sources like biogas, high-density pellets, and ethanol gas (Ministry of Energy and Petroleum, 2015). It is not enough for local people to merely be consulted on the integration and management of any chosen practices; they should be involved in the actual choosing of the practices to ensure their needs are met and uptake is successful (The Montpellier Panel, 2013).

Conclusion

The study revealed common challenges across the four villages relating mainly to land scarcity, decreased soil fertility, and pests and diseases in staple crops and fruit trees. However, each village had its own natural resource management issues and dynamics, thus requiring customized approaches to improving productivity of the existing farming systems. Farmers had detailed knowledge of the challenges faced in crop and livestock production but had significant knowledge gaps in terms of pest and disease identification and control.

Access to knowledge about integrated soil fertility management and integrated pest and disease control, along with better integration of farmers in market value chains, would be important interventions to increase agricultural productivity and income at farm and village level in order to improve smallholder livelihoods in the target area. The study demonstrates the importance of local knowledge research to better understand fine-scale variation in farming and community (here village) contexts and the needs and thinking of farmers in order to identify locally relevant entry points for sustainable intensification of farmer livelihoods (Coe et al., 2014). This study also reveals the trade-offs between on- and off-farm activities, e.g. in relation to labour, emphasizing the need for assessing the wider livelihood context and aspirations when agricultural innovations and interventions are negotiated with local communities. Any interventions should also be sensitive to gender roles within the household to have the greatest impact. Further research is needed to test which interventions are best suited and most likely to be adopted for sustainable intensification and improvements of farmer livelihoods in the study areas (Kiptot et al., 2007).

References

Abdelnour, S. and Branzei, O. (2010) 'Fuel-efficient stoves for Darfur: the social construction of subsistence marketplaces in post-conflict settings', *Journal of Business Research*, vol. 63, pp. 617–629, Richard Ivey School of Business, the University of Western Ontario.

Barrios, E., Coutinho, H.L.C. and Medeiros, C.A.B. (2012) *InPaC-S: Participatory Knowledge Integration on Indicators of Soil Quality – Methodological Guide*, World Agroforestry Centre (ICRAF), Embrapa, CIAT, Nairobi.

Berner, D.K., Kling, J.C. and Singh, B.B. (1995) 'Striga research and control, a perspective from Africa', *Plant Disease*, vol. 79, pp. 652–660.

Bloss, E., Wainaina, F. and Bailey, R. (2004) 'Prevalence and predictors of underweight, stunting, and wasting among children aged five and under in Western Kenya', *Journal of Pediatrics*, vol. 50, no. 5, pp. 260–270.

Bremner, J. (2012) 'Population and Food Security: Africa's Challenge', *Population Reference Bureau Policy Brief*, Washington, DC.

Chitere, P. and Omolo, B. (2008) 'Farmers' indigenous knowledge of crop pests and their damage in western Kenya', *International Journal of Pest Management*, vol. 39, no. 2, pp. 126–132.

Coe, R., Sinclair, F. and Barrios, E. (2014) 'Scaling up agroforestry requires research 'in' rather than 'for' development', *Current Opinion in Environmental Sustainability*, vol. 6, pp. 73–77.

Conelly, W. (1994) 'Population pressure, labour availability, and agricultural disintensification: the decline of farming on Rusinga Island, Kenya', *Human Ecology*, vol. 22, no. 2, pp. 145–170.

Conelly, W. and Chaiken, M. (2000) 'Intensive farming, agro-diversity, and food security under conditions of extreme population pressure in Western Kenya', *Human Ecology*, vol. 28, no. 1, pp. 19–51.

Crowley, E. and Carter, S. (2000) 'Agrarian change and the changing relationships between toil and soil in Maragoli, Western Kenya (1900–1994)', *Human Ecology*, vol. 28, no. 3, pp. 383–414.

Deininger, K., Jin, S., Gebre, S., Adenew, B. and Nega, B. (2003) 'Tenure security and land-related investment: evidence from Ethiopia', *World Bank Policy Research Working Paper*, no. 2991.

Dixon, H.J., Doores, J.W., Joshi, L. and Sinclair, F.L. (2001) *Agroecological Knowledge Toolkit for Windows: Methodological Guidelines, Computer Software and Manual for AKT5*, School of Agricultural and Forest Sciences, University of Wales, Bangor, UK.

Folke, C. (2006) 'Resilience: the emergence of a perspective for social–ecological systems analyses', *Global Environmental Change*, vol. 16, no. 3, pp. 253–267.

Folke, C., Carpenter, S., Walker, B., Scheffer, M., Chapin, T. and Rockstrom, J. (2010) 'Resilience thinking: integrating resilience, adaptability and transformability', *Ecology and Society*, vol. 12, no. 15, art. 20.

Francis, E. and Hoddinnot, J. (1993) 'Migration and differentiation in Western Kenya: a tale of two sub-locations', *The Journal of Development Studies*, vol. 30, no. 1, pp. 115–145.

Gebremedhin, B. and Swinton, S. (2003) 'Investment in soil conservation in northern Ethiopia: the role of land tenure security and public program', *Agricultural Economics*, vol. 29, no. 1, pp. 69–84.

Ginger, C. (2014) 'Integrating knowledge, interests and values through modeling in participatory processes: dimensions of legitimacy', *Journal of Environmental Planning and Management*, vol. 57, no. 5, pp. 643–659.

Glover, J., Reganold, J.P. and Cox, C.M. (2012) 'Plant perennials to save Africa's soils', *Nature*, vol. 489, pp. 359–361.

Godfray, H., Beddington, J., Crute, I., Haddad, L., Muir, J., Robinson, S., Thomas, S. and Toulman, C. (2010) 'Food security: the challenge of feeding 9 billion people', *Science*, vol. 327, no. 5967, pp. 812–818.

Hassanali, A., Herren, H., Khan, Z.R., Pickett, J.A. and Woodcock, C.M. (2008) 'Integrated pest management: the push–pull approach for controlling insect pests and weeds of cereals, and its potential for other agricultural systems including animal husbandry', *Philosophical Transactions of the Royal Society B*, vol. 363, pp. 611–621. doi:10.1098/rstb.2007.2173.

Helmreich, B. and Horn, H. (2009) 'Opportunities in rainwater harvesting', *Desalination*, vol. 248, no. 1–3, pp. 118–124.

HLPE (2013) *Investing in smallholder agriculture for food security*, A Report by the High Level Panel of Experts on Food Security and Nutrition of the Committee on World Food Security, Rome.

Hoffman, T.S. and O'Riain, M.J. (2012) 'Monkey management: using spatial ecology to understand the extent and severity of human–baboon conflict in the Cape Peninsula', South Africa', *Ecology and Society*, vol. 17, no. 3, p. 13. http://dx.doi.org/10.5751/ES-04882-170313.

Holden, S. and Shiferaw, B. (2004) 'Land degradation, drought and food security in less-favoured area in the Ethiopian highlands: a bio-economic model with market imperfections', *Agricultural Economics*, vol. 30, no. 1, pp. 31–49.

Iiyama, M., Neufeldt, H., Dobie, P., Hagen, R., Njenga, M., Ndegwa, G., Mowo, J., Kisoyan, P. and Jamnadass, R. (2015) 'Opportunities and challenges of landscape approaches for sustainable charcoal production and use', in Minang, P.A., van Noordwijk, M., Freeman, O.E., Mbow, C., de Leeuw, J. and Catacutan, D. (eds.) *Climate-Smart Landscapes: Multifunctionality in Practice*, World Agroforestry Centre (ICRAF), Nairobi, pp. 195–209.

Khan, Z.R., Midega, C.A.O., Wanyama, J.M., Amudavi, D.M., Hassanali, A., Pittchar, J. and Pickett, J.A. (2009) 'Integration of edible beans (*Phaseolus vulgaris* L.) into the push-pull technology developed for stemborer and *Striga* control in maize-based cropping systems', *Crop Protection*, vol. 28, pp. 997–1006.

Khan, Z.R., Pickett, J., Wadhams, L., Hassanali, A. and Midega, C.A.O. (2006) 'Combined control of *Striga hermonthica* and stemborers by maize-*Desmodium* spp. intercrops', *Crop Protection*, vol. 25, no. 9, pp. 989–995.

Kiptot, E., Hebinck, P., Franzel, S. and Richards, P. (2007) 'Adopters, testers or pseudo-adopters?' *Agricultural Systems*, vol. 94, no. 2, pp. 509–519.

Kongstad, P. and Monsted, M. (1980) 'Family labour, and trade in Western Kenya', Centre for Development Research no. 3, Scandinavian Institute of African Studies, Uppsala.

Lambin, E. and Meyfroidt, P. (2011) 'Global land use change, economic globalization, and the looming land scarcity', *Proceedings of the National Academy of Sciences of the United States of America*, vol. 108, no. 9, pp. 3465–3472.

Lasco, R., Delfino, R. and Espaldon, M. (2014) 'Agroforestry systems: helping smallholders adapt to climate risks while mitigating climate change', *Wiley Interdisciplinary Reviews: Climate Change*, vol. 5, no. 9, pp. 825–833.

Luedeling, E., Kindt, R., Huth, N. and Koenig, K. (2014) 'Agroforestry systems in a changing climate- challenges in projecting future performance', *Current Opinion in Environmental Sustainability*, vol. 6, pp. 1–7.

Malesu, M., Oduor, A. and Odhiambo, O. (2007) *Green Water Management Handbook: Rainwater Harvesting for Agricultural Production and Ecological Sustainability*, SearNet Secretariat, World Agroforestry Centre, Nairobi.

Mann, W., Lipper, L., Tenningkeit, T., MacCathy, N., Branca, G. and Paustia, K. (2009) *Food Security and Agricultural Mitigation in Developing Countries: Options for Capturing Synergies*, Food Agricultural Organization, Rome.

Masters, W., Anderson, D., Deehan, C., Hazel, P., Jayne, T., Jirstrom, T. and Reardon, T. (2013) 'Urbanization and farm size in Asia and Africa: implications for food security and agricultural research', *Global Food Security*, vol. 2, pp. 156–165.

May-Tobin, C. (2011) 'Chapter 8: Wood for Fuel', in Boucher, D., Elias, P., Lininger, K., May-Tobin, C., Roquemore, S. and Saxon, E. (eds.) *The Root of the Problem: What's Driving Tropical Deforestation Today?* Union of Concerned Scientists, Cambridge, MA, pp. 79–86.

Ministry of Energy and Petroleum (2015) *National Energy and Petroleum Policy*, Government of Kenya, Nairobi.

The Montpellier Panel (2013) *Sustainable Intensification: A New Paradigm for African Agriculture*, London. Agriculture for Impact.

Musemwa, L., Muchenje, V., Mushunje, A., Aghadasi, F. and Zhou, L. (2015) 'Household food insecurity in the poorest province of South Africa: level, cause, and coping strategies', *Food Security*, 7, vol. 647. doi:10.1007/s12571-015-0422-4.

Njenga, M., Karanja, N., Prain, G., Malii, J., Munyao, P., Gathuru, K. and Mwasi, B. (2009) 'Community-based energy briquette production from urban organic waste at Kahawa Soweto informal settlement, Nairobi', *Urban Harvest Working Paper Series*, no. 5 International Potato Center, Lima, Peru.

Öborn, I., Kuyah, S., Jonsson, M., Dahli, S., Mwangi, H. and Leew, J. (2015) 'Landscape-level constraints and opportunities for sustainable intensification in smallholder systems in the tropics', in Minang, P.A., van Noordwijk, M., Freeman, O.E., Mbow, C., de Leeuw, J. and Catacutan, D. (eds.) *Climate-Smart Landscapes: Multifunctionality in Practice*, World Agroforestry Centre (ICRAF), Nairobi, pp. 163–177.

Okello, M. (2005) 'Land use changes and human-wildlife conflicts in the amboseli area, Kenya', *Human Dimensions of Wildlife – An International Journal*, vol. 10, no. 1, pp. 19–28.

Palm, C.A., Myers, R.J. and Nandwa, S.M. (1997) 'Organic-inorganic nutrient interaction in soil fertility replenishment', in Buresh, R.J., Sanchez, P.A. and Calhoun, F. (eds) *Replenishing soil fertility in Africa*, Soil Science Society of America, Madison, WI, USA, Soil Science Society of America Special Publication 51, pp. 193–218.

Pickett, J.A., Woodcock, C.M., Midega, C.A.O. and Khan, Z.R. (2014) 'Push-pull farming systems', *Current Opinion in Biotechnology*, vol. 26, pp. 125–132.

Place, F., Barrett, C.B., Freeman, H.A., Ramisch, J.J. and Vanlauwe, B. (2003) 'Prospects for integrated soil fertility management using organic and inorganic inputs: evidence from smallholder African agricultural systems', *Food Policy*, vol. 28, no. 4, pp. 365–378.

Pretty, J. and Hine, R. (2001) *Reducing Food Poverty with Sustainable Agriculture: A Summary of New Evidence*, University of Essex, Colchester.

Pretty, J. and Smith, D. (2004) 'Social capital in biodiversity conservation and management', *Conservation Biology*, vol. 18, no 3, pp. 631–638.

Pretty, J., Toulim, C. and Williams, S. (2011) 'Sustainable intensification in African agriculture', *International Journal of Agricultural Sustainability*, vol. 9, no. 1, pp. 5–24.

Schmidhuber, J. and Tubiello, F. (2007) 'Global food security under climate change', *Proceedings of the National Academy of Sciences*, vol. 104, no. 50, pp. 19703–19708.

Sinclair, F.L. and Walker, D.H. (1998) 'Acquiring qualitative knowledge about complex agro ecosystems. Part 1: representation as natural language', *Agricultural Systems*, vol. 56, pp. 341–363.

Sturm, M., Zimmerman, M., Schutz, K., Urban, W. and Hartung, H. (2009) 'Rainwater harvesting as an alternative water resource in rural sites in central northern Namibia', *Physics and Chemistry of the Earth part S A/B/C*, vol. 34, no 13–16, pp. 776–785.

Terry, M.A. (2000) 'The emergence of human-wildlife conflict management: Turning challenges into opportunities', *International Biodeterioration and Biodegradation*, vol. 45, no 3–4, pp. 97–102.

Thorlakson, T., Neufeldt, H. and Dutilleul, F. (2012) 'Reducing subsistence farmers' vulnerability to climate change: evaluating the potential contributions of agroforestry in Western Kenya', *Agriculture and Food Security*, vol. 1, no. 15, pp. 1–13.

Vanlauwe, B., Coyne, D., Gockowski, J., Hauser, S., Huising, J., Masso, C., Nziguheba, G., Schut, M. and Van Asten, P. (2014) 'Sustainable Intensification and the African smallholder farmer', *Current Opinion in Environmental Sustainability*, vol. 8, pp. 15–22.

Waithaka, M.M., Thornton, P.K., Herrero, M. and Shepherd, K.D. (2006) 'Bio-economic evaluation of farmers' perceptions of viable farms in Western Kenya', *Agricultural Systems*, vol. 90, no.1, pp. 243–271.

Waithaka, M.M., Thornton, P.K., Shepherd, K.D. and Ndiwa, N.N. (2007) 'Factors affecting the use of fertilizers and manure by smallholders: the case of Vihiga, western Kenya', *Nutrient Cycling in Agroecosystems*, vol. 78, no. 3, pp. 211–224.

Walker, D.H. and Sinclair, F.L. (1998) 'Acquiring qualitative knowledge about complex agroecosystems, Part 2: formal representation', *Agricultural Systems*, vol. 56, pp. 365–386.

Zomer, R.J., Trabucco, A., Coe, R., Place, F., van Noordwijk, M. and Xu, J.C. (2014) 'Trees on farms: an update and reanalysis of agroforestry's global extent and socio-ecological characteristics', *Working Paper* 179. Bogor, Indonesia: World Agroforestry Centre (ICRAF) Southeast Asia Regional Program. DOI: 10.5716/WP14064.PDF.

13 Exploring options for sustainable intensification through legume integration in different farm types in Eastern Zambia

Carl Timler, Mirja Michalscheck, Stéphanie Alvarez, Katrien Descheemaeker and Jeroen C.J. Groot

Introduction

In Zambia maize is the main staple food crop and, with a share of 52% in the daily calorie intake of the local population, it is critical for ensuring the national food security (FAOSTAT, 2013). Of the total maize consumed in Zambia, smallholder farmers produce 80% in rain-fed systems under low soil fertility, frequent drought and with a limited use of high yielding varieties or inorganic fertiliser (Sitko et al., 2011). In eastern Zambia, the livelihoods of small-scale farmers depend largely on maize-legume mixed systems characterised by low productivity, extreme poverty and environmental degradation (Sitko et al., 2011). Thus, there seems to be a great need for sustainable intensification of these farming systems, for instance through promoting best practices in maize–legume integration. Maize–legume cropping provides protein-rich food for humans, residues for animal feed, composting and soil amendments and nitrogen inputs through symbiotic fixation by the legume. Sustainable intensification of farming systems can take place through changes in resource use and allocation that increase farm productivity while reducing pressure on local ecosystems and safeguarding social relations. According to Pretty et al. (2011), this entails the efficient use of all inputs to produce more outputs while reducing damage to the environment and building a resilient natural capital from which environmental services can be obtained. Sustainable intensification results from the application of technological and socio-economic approaches that may be categorised into genetic, ecological and socio-economic intensification (The Montpellier Panel, 2013).

Smallholder farming systems are often highly diverse in terms of biophysical and socio-economic characteristics. The diversity among systems stems *inter alia* from differences in soil fertility, in farmers' livelihood aspirations and the availability of resources such as land, labour as well as financial assets. Hence, instead of providing blanket recommendations for smallholder farmers, recognising and responding to the variability in local farm characteristics can lead to more appropriate, targeted and effective (design) recommendations to achieve improvements in agricultural production (Ojiem et al., 2006; Tittonell et al., 2010; Chikowo et al., 2014). Farm typologies aim at meaningful groupings of farms into subsets, homogenous according to specific criteria (Anderson et al., 2007; Alvarez et al., 2014), which can be used for technology targeting. Creating these typologies attempts to reach a useful compromise between analysing every single farm and assuming a broad category such as 'smallholders in general' based on average characteristics.

The main objective of this study was to perform an ex-ante evaluation of farm-type specific interventions for sustainable intensification and innovation at the farm level. Subsidiary objectives were to: (i) characterise the diversity of farming systems within the action sites in terms of resource endowment and legume cultivation practices; (ii) diagnose the systems in terms of productive, environmental and economic performance; (iii) explore trade-offs and synergies among various farm performance indicators across farm types; and (iv) identify potential points of improvement based on farm interviews and model explorations.

Methodology

A baseline survey was conducted in 2011/2012 in Eastern Zambia (Chipata, Katete and Lundazi districts) to obtain an initial description of the local farming systems and their diversity, and to derive a statistical farm typology. The resulting typology allowed selection of representative farms per type for the detailed characterisation (DC) survey. The DC survey, conducted during June 2014, provided the basis for a complete farming system diagnosis and an exploration of innovations using the whole-farm model FarmDESIGN. The exploration with the computer model yielded suggestions for system redesign, aiming at an improvement in the economic, social and environmental performance as compared to the current farm situation.

Typology

The farm types for this research were generated by two multivariate analyses, a principal component analysis (PCA) and a hierarchical clustering analysis (performed with the statistical software R, package *ade4*) on the surveyed baseline farms (n = 746). An early expert consultation served to develop a hypothesis on important farm characteristics to use to distinguish between farm types: 'farms differ in terms of their farming resources (land and labour) and their current application of integration of grain legumes'. This hypothesis was used to support the selection of variables for PCA: variables related to farm structure (operated area, tropical livestock units), to labour resource and constraints (total labour inputs, cost of hiring labour, proportion of total labour input used for land preparation and for weeding), to income source (crop, livestock and off-farm incomes) and to legume[1] practices (proportion of total operated area cultivated with legumes, years of experience in growing legumes and farmer's legume evaluation) were used. The hierarchical cluster analysis allowed classification of the farms into different farm types. The typology method was based on the guidelines set out in Alvarez et al. (2014).

Detailed characterisation

To perform the DC, for each farm type, a representative farm was selected from each of the three districts (Chipata, Katete and Lundazi) in the Eastern Province of Zambia (n = 15). The DC survey tool was developed for the data needs of the FarmDESIGN model. The captured data was used for the parameterisation of the model. The DC was complemented by secondary data (results of trials conducted at Msekera Research station in Chipata, project reports and external literature).

Model analysis

FarmDESIGN is a bio-economic static model, capturing structural as well as functional farm characteristics (Groot et al., 2012). It uses field crop information (e.g. plot sizes, crop types, intercrops and crop products) and cropping management practices such as manure, inorganic fertiliser, and pesticide use. The model also uses information on livestock (types, numbers and products) and on livestock management practices (e.g. animal feeding, livestock allotment, manure storage and herd replacement strategy). FarmDESIGN further assesses the destinations of crop and animal products such as household consumption, market sales or incorporation of residues into the fields. Also, soil and climate characteristics are integrated in the model. The FarmDESIGN model hence captures biophysical and economic features as well as management aspects of the particular farming system.

Based on these inputs, FarmDESIGN determines detailed nutrient cycles and annual feed balances, soil organic matter status, operating profit and labour balances. Beyond displaying the current farm situation, FarmDESIGN allows the exploration and evaluation of the impacts of different management decisions, changes in input use and production priorities. Based on available resources, the model is given a delimited room to reallocate these resources aiming towards defined farm objectives (desired outputs). The multi-objective optimisation algorithms generate diverse sets of alternative farm configurations that represent windows of opportunities or solution spaces for the case study farm (Groot and Rossing, 2011). The model aims to find alternative farm configurations using different decision variables to find configurations that achieve the objectives and that are within the constraints that have been set.

In this study, the decision variables used were the areas of the currently grown crops and five new 'intervention crops' suggested by project partners: maize–cowpea intercrop, sole soybean crop, sole cowpea crop, maize after cowpea and maize after soybean. The explorations used three objectives: (i) to maximise farm operating profit, (ii) to maximise the organic matter added to the soil, and (iii) to minimise the farm labour requirements. The ranges of non-maize crops were restricted between 0 and 70% of the total area and the range of maize and maize intercrops between 0 and 100% of the total area. As the total farm area remains unchanged, a reduction in area of one crop will be reflected by an increase in area of a crop that is more favourable in terms of achieving the objectives. Constraints were set on the total farm area and the ruminant feed balance (animals must always be sufficiently fed in all configurations). The frontier of the resulting graphical solution cloud represents the possible Pareto-optimal farming systems alternatives according to the model and makes the trade-offs and synergies between objectives visible and able to be evaluated.

From the 15 farms surveyed in the DC, one farm of each type was chosen, based on its representativeness to its type, to be used for the final model analysis; one farm from Chipata and two farms each from Katete and Lundazi districts.

The information derived from the modelling is important in guiding discussions between farmers and stakeholders towards the selection of farm designs that are likely to be adopted by target farmers. The systems approach allows assessing the combined effects of changes in farm configuration on all other system components. Revealing the impacts of these system component changes provides information as to their suitability for that specific farm and for the type they represent.

Results

Typology

The local farming systems were grouped into five farm types mainly according to their resource endowment, their income source and their labour constraints (Table 13.1).

Type L-LEGU: Low resource endowed, most labour for land preparation, legume growers

L-LEGU farms tend to have the least cultivated land area and the lowest number of tropical livestock units (TLUs) with on average only one cattle and one goat (Table 13.1). On average, this farm type has the lowest share of farmers growing cash crops (62%) and the highest proportion of households reporting food insecurity (35%). L-LEGU farmers tend to cultivate a relatively large proportion of their fields with legumes and due to the low number of cattle available for draft power, spend the most labour on land preparation. They tend to spend the least proportion of labour on weeding compared to all types, probably due to their highest cost per hectare of herbicides.

Table 13.1 Average characteristics per farm type for rain-fed smallholder systems in the Eastern Province of Zambia.

Farm types[1]	L-LEGU	L-WEED	M-LEGU	MH-OFI	H-LVST
Household characteristics					
Number of people in household	6	6	7	8	9
Land use					
Cultivated land area (ha)	2.8	2.9	3.4	5.9	14
No. of crops grown	3	3	4	4	5
% of farmers growing cash crop(s)	62	70	72	74	82
Livestock					
Number of cattle	1	2	2	4	13
Number of goats	1	1	2	2	4
Number of sheep	0	0	0	0	1
Number of pigs	2	3	3	4	6
Number of chickens	9	7	12	17	16
Tropical livestock units (TLU)	1	1.6	2.4	4.1	10.7
Animal income per TLU (US$)[2]	20.3	17.7	25.2	24.4	22.3
Food security					
% of farms facing food shortage throughout the year or occasionally	35	29	25	17	8
Residue use					
% of all residues used as green manure	52	58	52	57	57
% of all residues fed to livestock	23	21	24	20	24
Income sources and amounts					
Off-farm income as % of total income	32	26	23	44	8
Crop income as % of total income	64	69	70	53	87
Animal income as % of total income	4	5	7	3	5
Total revenues (US$)[2]	508	567	865	3339	4762
Revenues per hh. member (US$)[2]	83.0	89.9	128.9	428.2	555.7
Herbicide costs per hectare (US$)[2]	0.68	0.13	0.17	0.16	0.45
Labour allocation					
Total labour (person-days year^{-1})	334	334	637	774	1 031
Labour days per hectare	119	115	185	131	73
Labour for land preparation (%)	32	11	15	13	15
Labour for weeding (%)	24	46	34	29	27
Labour for harvesting (%)	29	31	34	36	36
Labour for shelling & threshing (%)	15	12	17	22	23
Legume related information					
% of total area cultivated to legumes	24	14	27	15	15
Years of experience growing legumes	4.5	3.9	8.7	4.7	8.9

[1] L-LEGU: Low resource endowed, most labour for land preparation, legume growers; L-WEED: Low resource endowed, most labour for weeding, few legumes grown; M-LEGU: Medium resource endowed, legume growers, highest relative animal income; MH-OFI: Medium to high resource endowed, highest off-farm income; H-LVST: High resource endowed, high crop and animal income.
[2] 1 US$ = 5115 ZMK as at 31 December 2011 (www.xe.com). ZMK is an obsolete currency since 1 January 2013. New currency is ZMW

Type L-WEED: Low resource endowed, most labour for weeding,
few legumes grown

L-WEED farms tend to be relatively small in family size, cultivated land area and animal numbers (Table 13.1). L-LEGU and L-WEED types are quite similar in household size, operated area, crop diversity, per head income and total labour inputs, but a striking difference can be observed in their labour allocation. While L-LEGU farmers tend to spend most labour on land preparation, L-WEED farmers allocate the least labour to it and more to weeding. This might be associated with a higher number of cattle owned by L-WEED farms, which can assist with land preparation. Among all farm types, L-WEED farmers spend the largest share of labour on weeding and the smallest share on land preparation. More weeding labour was associated to low herbicide costs. Farmers of this type tend to be more food insecure than other farm types (except L-LEGU). On average, L-WEED farmers assign the least area of land to the cultivation of legumes.

Type M-LEGU: Medium resource endowed, legume growers, highest relative
animal income

M-LEGU farms tend to have a medium farm and family size, intermediary animal numbers as well as an intermediary income compared to the other farm types (Table 13.1). On average this type cultivates the greatest share of their land with legumes. They tend to have long term experience in growing legumes, and this farm type could potentially provide useful information about farmers' reasons for adopting legumes, about best practices and how to overcome constraints reported by other types of farmers. They have the highest total labour inputs per hectare (185 person-days ha^{-1}).

Type MH-OFI: Medium to high resource endowed, highest off-farm income

MH-OFI farms tend to have, by far, the highest off-farm income. Whilst having on average a relatively large family size, farm area, animal number, crop diversity and a high food security, this farm type has the lowest shares of crop and animal incomes among all farm types (Table 13.1). Despite the small share of animal income compared to total income, the animal income per TLU is the second largest among all types indicating a large share of the TLU sales. MH-OFI farms are inclined to allocate relatively little labour to land preparation, which is possibly associated with the high number of cattle (on average four per farm) available for traction.

Type H-LVST: High resource endowed, high crop and animal income

H-LVST farms tend to have the highest overall revenues, attributable to their significantly higher resource endowment in terms of operated area as well as TLUs. The numbers of animals are the highest among all farm types (Table 13.1).

H-LVST farms also have on average the largest share of farmers growing cash crops. They allocate more labour than other types to harvesting and shelling and threshing, indicating greater efforts in collecting and processing, adding market value to their farm products. This farm type has the greatest number of family members who contribute most of their labour to on-farm activities (concluded from comparatively low off-farm income). They are inclined to have the lowest amount of labour inputs per hectare (on average 2.5 times less than farm type M-LEGU), quite possibly due to their highest absolute expenses on herbicides when compared to other farm types. The high crop diversity makes farm households of this type resilient against climate and market price fluctuations, shown by the lowest share of households with food shortages. H-LVST farms tend to have the most experience in growing legumes among all farm types, but they allocate a relatively low share of their cultivated area to legumes.

In conclusion, from the types L-LEGU to H-LVST, an increasing gradient of revenues per household member, TLUs, land area and total labour is highlighted while food shortage decreases (Table 13.1). L-LEGU and M-LEGU farm types crop more legumes, and MH-OFI and H-LVST farm types have respectively an off-farm income generation or livestock activities orientation.

Model-based exploration

Model-based explorations were performed for five representative farms selected from each farm type (based on average features presented in Table 13.1). The scenario used entailed variable areas of the five new 'intervention crops'. The results from the explorations are presented below. The current situation of each farm is presented in Table 13.2. The results of the explorations (i.e. the solution space, with each dot representing an alternative farm configuration) are visualised in Figures 13.1, 13.2 and 13.3.

Trade-offs were identified between increasing operating profit and the other two objectives (increasing organic matter inputs and reducing labour requirements) for the five farm types, with only a few exceptions. In general, increasing the operating profit would require an increase in labour input (except for farm L-WEED; Figure 13.1c), and farm configurations with larger amounts of organic matter inputs into the soil would have lower operating profit (except for farm H-LVST; Figure 13.1a). There was a synergy between increasing organic matter inputs and reducing the labour requirements for farm L-LEGU and M-LEGU (Figure 13.1b).

H-LVST farm had the highest operating profit for all alternative configurations and the M-LEGU farm reaches the highest organic matter added to the soil (Figure 13.1a). The distance between the alternative farm configuration points and the current situation (horizontally or vertically) indicates the magnitude of the increase or decrease that can be reached in each objective. It can be seen that the L-WEED farm had relatively little room for increases in operating profit, yet has a large range for improvement in soil organic matter inputs. The result of this small range in operating profit probably stems from the

Table 13.2 The current situation of the five representative rain-fed smallholder systems in the Eastern Province of Zambia, chosen as representative of their type for exploration in FarmDESIGN.

Farm types[1]	L-LEGU	L-WEED	M-LEGU	MH-OFI	H-LVST
Farm area (ha)	3.2	2.0	6.7	13.4	23.0
Crops currently grown	Maize Groundnut Cowpeas Tobacco Pumpkin	Maize Groundnut Sunflower Cotton Sw. Potato Sugarcane Pumpkin Vegetables	Maize Groundnut Sunflower Soybean Sw. Potato Cassava	Maize Groundnut Sunflower Pumpkin Cowpea Vegetables	Maize Groundnut Sunflower Cotton Vegetables
Animals currently owned	Pigs Chickens	Cattle Pigs	Cattle Chickens Goats	Cattle Chickens Goats Sheep Pigs Ducks	Cattle Chickens Goats Pigs Doves
Operating profit (US$ year⁻¹)[2]	1 299	101	939	2 625	5 604
Organic matter added (kg ha⁻¹ year⁻¹)	1 229	1 147	1 451	1 222	710
Labour required (hours year⁻¹)[3]	0	50	3 027	5 503	360

[1] L-LEGU: Low resource endowed, most labour for land preparation, legume growers; L-WEED: Low resource endowed, most labour for weeding, few legumes grown; M-LEGU: Medium resource endowed, legume growers, highest relative animal income; MH-OFI: Medium to high resource endowed, highest off-farm income; H-LVST: High resource endowed, high crop and animal income.
[2] 1 US$ = 6.259 ZMW as at 1 July 2014 (www.xe.com)
[3] Additional hours over and above family labour required to manage crops and animals; represents labour hours that will have to be hired

fact that this farm's yields for maize are low (using local maize variety with low yield and possibly poor management) and hence the predicted yields used for intervention crops were consequently low too. The reason that the points for the H-LVST farm have a different shape to that of the other types is due to the fact that this farm with its large area (23 ha) has a larger room to manoeuver to find different configurations and thus the trade-offs between operating profit and the other objectives were less pronounced than for the smaller farms.

For each alternative configuration, it is also possible to examine the corresponding changes in crop areas, i.e. decision variables, according to the three

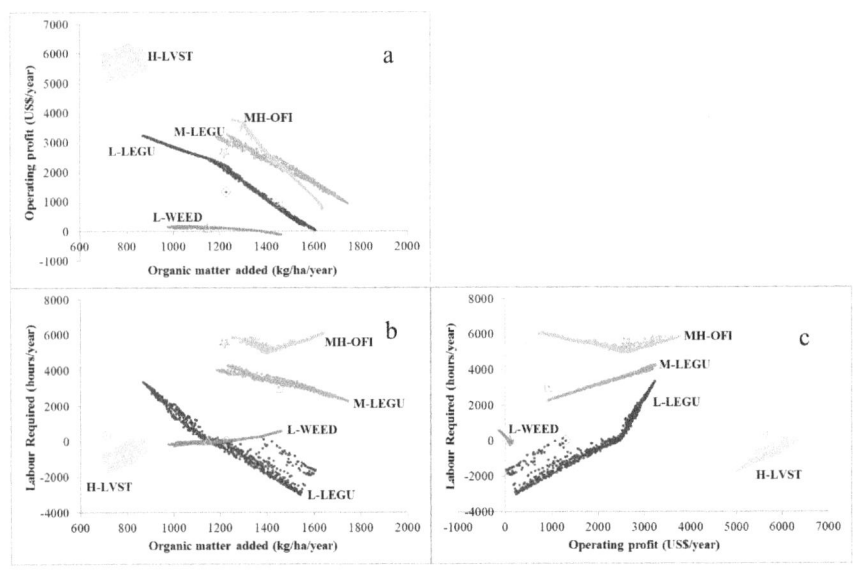

Figure 13.1 Performance of alternative farm configurations in terms of three farmer objectives, for five farm types in Eastern Zambia. The symbols indicate the performance of the original farm configurations (L-LEGU: ◊, L-WEED: Δ, M-LEGU: □, MH-OFI: and H-LVST: ○)

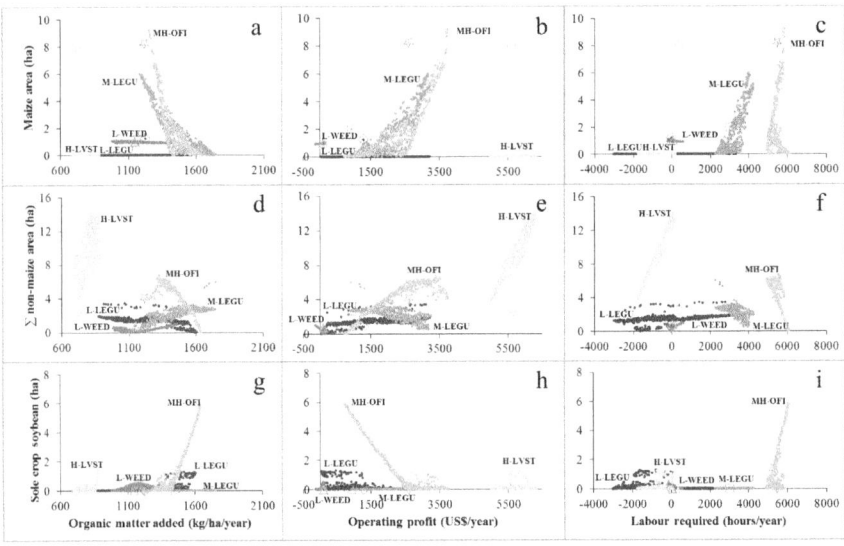

Figure 13.2 Performance of alternative farm configurations with different selected decision variables affecting changes in three farmer objectives, for five farm types in Eastern Zambia. Different points refer to the different farm types. Each point refers to an alternative farm configuration for that farm type. Maize area, ∑ non-maize area and sole crop soybean are decision variables related to allocation of area to these crops by FarmDESIGN. The following symbols indicate the performance of the original farm configurations (L-LEGU: ◊, L-WEED: Δ, M-LEGU: □, MH-OFI: and H-LVST: ○)

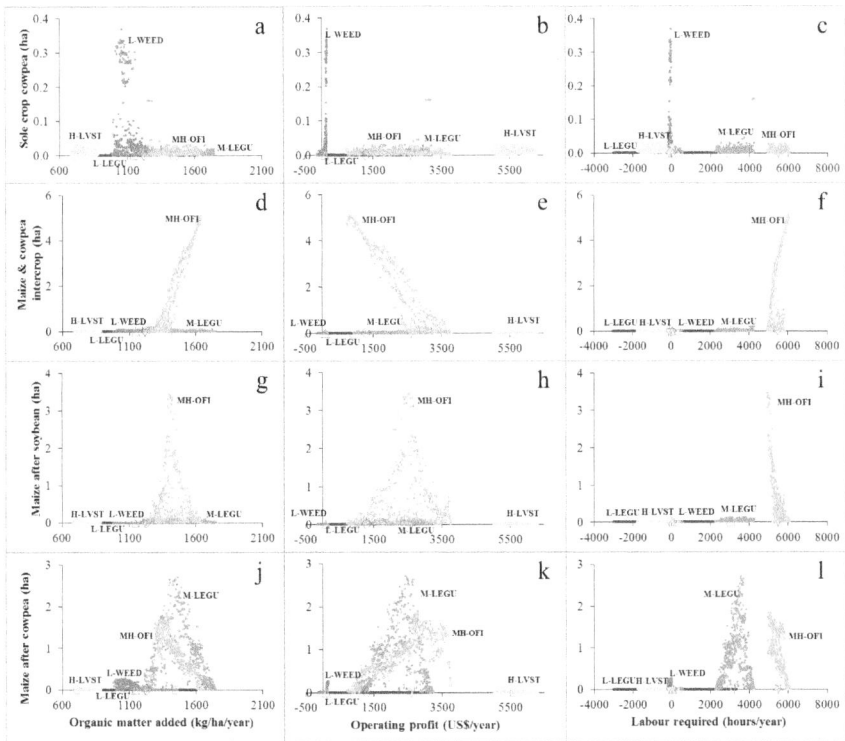

Figure 13.3 Performance of alternative farm configurations with different selected decision variables affecting changes in three farmer objectives, for five farm types in Eastern Zambia. Different points refer to the different farm types. Each point refers to an alternative farm configuration for that farm type. Sole crop cowpea, maize & cowpea intercrop, maize after soybean and maize after cowpea are decision variables related to allocation of area to these crops by FarmDESIGN. The following symbols indicate the performance of the original farm configurations (L-LEGU: ◊, L-WEED: ∆, M-LEGU: □, MH-OFI: and H-LVST: ○)

objectives (Figures 13.2 and 13.3). Figures 13.2a–c and 13.2d–f show the maize area and the sum of the areas of other currently grown non-maize crops respectively. Almost all alternative configurations for the five farm types had less maize area than are currently allocated to that crop; thus it seems to be more advantageous in terms of profit, organic matter additions and labour reduction to replace (at least partly) the currently grown maize crop with either a currently grown non-maize crop or a new 'intervention crop'. In the Figure 13.2a–c, it can be seen for the L-LEGU and H-LVST farms that the model chose to replace the entire current maize crop area for another crop; all the points for these two types are at or near zero.

From Figures 13.2g–i and 13.3a–l it is apparent that the model chose to create alternative configurations with specific intervention crops for specific farm types. This indicates the suitability of that intervention for that type showing the potential adoption of these interventions under the constraints faced by the farm. Some intervention crops, such as the maize and cowpea intercrop and the maize after soybean crop, were only chosen for one type, the MH-OFI. The figures also show that in some cases only one intervention crop was chosen for a type: the sole soybean for the L-LEGU type and maize after cowpea for the M-LEGU type. For the L-WEED type, the model chose the sole cowpea intervention in a greater amount, although the absolute area was relatively small. In addition, the model also chose to allocate land to maize after cowpea for the L-WEED type; thus for this type the combination of cowpeas in rotation with maize could prove to be a successful intervention. The MH-OFI type was allocated area by the model to almost all of the tested interventions. The increases of sole soybean area and maize and cowpea intercrop area for this type correspond with the trade-off trend of increases in the organic matter added and decreases in operating profit (Figures 13.2g–h and 13.3d–e). For the H-LVST type the model allocates relatively little land to intervention crops; the only intervention crop chosen by the model is sole soybean (Figures 13.2g–i and 13.3a–l) but the area is quite small (just over 1 ha out of the 23 ha that are available).

Finally, sole soybean should be suitable for all types except M-LEGU, sole cowpea for all types except L-LEGU. A maize and cowpea intercrop is suited to MH-OFI, but only for small areas as for larger areas the added labour and hence lower profit would make this intervention less attractive. Maize after soybean would be a better intervention for MH-OFI farms, as there are synergies with labour required and profit. Maize after cowpea would be suitable for M-LEGU, MH-OFI and L-WEED types.

Discussion

Practices such as integrating legume as an intercrop or in rotation are viewed as a means for sustainable intensification. Indeed, intercropping maize with legumes such as cowpea or soybean may lead to increased land use efficiency, crop diversity, soil fertility and farm household income if competition between component crops is minimised while beneficial interactions are maximised (Giller et al., 2009; Baudron et al., 2012; Rusinamhodzi et al., 2012). We can see from the results that, at least for the MH-OFI type, an increase of intercropped maize and cowpea does not necessarily increase operating profit (Figure 13.3e). Including legumes in a rotation appears to have more potential to improve operating profit (Figure 13.3h, k) and it has slight synergies with labour (Figure 13.3i, l) for M-LEGU and MH-OFI.

The surveyed farmers for the DC may have similar structural characteristics and farming orientation to the averages of the farm types; however their personal motivations, desires and fears could diverge from others of the same farm type. For instance, the H-LVST farmer we surveyed had already been exposed to legume diversification intervention activities in his region, and explained he was keen to integrate legumes in his system, yet the model did not choose to include any large areas to intervention crops for his farm, the greatest area being to sole soybean (Figure 13.2g–i). In FarmDESIGN, the windows of opportunities are defined using fixed assumptions on the achievable yields and market prices. However, in real conditions, farmers have to make decisions early in the cropping season under uncertainties on the future production and market situations. This decision making process may be influenced by the farmer's personal background and socio-cultural factors. Moreover, it should be noted that a typology captures a 'snapshot in time' of a farming community (Kostrowicki, 1977). As farms are highly dynamic, farmers may change over time from one farm type to another. Interventions encouraging to improve farming systems by increasing the legume cultivation could be a driver for the change from one farm type to another (e.g. from L-WEED to L-LEGU and then M-LEGU) or even for the creation of a new type (e.g. H-LVST with legume).

Conclusion

The model exploration showed which intervention crops would be most suitable to which farm types taking into account their structural constraints and their objectives to maximise operating profit and organic matter added and to minimise their labour requirements. Sole legume crops like soybeans were found beneficial (i.e. higher profit, organic matter added to the soil and lower labour requirement) to L-LEGU, MH-OFI and H-LVST types, whereas L-WEED and M-LEGU types benefitted more from sole cowpea. For types L-WEED, M-LEGU and MH-OFI, including maize after the legume crop was found to be beneficial. Only the MH-OFI type was shown to have some benefit from an intercrop of maize and cowpea. The results show the need for differentiated solutions for different farm types in the Eastern province of Zambia and can act as a guideline for improved targeting of novel innovations for sustainable intensification that can possibly lead to improved adoption and hence enhanced livelihoods of smallholder farmers.

Future research can focus on a feedback of this data to the farmers aiming to gauge their opinions of the suitability of the interventions targeted at their farm type, and thereafter mapping their trajectories over time. Whether they adopt or reject the interventions, and the effect that this has on their farm type, would be of interest.

Acknowledgements

The research was part of farming systems analysis within the project-alliance between SIMLEZA (Sustainable Intensification of Maize-Legume Systems in Eastern Province of Zambia) and Africa RISING (Africa Research in Sustainable Intensification for the Next Generation), a research program of the

Feed-the-Future initiative of the USA government. We would like to acknowledge assistance from the following people and organisations: Peter Setimela (CIMMYT – International Maize and Wheat Improvement Centre), Julius Manda (IITA – International Institute for Tropical Agriculture) and Munyaradzi Mutenje (CIMMYT) for supplying us with the baseline survey dataset from 2011/2012; CIMMYT experts in the field Jens Andersson and Christian Thierfelder for discussions on the typology; CIMMYT and IITA for data from legume integration trials at Msekera Research station in Chipata; Mulundu Mwila and Abell Mwale who translated and assisted in the DC fieldwork; and finally our thanks go to the farmers who willingly gave us their valuable time to answer our many questions.

Note

1 Legumes included common bean, soybean, pigeon pea, groundnut and cowpea.

References

Alvarez, S., Paas, W., Descheemaeker, K., Tittonell, P. and Groot, J.C.J. (2014) *Constructing typologies, a way to deal with farm diversity: General guidelines for the Humid Tropics*. Report for the CGIAR Research Program on Integrated Systems for the Humid Tropics. Plant Sciences Group, Wageningen University, The Netherlands. http://humidtropics.cgiar.org/wp-content/uploads/downloads/2015/04/Typology-guidelines_v2.pdf. Accessed 20 October 2015.

Andersen, E., Elbersen, B., Godeschalk, F. and Verhoog, D. (2007) 'Farm management indicators and farm typologies as a basis for assessments in a changing policy environment', *Journal of Environmental Management* 82, pp. 353–362.

Baudron, F., Tittonell, P., Corbeels, M., Letourmy, P. and Giller, K.E. (2012) 'Comparative performance of conservation agriculture and current smallholder farming practices in semi-arid Zimbabwe', *Field Crops Research* 132, pp. 117–128.

Chikowo, R., Zingore, S., Snapp, S. and Johnston, A. (2014) 'Farm typologies, soil fertility variability and nutrient management in smallholder farming in sub-Saharan Africa', *Nutrient Cycling in Agroecosystems* 100, pp. 1–18.

FAOSTAT (2013) *Country statistics for food consumption patterns in Zambia for 2013*. http://faostat3.fao.org/download/FB/FBS/E. Accessed 10 November 2015.

Giller, K.E., Witter, E., Corbeels, M. and Tittonell, P. (2009) 'Conservation agriculture and smallholder farming in Africa: the heretics' view', *Field Crops Research* 114, pp. 23–34.

Groot, J.C.J., Oomen, G.J.M. and Rossing, W.A.H. (2012) 'Multi-objective optimisation and design of farming systems', *Agricultural Systems* 110, pp. 63–77.

Groot, J.C.J. and Rossing, W.A.H. (2011) 'Model-aided learning for adaptive management of natural resources: An evolutionary design perspective', *Methods in Ecology and Evolution* 2, pp. 643–650.

Kostrowicki, J. (1977) 'Agricultural typology concept and method', *Agricultural Systems* 2, pp. 33–45.

The Montpellier Panel (2013) *Sustainable intensification: A new paradigm for African agriculture*. London. http://ag4impact.org/publications/montpellier-panel-report2013/ Accessed 20 October 2015.

Ojiem, J.O., Ridder, N. de, Vanlauwe, B. and Giller, K.E. (2006) 'Socio-ecological niche: a conceptual framework for integration of legumes in smallholder farming systems', *International Journal of Agricultural Sustainability* 4, pp. 79–93.

Pretty, J., Toulmin, C., and William, S. (2011) 'Sustainable intensification in African agriculture', *International Journal of Agricultural Sustainability* 9, pp. 5–24.

Rusinamhodzi, L., Corbeels, M., Nyamangara, J. and Giller, K.E. (2012) 'Maize–grain legume intercropping is an attractive option for ecological intensification that reduces climatic risk for smallholder farmers in central Mozambique', *Field Crops Research* 136, pp. 12–22.

Sitko, N.J., Chapoto, A., Kabwe, S., Tembo, S., Hichaambwa, M., Lubinda, R., Chiwawa, H., Mataa, M., Heck, S. and Nthani, D. (2011) *Food security research project technical compendium: descriptive agricultural statistics and analysis for Zambia in support of the USAID mission's feed the future strategic review.* http://fsg.afre.msu.edu/zambia/wp52.pdf Accessed 10 November 2015.

Tittonell, P., Muriuki, A., Shepherd, K.D., Mugendi, D., Kaizzi, K.C., Okeyo, J., Verchot, L., Coe, R. and Vanlauwe, B. (2010) 'The diversity of rural livelihoods and their influence on soil fertility in agricultural systems of East Africa – a typology of smallholder farms', *Agricultural Systems* 103, pp. 83–97.

14 Sustainable intensification of smallholder agriculture in Northwest Vietnam

Exploring the potential of integrating vegetables

*To Thi Thu Ha, Pepijn Schreinemachers, Fenton Beed,
Jaw-Fen Wang, Nguyen Thi Tan Loc, Le Thi Thuy,
Dang Thi Van, Ramasamy Srinivasan, Peter Hanson
and Victor Afari-Sefa*

Introduction

Northwest Vietnam, one of seven regions of Vietnam, comprises six provinces. The region is of strategic importance to Vietnam as it borders with Laos and China and provides considerable amounts of hydroelectric power. The area covers 10 million hectares accounting for 30% of Vietnam's land area with a population of 12 million inhabitants, consisting mainly of ethnic minorities such as Thai, Muong, Kinh, Hmong, etc. The prevalence of poverty due to the poor infrastructure base and rudimentary farming practices is fairly high, particularly in comparison to other regions of Vietnam. The region is mostly mountainous with 80% of the land being sloping. Until recently, the area was difficult to access by road and farmers were poorly integrated into markets.

Agricultural land use in Northwest Vietnam is dominated by the production of rice in the valleys and maize on sloping lands. Maize cultivated area has increased rapidly in recent years, sometimes at the expense of forest lands through crop expansion. The extension of maize into poor and marginal land areas has led to decreased average yields. Widespread soil erosion has further exacerbated this downward trend and poses a significant threat to the sustainability of agriculture and livelihoods (Ha et al., 2004). There is, therefore, an urgent need to diversify the current land use, particularly by growing fruit trees on sloping lands while at the same time intensifying production systems on existing cropland through the inclusion of high-value crops such as vegetables.

This study explores the potential of Northwest Vietnam to diversify its land use through vegetable production systems. Temperate vegetables such as tomato (*Solanum lycopersicum*) and common cabbage (*Brassica oleracea* var

Capitata) have a high potential for Northwest Vietnam because of its relatively cool climate and these products offer the opportunity to reduce reliance on imported temperate vegetables. Constraints and opportunities for Northwest Vietnam to expand vegetable production were evaluated by focusing on the representative provinces of Son La and Dien Bien and their potential to supply fresh produce to the urban markets of Hanoi. Data from producers and consumers was analyzed for changing trends. In addition, on-farm trials were conducted to determine the suitability of introducing elite vegetable varieties and integrated pest management (IPM) options into existing farming systems.

Materials and methods

The study collected primary data through survey questionnaires from consumers and retailers in Hanoi and Son La and from vegetable producers in Son La and Dien Bien provinces, which are known to have high potential for agricultural production. Focus group discussions and in-depth surveys were conducted in 2013 with selected vegetable producers and consumers in Son La province and in Hanoi. Twelve consumer focus group discussions were carried out in three representative wards in the urban and peri-urban districts of Hanoi and from one ward in Son La city. The wards were selected to represent high-, medium- and low-income neighbourhoods. From each ward, representatives from 7–10 households were interviewed in a group. Questions that were posed to respondents included: frequency of vegetable consumption, origin of purchased produce and their market preferences for vegetables.

The qualitative focus group discussions were augmented with quantitative primary data collected via in-depth one-on-one interviews with 29 selected retailers operating at open markets, stores and supermarkets in Hanoi and Son La. They were questioned on types of traded vegetables, consumer preferences, challenges and opportunities to expand the sale of vegetables, with a particular focus on vegetables sourced at farm gates from Northwest Vietnam.

Six communes in Son La and Dien Bien provinces were chosen as focus discussion groups specifically for producers. Each discussion group had 10 households. These were noted for producing vegetables for commercial sale in popular markets. In addition, selected commune and district-level government officers were interviewed to determine the status, opportunities and challenges of vegetable production in their locations.

Secondary data were obtained from government offices in Hanoi, Son La and Dien Bien provinces. Data collected included agricultural production as well as socio-economic situation analyses and other relevant information.

On-farm testing of suitable methods for safe vegetable production was conducted in Mai Son district, Son La province in the 2014 off-season (i.e., in summer and autumn season). The trials evaluated introduced and existing varieties using a randomized complete block design on three farms for each crop of tomato, French bean and radish. Each farm with plot measured 10 m² and there were three replications for each treatment. The varieties were assessed based on performance, and likely adoption as influenced by farmer's preferences (Tables 14.1a, b, c).

Trials to assess the efficacy of integrated pest management (IPM) options for control of common pests, including fruit borer on tomato, were performed in Mai Son district, Son La province in 2014. The popular variety VL2910 (SEMI-NIS) was used for the purpose of the assessment. Two treatments – IPM and farmers' standard practice, each with 11 replications with a plot size of 300 m² – were implemented for the purpose of the assessment. The IPM treatment included pheromone lures to remove male adults of *Helicoverpa armigera* and *Spodoptera exigua*, and three kinds of biopesticides – Xentari® (*Bacillus thuringiensis* subsp. *aizawai*, Serotype H-7, Strain ABTS-1857), BT911 (*B. thuringiensis*

Table 14.1a List of tomato entries of on-farm trials in Son La, 2014

No	Summer season			Autumn season		
	Name	Source	Type	Name	Source	Type
1	FMTT1733A	WorldVeg	OP	FMTT1733A	WorldVeg	OP
2	CLN3670G	WorldVeg	OP	CLN3670G	WorldVeg	OP
3	CLN3682D	WorldVeg	OP	CLN3682D	WorldVeg	OP
4	CLN3682C	WorldVeg	OP	CLN3682C	WorldVeg	OP
5	CLN3643A	WorldVeg	OP	CLN3643A	WorldVeg	OP
6	CLN3552C	WorldVeg	OP	CLN3552C	WorldVeg	OP
7	CLN3125 L	WorldVeg	OP	CLN3125 L	WorldVeg	OP
8	CLN3984	WorldVeg	F1	CLN3241H-27	WorldVeg	OP
9	CLN3979	WorldVeg	F1	CLN3670E	WorldVeg	OP
10	CLN4000	WorldVeg	F1	CLN3984	WorldVeg	F1
11	CLN3078C	WorldVeg	F1	CLN3979	WorldVeg	F1
12	Savior (control)	SYNGENTA	F1	CLN4000	WorldVeg	F1
13	FM 59	FAVRI	F1	CLN3946	WorldVeg	F1
14	ANNA (control)	SEMINIS	F1	CLN3940	WorldVeg	F1
15				CLN3941	WorldVeg	F1
16				CLN3953	WorldVeg	F1
17				CLN3976	WorldVeg	F1
18				CLN3948	WorldVeg	F1
19				VNS390 (control)	Southern Seed Company	F1
20				CN3500 (control)	SEMINIS	F1
21				FM29	FAVRI	F1
22				FM1080	FAVRI	F1

Table 14.1b List of French bean entries of on-farm trials in autumn and winter season in Son La, 2014

No	Name of varieties	Source
1	Tu Quy No.1	VEGESEED
2	Tu Quy No. 2	VEGESEED
3	Khang Binh No. 1	VEGESEED
4	Cao San Hat Den 1	Hung Nong Company
5	Cao San Hat Den 2	Giong Moi Company
6	Cao San Hat Den 3	SSC
7	Trach Lai (Check)	VEGESEED

Table 14.1c List of radish entries of on-farm trials in autumn and winter season in Son La, 2014

No	Name of varieties	Source/Note
1	TN 48	Trang Nong Seed Company
2	TN 45	Trang Nong Seed Company
3	White King RA 50	Nishi – Nihon Miyairi Hanbai – Japan
4	Hanoi radish	VEGESEED
5	Radish No.13 (Control)	VEGESEED
6	NP – 04	Tan Nong Phat Seed Company

subsp. *kurstaki*) and Metarhizium (*Metarhizium anisopliae*) – were sprayed at nine-day intervals on fields. The IPM treatment was compared against the control, in which farmers' standard practice of using locally available pesticides was observed. Most farmers traditionally used an insecticide known commercially as Coc Chua (Emamectin Benzoate at 1.43 g/kg, Matrine at 2 g/kg and special additives at 850g/kg). Farmers sprayed insecticides whenever they detected tomato fruit borer in their field, normally at 7–10-day intervals. The number of male adults caught in traps at 20-day intervals in each plot was assessed and crop yield at harvest determined.

The data were analyzed using analysis of variance (ANOVA). The significant treatment differences were indicated, and means were separated by Tukey's HSD Test.

Results

Vegetable consumption in Hanoi and Son La

Surveys of consumer groups in Hanoi and Son La showed that household diets contained a wide variety of vegetables. Consumers in Hanoi city used a total of 40 different vegetables while the consumers in Son La city used 47 kinds of vegetables, including several traditional species. Consumers in both locations were specifically concerned about vegetable safety, particularly the contamination with pesticide residues.

The frequency of vegetable consumption in households varied for the different crops and seasons. Consumers often bought tomatoes, green peas, potatoes and kohlrabi from January to March. Tropical vegetables including kangkong (*Ipomoea aquatic*), Ceylon spinach (*Basella alba*) and pumpkin (*Cucurbita pepo* L.) were consumed in Hanoi from April to September and bamboo shoot and loofah (*Luffa cylindrica*) were consumed in Son La during the same period. Consumers preferred using temperate vegetables, particularly cabbage, kohlrabi (*Brassica oleracea* var Gongylodes) and sweet pepper (*Capsicum annum*) in the summer off-season. However, they were very much concerned about vegetable safety because imported off-season vegetables such as tomato and common cabbage can contain potential toxic chemical residues (Hoi et al., 2009). From October to December, consumers ate more local vegetables because it is the main vegetable production season in lowland areas with the perception that less chemicals were applied during this period because of the low incidence of pest damaged during this time.

During the off-season period, 10–50% of tomatoes and 31–50% of common cabbage were supplied to Hanoi from Lam Dong province (southern part of Vietnam) supplemented by peri–urban vegetables from Hanoi, with the rest being imported from China (An, 2005). Other vegetables such as Hmong mustard, green peas, pak choi and chayote come from the Northwest region and were only recently introduced to consumers in Hanoi.

The results of consumer focus group discussions showed that consumers prefer vegetables which are of known origin (preferably from Vietnam), fresh, purchased from known acquaintances and considered to be safe and not contaminated with pesticide residues or bacteria. There was interest in sourcing safe or organically certified vegetables such as Vietnam Good Agricultural Practice (VietGAP). Unlike Hanoi, consumers in Son La were more interested in colour and appearance and perceived that pesticide residues was not a major concern as the local climate is favourable for production and farmers spray less pesticide (Loc et al., 2013).

The comparison of the origin of vegetables – the Northwest region, Hanoi, Lam Dong or China – showed that consumers most strongly preferred vegetables from the Northwest. Consumers explained that some kinds of vegetables such as chayote, Hmong mustard, taro, pumpkin, celery cabbage and bamboo shoots remain uncommon and are not always available. However, these vegetables are routinely grown in the Northwest region and thus offer the potential for new market introduction development. They explained that such vegetables are of better quality, are fresher, greener and confer good taste and aroma while containing fewer chemicals.

Respondents interviewed at supermarkets and shops also claimed that vegetables from the Northwest region were of better quality and higher value compared to those from other regions. Results from shop owner and street vendor interviews showed that the quality and price of vegetables from the Northwest region was better than those of other regions as confirmed by 60% of respondents.

The observations from supermarkets, vegetable shops, retailers and street vendors showed that vegetables are often sold in large quantities. Besides the main season, off-season vegetables were available at Hanoi markets. This demonstrates a good opportunity for expanding vegetable production in the Northwest region to satisfy the demand for safe vegetables during the off-season, i.e. hot-wet season when production in the Red River Delta around Hanoi is problematic (due to increased pest and disease pressure and resulting high pesticide applications). It will also help improve self-sufficiency and reduce the supply of vegetables from distant provinces and imports from China.

Commercial vegetable production in Son La and Dien Bien province

The results of the focus group discussions and in-depth interviews conducted in three main vegetable areas showed that land for growing vegetables is typically located near farmers' residences. The average area under vegetables was 0.17 hectares against a total average agricultural land area of 0.66 hectares per household. Soils were mainly alluvial. Water for irrigation of vegetables was mainly from streams, canals, ditches and wells. Household members indicated that in many locations there is a critical shortage of irrigation water for vegetables from February to April.

Various vegetables can be grown in different seasons of the year. Chayote and Hmong mustard can be grown throughout the year. Tomato, cabbage, kohlrabi and green peas can be harvested from May to October when the market supply in urban areas such as Hanoi is otherwise low. Vegetable prices fluctuated over the year due to variations in market supply. Vegetable retail prices in Hanoi markets were higher than in the Son La and Dien Bien markets (Table 14.2). There is therefore an opportunity for producers and traders to supply the Hanoi market and obtain higher returns during summer, provided that transport can be organized cheaply, and that produce does not get damaged. However, there is a lack of planning among vegetable farmers; they do not usually plan their production or distribution and simply try to sell on spot markets when they harvest and without linkages to markets that can offer improved returns.

The results from group discussions confirmed vegetable seed is sold through local village shops, but is of low quality and there is no variety label information. Seed sold in the central district is more expensive and available only in large packets while there is not much choice in terms of varieties. Many

Table 14.2 Retail price of some vegetables in Hanoi, Son La and Dien Bien markets in off-season, in 1,000 VND kg⁻¹

	Tomato	Cabbage	Hmong mustard	Zucchini	French bean	Lettuce	Taro
Hanoi	19*–25**	15–24	15–25	10–30	10–25	15–50	19–29
Son La	7–13	5–10	7–12	8–10	8–15	8–12	8–10
Dien Bien	5–15	5–12	7–10	8–10	8–10	6–12	7–10

* Minimum price
** Maximum price
Source: Loc et al., 2013.

farmers, therefore, produce their own vegetable seed. However, this often has a low quality resulting in poor yields.

Common pests and diseases affecting vegetable production include bacterial diseases on tomatoes and fruit worm, caterpillar and triple flea beetle on leafy vegetables. The use of available chemical pesticides is common (Schreinemachers et al., 2015). However, pesticide use efficiency is low due to the lack of knowledge among farmers. Knowledge is particularly limited about biopesticides. In general, not many farmers have received training in vegetable cultivation methods including agronomic practices to reduce introduction and spread of pests and diseases, removal of diseased plants, use of disease-free seed, soil and water, and crop rotation, among others.

Vegetable variety trials in Son La province

Tomato variety trial

The results from the tomato trial in the summer season included analysis based on fruit setting, average fruit weight and disease damage data. The varieties CLN3979, CLN3984 and CLN4000 performed similarly to the Savior check. These varieties also produced relatively higher yields. The yield of CLN3984 at 41.7 ton ha⁻¹ was not significantly different from Savior at 37.0 ton ha⁻¹. Tomatoes were attacked by black cutworm (*Agrotis ipsilon*) and bacterial wilt (*Ralstonia solanacearum*). Particularly the Anna control variety completely failed because of the incidence of bacterial wilt disease, followed by CLN3670G and Savior control variety at 33% and 18% of plants, respectively. Southern blight (*Sclerotium rolfsii*) occurred in most of the lines and varieties and infected 5–15% of the plants on average. Two varieties, CLN3984 and CLN3078C, were not affected by Southern blight disease.

Among the 22 lines/varieties tested in the autumn season, four F1 varieties (CLN3953, CLN3948, FM29 and FM1080) performed rather well in terms of fruit setting rate at 60–70%, fruit weight and yield. The control variety (CN3500) had the highest yield of 46.1 ton ha⁻¹. The yields of other varieties (CLN3953, CLN3948, FM29 and FM1080) were not significantly different from the control (CN3500), ranging from 40.7 to 43.9 ton ha⁻¹. Yields from all open pollinated lines were modest and ranged from 8.1–27.5 ton ha⁻¹. In

the reference season, bacterial wilt affected 3–38% of the plant of most varieties, depending on soil conditions and location. As this bacterium is soil borne,
it is likely that much of the variability was due to the differential levels of soil
contamination between farms. Other varieties (CLN3953, FM29 and FM1080)
were not damaged by bacterial wilt, followed by CLN3948, which was affected
at 2%.

French bean variety trial

There were no marked differences across growth duration between the two
seasons. Among the seven evaluated varieties, Tu Quy 1 and Tu Quy 2 had a
longer growing period of 101–102 days while the other varieties ranged from
93–96 days after sowing.

Tu Quy 1 and Cao San Hat Den 3 varieties achieved more vigorous growth
and attained high fruit setting at 46–49%, compared to the Trach Lai control
variety at 44–46% in both seasons. The highest average yields were obtained for
Tu Quy 1 and Cao San Hat Den 3 at 29.6 ton ha^{-1} while the Trach Lai control
variety produced 22.5 ton ha^{-1} (Table 14.3). Evaluated French bean varieties
performed quite well across the two seasons and produced similar observed
yields. In terms of customers' preference, Cao San Hat Den 3 produced green
coloured and round shaped pods that were not suitable for the local market,
but are preferred in the Hanoi market. Tu Quy 1 and Quy Tu 2 produced more
flattened and lighter green fruits and were more preferred by local consumers.

RADISH VARIETY TRIAL

Most of the radish varieties had a short growing duration, ranging from 45–48
days and 25–46 days in the autumn and winter seasons, respectively. Among
these, Hanoi radish showed the shortest growth duration and was ready for
harvesting at 25 and 32 days after sowing in winter and autumn, respectively.

Table 14.3 Yield of French bean varieties in Mai Son, Son La, in ton ha^{-1}

No	Variety	Autumn Farm 1	Autumn Farm 2	Winter Farm 3	Average yield
1	Tu Quy 1	29.8	29.2	30.2	29.7
2	Tu Quy 2	25.2	25.6	26.4	25.8
3	Khang Binh 1	24.2	23.8	25.0	24.3
4	Cao San Hat Den 1	23.8	23.4	24.2	23.8
5	Cao San Hat Den 2	23.0	22.6	24.0	23.2
6	Cao San Hat Den 3	29.2	29.3	30.4	29.6
7	Trach Lai (Check)	22.3	22.1	23.3	22.5
	Average	24.6	24.5	25.5	
	CV%	5.9	7.1	5.0	
	LSD$_{0.05}$	2.64	3.18	2.35	

Source: Research results by WorldVeg and FAVRI, 2014.

Shorter duration varieties were better suited to production between crop seasons when vegetable market supply is limited. However, Hanoi radish produced low yields of 13.5 ton ha^{-1}. The highest yield was 44 ton ha^{-1} produced by TN48. The yield of the six evaluated varieties varied depending on the season of cultivation. Sowing in winter was favourable to the growth and development of radish with limited infection with pests and diseases, and therefore gave a higher number of surviving plants at harvesting time (22.7 plants m^{-2}) than when sowing in summer (14.2 plants m^{-2}). The radish roots in winter season were also of relatively larger sizes. The greatest average root weight was produced by TN48 and TN45 at 260 and 241 g root^{-1}, respectively. As a consequence, TN48 gave the highest yield of 44.4 ton ha^{-1}, followed by TN45 at 37.3 ton ha^{-1}; Hanoi radish produced the lowest yield of 9.1–14.0 ton ha^{-1} at all three farms in both seasons.

Integrated pest management tomato trial

Tomato fruit borers (*Spodoptera exigua* and *Helicoverpa armigera*) are the most common insects damaging tomatoes, especially under warm climatic conditions during the off-season (Kashyap and Batra, 1987; Venette et al., 2003; Srivastava et al., 2010).

The results showed that tomato fruit borer damage in Mai Son district was mainly due to *S. exigua*. The average number of *S. exigua* caught was 26.2 insects time^{-1} plot^{-1}. It was 2.4 times higher than *H. armigera* (Table 14.4). This may have been due to *S. exigua* infestations from infested maize plants from the surrounding fields in Son La.

The results of the trial showed that there were no significant differences in total marketable and non-marketable tomato yields between the plots with IPM treatment and standard farmers' practice. In the IPM treatment, the use of pesticides was minimized and largely included non-harmful biopesticides. Thus, environment and human health were better protected. Tomato growers

Table 14.4 Effect of integrated pest management (IPM) on tomato fruit borer (*Spodoptera exigua* and *Helicoverpa armigera*) control in Son La, Vietnam

Treatment	Average number adults caught time^{-1} plot^{-1}		Non-marketable yield		Marketable yield (ton ha^{-1})
	S. exigua	*H. armigera*	*Number of fruit ha^{-1}*	*Fruit weight (ton ha^{-1})*	
IPM	26.18	10.84	5,640	4.5	50.0
Farmer practice	–	–	4,463	3.2	49.8
					P = 0.0005
					t = 0.12

Source: Research results by WorldVeg and FAVRI, 2014.

were interested in the results and willing to apply the IPM package to a larger production area.

Discussion

Cities like Hanoi and Son La have rapidly growing demand for fresh and safe vegetables. Hanoi consumers were found to prefer vegetables from the Northwest region because they associate vegetables from there with high quality (freshness and low pesticide contamination). They are particularly interested in new kinds of vegetables (e.g., Hmong mustard, Hmong cucumber) and temperate vegetables in the off-season. Consumers are seriously concerned about food safety (Hoai et al., 2011) and strongly prefer domestically produced vegetables. But it remains difficult to recognize domestically produced vegetables in terms of quality and origin in the market. There is, therefore, much potential to increase the quantity and quality of vegetables from the Northwest region (Loc, 2013).

Climatic conditions in the Northwest region have shown to be suitable for growing vegetables all year round, especially for temperate crops in the off-season. Although farmers in this region already have some experience in vegetable production, they lack appropriate training in agronomy and pest and disease management.

The supply of quality vegetable seeds of local and traditional vegetable varieties is limited, which can result in low vegetable productivity. The seeds of high-value crops such as tomato, cauliflower, sweet pepper and cabbage are also not available or are not suitable for local conditions into specific varieties.

Pests and diseases are another major constraint and farmers should be trained in pest and disease diagnosis as a first and critical step towards deployment of appropriate management technologies. These should include agronomic practices to remove diseased plants and to use clean seeds, soil and water, and through rotating crops and the use of resistant varieties combined with biological solutions such as pheromone traps, botanicals and natural enemies. Participatory training and supply of appropriate inputs will be quite helpful in this regard. However, the above highlighted problems should be seen not only in the context of Northwest Vietnam but more generally for the whole of Vietnam. Schreinemachers et al. (2015), for example, reported a rapid growth in pesticide use for Vietnam and an average use of 16.5 kg of commercial product ha^{-1}. Research and on-farm trials of sex pheromones and biopesticides started in the 1990s in Vietnam and some biocontrol methods are already commercially available (Uyen, 2005). However, their use remains very limited; farmers have little knowledge about them and others feel they are slow to react and complicated to install. The sex pheromones used in this study were found to be effective in controlling tomato fruit borer and yields were not significantly different from standard farmers' practice, which shows there is much potential in using these methods to produce similar yields that are safe for consumers while mitigating harmful effects of pesticide applications.

Income from vegetable production is three or four times higher than from crops such as rice and maize (Ha, 2008; Mai Son Division of Agriculture, 2013). Therefore, there is great potential to increase household incomes through vegetable production from the poverty-stricken Northwest region. Besides supplying vegetables to Hanoi, there are opportunities to export to Laos and China and to regional and local markets in towns such as Son La. Market access has been a constraint, but recent improvements in road infrastructure make it possible to transport large quantities of vegetables quickly to Hanoi and other parts of the region. The key is to link the rural producers to urban markets to ensure equitable revenue returns.

Vegetables have been grown in large areas of Mai Son district of Son La and there have been projects promoting safe vegetable production. However, growers in Mai Son district continue to lack basic production skills (Mai Son Division of Agriculture, 2013). Integrating vegetable cultivation with other crops can contribute to a more sustainable agricultural production system to protect the environment and diversify income sources, diet diversity and nutrition. Vegetable cultivation provides a more diversified income and optimal use of land for farmers than sole maize or rice cultivation. However, there is a need to develop and introduce new high-value vegetable crops and varieties suited to environmental conditions and consumer demands and IPM methods (Loc et al., 2013). Currently, there is not much choice in vegetable crops and varieties because of very limited supply and high price of vegetable seed. The trial results show that the available tomato, French bean and radish varieties are well adapted to local conditions, especially in the off-seasons.

Conclusion

This study shows that there is much potential to diversify and intensify agricultural production systems in the Northwest region of Vietnam through cultivation of vegetables. This is because of high and rapidly rising market demand in urban areas, especially Hanoi; favourable perceptions of consumers about vegetables from the Northwest region; suitable agro-climatic conditions, particularly during the off-season; and a substantial price differential between vegetables sold at local markets in the Northwest and the same vegetables sold in Hanoi. However, basic production constraints will need to be addressed, including the supply of quality seeds of a diverse range of vegetable varieties and knowledge on integrated pest management. The use of integrated pest management is particularly important because Hanoi consumers associate vegetables from the Northwest with low levels of pesticide use. Most current vegetable farmers in the Northwest have not received any training in vegetable production methods. Providing such training, including introducing new vegetables and integrated pest management tools, while simultaneously addressing the production constraints, could have a large impact on rural livelihoods in Northwest Vietnam.

Acknowledgments

The financial support from the CGIAR Research Program on Integrated Systems for the Humid Tropics (Humidtropics), through AVRDC – The World Vegetable Center – to this research is gratefully acknowledged.

References

An, H.B. and Moustier, P. (2005) *Vegetable market information and consultation systems in the Mekong region*, SUSPER Project Report, Hanoi, Vietnam.

Fruit and Vegetable Research Institute (FAVRI). (2014) 'Testing of sustainable methods of safe off-season commercial vegetable production and home-based food production in Son La, Vietnam', *Humidtropics Technical Report.*, Submitted to World Vegetable Center, Hanoi, Vietnam.

Ha, D.T, Thao, T.D, Khiem, N.T, Trieu, M.X., Gerpacio, R.V. and Pingali, P.L. (2004) *Maize in Vietnam: Production Systems, Constraints, and Research Priorities*, CIMMYT, Mexico, DF.

Ha, T.T.T. (2008) 'Sustainability of Peri-Urban Agriculture of Hanoi: The Case of Vegetable Production', PhD thesis. AgroParisTech, Paris, France.

Hoai, P.M., Sebesvari, Z., Minh, T.B., Viet, P.H. and Renaud, F.G. (2011) 'Pesticide pollution in agricultural areas of Northern Vietnam: case study in Hoang Liet and Minh Dai communes', *Environmental Pollution* 159 (12), pp. 3344–3350.

Hoi, P.V., Mol, A.P.J., Oosterveer, P. and van den Brink, P.J. (2009) 'Pesticide distribution and use in vegetable production in the Red River Delta of Vietnam', *Renewable Agriculture and Food System* 24 (3), pp. 174–185.

Kashyap, R.K. and Batra, B.R. (1987) 'Influence of some crop management practices on the incidence of *Heliothis armigera* (Hübner) and yield of tomato (*Lycopersicon esculentum* Mill) in India', *Trop Pest Manage*, 33, pp. 166–169.

Loc, N.T.T. (2013) 'Supporting activities for farmer groups in Northern mountainous area to connect to Hanoi market', Technical Report on Project Improvement Linkage between Market and Cross-Season Vegetable Producers in Northwest Vietnam, Hanoi, Vietnam.

Loc, N.T.T, Thinh, L.N., Thuy, N.T.T. and Anh, H.V. (2013) 'Scoping study on commercial vegetable production in Son La and Dien Bien', Humidtropics Program Technical Report 2013, AVRDC, Hanoi, Vietnam.

Mai Son Division of Agriculture (2013) *Agriculture Production Annual Report of 2012*, Mai Son, Son La, Vietnam.

Schreinemachers, P., Afari-Sefa, V., Heng, C.H., Dung, P.T.M., Praneetvatakul, S. and Srinivasan, R. (2015) 'Safe and sustainable crop protection in Southeast Asia: status, challenges and policy options', *Environmental Science & Policy* 54, pp. 357–366.

Srivastava, C.P., Nitin, J. and Trivedi, T.P. (2010). 'Forecasting of Helicoverpa armigera populations and impact of climate change', *Indian Journal Agricultural Science* 80, pp. 3–10.

Uyen, N.V. (2005) *Pest and Disease Control Using Biological Solutions*, Agricultural Publishing House, Hanoi, Vietnam.

Venette, R.C., Davis, E.E., Zaspel, J., Heisler, H. and Larson, M. (2003) 'Mini Risk Assessment: Old World Bollworm, *Helicoverpa armigera* Hübner [Lepidoptera: Noctuidae]', University of Minnesota, St Paul, MN, USA. http://www.aphis.usda.gov/plant_health/plant_pest_info/pest_detection/downloads/pra/harmigerapra.pdf (downloaded: Feb 3, 2016).

15 Improved grain legumes for smallholder maize-based systems in Western Kenya

Paul L. Woomer, Bonface Omondi, Celister Kaleha and Moses Chamwada

Introduction

Western Kenya is dominated by small-scale, maize-based farming systems that are undergoing diversification toward market-oriented agriculture as households raise their expected living standards. Two legumes offer potential for both income generation and improved household nutrition within these systems: soybean and climbing bean. Soybean (*Glycine max* (L) Merr) is an important source of oil and protein throughout the world and is growing in importance in Africa (Tefera, 2011). It is a crop new to small-scale farmers in Western Kenya but its demand is rapidly growing (Sinclair et al., 2014) and soybean self-sufficiency and export by Africa are important regional development targets (Chianu et al., 2010). Climbing bean (*Phaseolus vulgaris* L.) originates from tropical America but has spread to the Great Lakes region of Africa, including Western Kenya, and has greater biological nitrogen fixation (BNF) than bush varieties (Graham and Rosas, 1977). In contrast to intercropped bush beans commonly grown by farmers, climbing beans are higher yielding, have a longer growing period and require support from stakes or trellises. The N2Africa Project is actively pursuing management strategies that allow small-scale farmers in Western Kenya and elsewhere to adopt and prosper from these two grain legumes (Woomer at al., 2014) and this work is part of that effort.

Grain legumes occupy a special role in better integration of small-scale farming systems in Africa. Their grain and edible leaves are key to balancing cereal-based diets (Chianu et al., 2010). So too their residues are useful as animal feeds and organic inputs to soils. These dual benefits are based in part on symbiotic nitrogen fixation by these legumes that redirects inert atmospheric nitrogen into the diets of humans and livestock, and the mineral pools of soils (Sanginga et al., 2001). Soybean is particularly important in this regard as it is also a cash crop used in the manufacture of textured vegetable proteins and blending animal feeds, and greater domestic production results in savings of foreign reserves otherwise spent on commodity importation. At the same time, legume intensification is built upon the availability and affordability of key production inputs, particularly improved, disease-resistant seed, specially blended fertilizer and rhizobial inoculants; and their placement within value chains in a demand-driven fashion. This paper not only examines the testing of promising new legumes, but subsequent developments that promote their adoption by small-scale farmers in Western Kenya (Woomer at al., 2016).

Materials and methods

The WeRATE N2Africa Outreach Network in Western Kenya (Woomer et al., 2016) assembled and distributed 27 legume technology packages and accompanying field protocols exploring BNF technologies during the 2013–2014 short rains. Each demonstration consisted of eight managements in a 2 x 2 x 2 arrangement (two varieties ± fertilizer and ± inoculant) with either improved soybean or climbing bean varieties. The fertilizer evaluated was the new SYM-PAL (0–23–16+) blend applied at 276 kg per ha. The inoculant was BIOFIX, a commercially available product for either soybean or bean, applied at 10 g inoculant per kg of seed. The two soybean varieties were SB19 and SC Squire, and the climbing beans were Kenya Tamu and Rwanda Red. Cooperators ranked root nodulation on a 0–5 scale and collected yield data. The trials required an area of 13 m x 13 m and were established in prominent roadside locations but not too close to dusty areas. Cooperators completed the data report forms accompanying the field protocols and submitted them to the Kenya Country Coordinator through their respective group leaders. A data base was assembled from these results and summary statistics generated. As the managements were systematically arrayed to assist farmer understanding (non-randomized) and the sites were based upon voluntary farmer group subscription to the field protocol rather than a balanced array of locations across agro-ecological zones, the assumptions underlying mixed ANOVA were considered lacking, and errors were rather expressed in a more conservative manner as SEM = (SD/(SQRT n).

Yield results were also combined with previous season information to perform economic analysis using the N2Africa EZ Cost and Return Utility. This utility is constructed in MS Excel and assists in complete, as opposed to partial, economic analysis. Users enter rates and prices of inputs including seed, inoculant, fertilizers, pest control products and any other materials. Labour includes land preparation, tillage, planting, weeding, spraying and harvest, and requires that duration (days) and pay (daily wages) be entered for each operation. Outputs include Total Costs, Gross Return, Net Return and Benefit to Cost Ratio (= Gross Return/Total Costs). In addition, many of these field tests provided the focus of farmer field days later in the season and farmer impressions of different varieties and their management were assessed in an informal setting by WeRATE farm liaison staff.

Results

Field reports were returned from 25 test sites with a majority (68%) of those using soybean as a test crop. Results were submitted from seven counties and three different agro-ecological zones, although one county (Migori) and the Lake Victoria Basin were under-represented. Results for climbing bean nodulation, grain yield and economic return appear in Table 15.1. BIOFIX for bean inoculant contained *Rhizobium tropici* strain CIAT 899 at approximately

Table 15.1 Nodulation characteristics, yield and economic returns of two climbing bean varieties under four different managements in Western Kenya (± SEM) (n = 8).

Management	Climbing bean variety	Nodule number plant⁻¹	Grain yield (kg ha⁻¹)	Return Net (US $)	Ratio
No inputs	Kenya Tamu	13 ± 2	1141 ± 340	348	2.0
No inputs	Rwanda Red	13 ± 2	1141 ± 302	348	2.0
BIOFIX inoculant	Kenya Tamu	21 ± 3	1524 ± 423	569	2.6
BIOFIX inoculant	Rwanda Red	24 ± 4	1730 ± 499	691	3.0
SYMPAL fertilizer	Kenya Tamu	16 ± 2	1843 ± 603	549	2.0
SYMPAL fertilizer	Rwanda Red	17 ± 2	1470 ± 440	328	1.6
BIOFIX & SYMPAL	Kenya Tamu	29 ± 4	2165 ± 644	734	2.3
BIOFIX & SYMPAL	Rwanda Red	32 ± 3	1780 ± 422	505	1.9

1.3 x 10^9 cells per gram (data from Nairobi MIRCEN), or about 8.6 x 10^6 rhizobia per seed.

Improved management increased resulting root nodule number over two-fold with marked increase in crown nodulation and red interior pigmentation (associated with more effective BNF symbiosis). Inoculation with BIOFIX alone improved nodulation, but more so in conjunction with applied SYMPAL fertilizer. Trends suggest that Rwanda Red has greater nodulation capacity. Under best management it formed 32 nodules per plant with 48% crown nodulation and 100% red interior pigmentation. The average yield of climbing bean increased by 831 kg ha⁻¹ in response to applied BIOFIX and SYMPAL. Kenya Tamu outperformed Rwanda Red because of its longer pods and larger seeds, and its apparent resistance to aphids. The cost of establishing one ha of climbing bean Kenya Tamu under full management (inoculant and fertilizer) is $664 (data not presented). The net return on investment is greatest with Kenya Tamu ($863), offering a return ratio of 2.3:1.

Nodulation, yield and economic returns for soybean appear in Table 15.2. BIOFIX for soybean inoculant contained *Bradyrhizobium japonicum* strain USDA110 at approximately 2.7 x 10^9 cells per gram (data from Nairobi MIRCEN) or about 3.6 x 10^6 rhizobia per seed. Improved management increased nodule number 2.5-fold, again with marked increase in crown nodulation and red interior pigmentation.

Inoculation with BIOFIX alone improved nodulation, but 48% more so in conjunction with SYMPAL fertilizer. Soybean yield increased by 1067 kg ha⁻¹ in response to inputs. Squire consistently outperformed SB19 in part because of its much larger seed size and tolerance to Asian rust disease. The cost of

Table 15.2 Nodulation characteristics, yield and economic returns of two soybean varieties in Western Kenya (± SEM) (n = 19)

Management	Soybean variety	Nodule number plant⁻¹	Grain yield kg ha⁻¹	Return Net ($ ha⁻¹)	Ratio
No inputs	SB19	10 ± 1	1046 ± 174	405	3.4
No inputs	SC Squire	11 ± 2	1250 ± 191	430	2.7
BIOFIX inoculant	SB19	17 ± 2	1427 ± 244	602	4.3
BIOFIX inoculant	SC Squire	18 ± 2	1699 ± 236	664	3.4
SYMPAL fertilizer	SB19	15 ± 2	1656 ± 320	519	2.3
SYMPAL fertilizer	SC Squire	15 ± 2	1790 ± 258	506	2.1
BIOFIX & SYMPAL	SB19	27 ± 2	2031 ± 303	712	2.8
BIOFIX & SYMPAL	SC Squire	24 ± 3	2399 ± 348	827	2.7

establishing one ha of the better performing soybean variety, Squire, under full management is $492 (data not presented). The net return is $827 per ha with a return ratio of 2.7:1. Higher benefit:cost ratios are observed under less intensive managements, suggesting that the crop is better suited to lower input regimes. Nonetheless, soybean responds well to management but inoculation has a reduced effect in the absence of fertilizer. Squire planted in conjunction with inoculant and fertilizer offers 14% greater returns than SB19.

Note that the best performing varieties of both climbing bean and soybean demonstrate a strong trend of stepwise yield improvement in response to BIOFIX, then SYMPAL and then both where the other varieties respond strongly to inoculation but not fertilizer, likely due to interference by pest and disease. The yield of Squire signals a breakthrough as it is the first time crop yield has greatly exceeded the two tons per ha. This is possibly due to a combination of the increased rate of SYMPAL addition, greater compliance with recommended plant populations and rust resistance. This trend is not, however, uniform across all counties and agro-ecological zones as yield declines in the Upper Midland counties of Kakamega and Vihiga (Figure 15.1) even though the BNF technology package of BIOFIX and SYMPAL continues to result in increased yield. Climbing bean performs best in the Upper Midlands (data not presented).

Economic analyses for soybean were also conducted at the county level based upon yields presented in Figure 15.1 and the calculated costs. Outputs were compiled both as Net Return and Benefit-to-Cost Ratios (data not presented). Maximum Net Return per ha of intensively managed soybean ranged from $317 in Vihiga to $1898 in Migori Counties, with a strong trend of greater returns in warmer agro-ecologies. In the Upper Midlands, higher returns were obtained without inoculation suggesting that effective native rhizobia (or colonizing exotics) inhabit the soil. These same differences are reflected in the Benefit-to-Cost Ratios even though fertilizer is far more expensive than inoculant. Many management combinations in Kakamega and Vihiga are non-economic

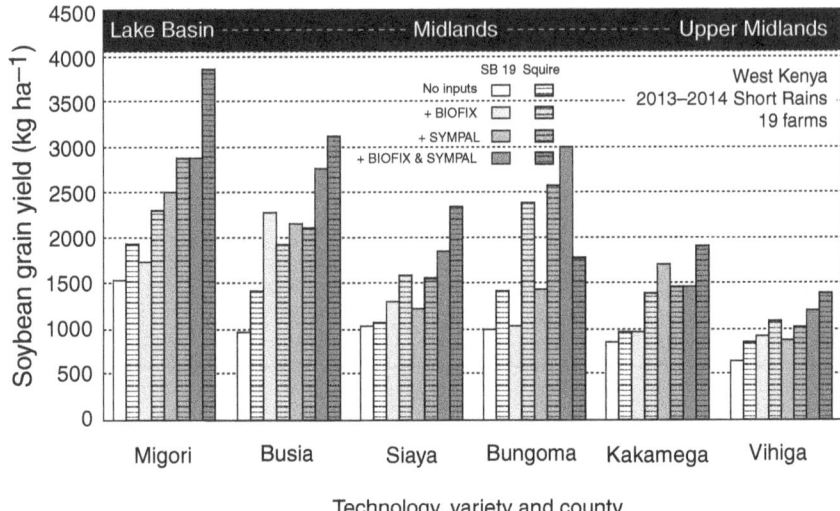

Figure 15.1 Yield of two soybean varieties receiving BIOFIX inoculant and/or SYMPAL
fertilizer in three agro-ecological zones and five counties of Western Kenya
during the 2013–2014 short rains growing season (n = 19: Migori 1, Busia 3,
Siaya 4, Bungoma 2, Kakamega 6, Vihiga 3)

based upon a threshold value of 2:1. Clearly, soybean production is far more
profitable in Migori, Busia and Siaya than in counties in higher elevations (and
cooler temperatures). These observations should be tempered with observations that the Upper Midlands suffered short-term mid-season drought and
that yield potential is usually higher during the following long rains.

Discussion

These findings suggest that our current combination of BNF technologies are
on-target but that finer, more site-specific adjustments are necessary (Woomer
et al., 1999), particularly among Western Kenya's diverse agro-ecological zones
(Ojiem et al., 2007). Clearly, the iterative process of testing soybean and climbing bean varieties over the past several seasons has resulted in farmers' access
to well-performing varieties. SC Squire now appears the best variety but two
other related lines, SC Saga and Salama, were recently released and warrant
further on-farm testing. A weakness in our trials is the inconsistent rate of
SYMPAL fertilizer addition over several seasons. Initially, we applied SYMPAL
at 125 kg ha⁻¹ as the formulation was refined, and this rate is still recommended
within our best practice guidelines. This season's addition of over twice that
amount was the result of applying a 2 kg bag of fertilizer to the two plots
receiving SYMPAL, in part because we needed to use factory pre-packaged

material and 2 kg was the smallest size available. The excellent results this season may well be a result of this higher, but evidently still economical, fertilizer application rate. It is clear that future studies should examine the effects of SYMPAL fertilizer addition rates in different agro-ecological zones on multiple farms. Similarly, recent findings suggest that a locally obtained strain, NAK128, outperforms USDA110 on SB19 in Western Kenya, and there is a strong likelihood that changing the BIOFIX formulation may result in increased yield of soybean (Woomer et al., 2014). This logic leads to four elements for on-farm trials in the future, comparing short- and long-rains performances, evaluating additional new, rust-tolerant varieties, examining different rates of SYMPAL fertilizer and comparing standard and experimental formulations of BIOFIX.

One facet of our expanded adaptive research agenda that is not well developed is the understanding of how improvements in legume enterprise interact with other components of the small-scale maize-based farming systems common throughout Western Kenya (Woomer et al., 2002). Efforts are underway to interpret grain legume enterprise within the fuller farming system context including interactions with three other key entry points; Striga Elimination, Crop Diversification and Animal Enterprise (see Figure 15.2).

We understand that these legumes provide residual benefits to following crops and feed to livestock, and that organic inputs from livestock manures can improve legume production in less responsive soils (Mpepereki et al., 2000), but other more nuanced interactions require further insights and all positive interactions and tradeoffs require quantification (Sanginga et al., 2001;

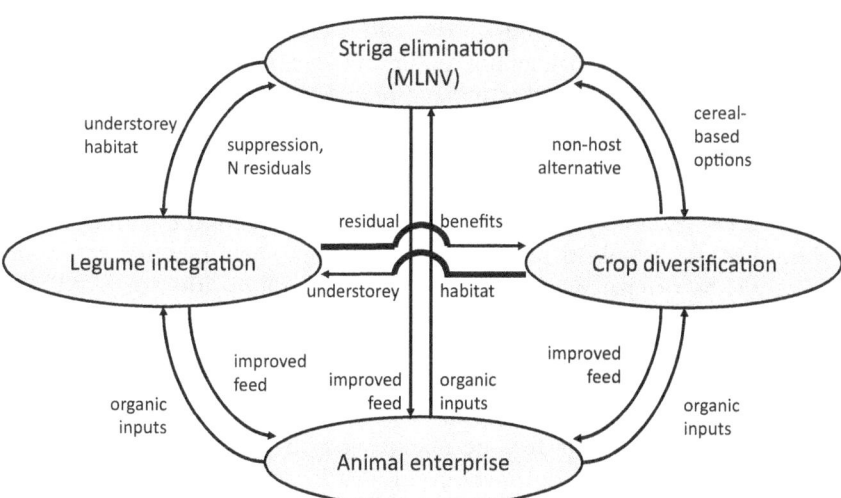

Figure 15.2 Examining legume integration and its interactions with other key farming operations within the Western Kenya Action Site as mandated by Humidtropics will require a new suite of field activities and cooperator skills

Kaizzi et al., 2012). How these will be achieved will certainly be an important aspect of future outreach activities and it is likely that the skill sets of Master Farmers and County Officers working in Western Kenya must be expanded to benefit from a more holistic approach (Lacy, 1996).

Following the on-farm technology testing described in this paper, the N2Africa Project entered into a more development-oriented phase intended for farmers in Western Kenya to enjoy more ready access to commercialized BNF technologies (Woomer et al., 2014). This shift signalled strategic partnership with both the private sector and the WeRATE network of farmer associations. Commercial partnership with MEA Fertilizers Ltd. also led to the wider distribution of BIOFIX legume inoculant and blending of SYMPAL, a fertilizer specifically designed for symbiotic legumes.

BIOFIX legume inoculant resulted from product licensing by the University of Nairobi MIRCEN laboratory and it is now produced at MEA's factory in Nakuru for several legume hosts (soybean, bean, pea, green gram, lucerne and others) in a variety of packaged quantities (10, 20, 50, 100 and 150 g). MIRCEN continues to offer quality control inspection with the most recent tests averaging over 5×10^9 colony-forming units (CFU) per gram. N2Africa designed software used in the analysis and distribution of these quality control results and worked with "last mile" stockists to display BIOFIX in glass-fronted refrigerators. It also worked with MEA to develop a product return policy where unsold stock at the end of each season is replaced with fresh stock the following one. Production of BIOFIX exceeded 10 tons in 2015, sufficient to treat over 20,000 ha, but not all is being used in Kenya as the product is also exported to Rwanda, Tanzania, Uganda and Zambia.

SYMPAL fertilizer blend is also commercially produced and distributed by MEA. It contains no mineral nitrogen but offers a balanced supply of phosphorus, potassium, calcium, magnesium, sulfur and zinc. This fertilization strategy optimizes BNF by assuring that mineral nutrient supply remains non-limiting. This blend was formulated by N2Africa, proto-types packaged by MEA and distributed free-of-charge, tested and refined through on-farm testing by WeRATE and the final product then produced at MEA's blending facility in Nakuru. SYMPAL enjoys growing popularity, especially for use on inoculated soybean because the plants become dark "blue" and yield increases by about 700 kg per ha compared to other managements. Appreciation of SYMPAL extends well beyond our network as over 128 tons of this product were blended and marketed by MEA over the past few years. It is available in 2, 10 and 50 kg plastic-lined woven polythene bags.

Grain legume seed is widely produced and marketed throughout Kenya by several seed companies, but few are marketing soybean. N2Africa conducted widespread on-farm testing of soybeans in West Kenya (Woomer et al., 2014), examining several traits such as seed size, protein and oil content and "dual purpose" growth habit, but ultimately varietal choice was largely determined by tolerance to Asian Rust, a foliar fungal disease outbreak that occurred after soybean production grew in popularity. Two soybean varieties developed by SeedCo are extremely resistant to rust SC Saga and SC Squire. SC Saga was licensed for distribution in Kenya in 2014 and is first appearing on stockists shelves in Western Kenya. Our work with climbing bean also led to the recognition and commercial release of climbing bean cultivar "Kenya Muvano", a very aggressive variety similar to Kenya Tamu, first sourced from Rwanda and now distributed by Kenya Seed Company.

Last-mile input supply is supported through several mechanisms. N2Africa organized agrodealer training of 32 members belonging to the Western Kenya Chapter of the Kenya AgroDealer Association (KENADA), sensitizing many stockists to BNF technology products for the first time. Next, 12 "One-Stop Shops" were initiated among partner farmer associations, allowing for BNF technologies to be directly marketed to farmers who participated in grain legume outreach. The Agricultural Technology Clearinghouse was formed as a semi-annual event bringing together representatives from farmer associations, input manufactures and distributors, and the development community to discuss which input products are performing best and which new products are becoming available (Woomer et al., 2016). This mechanism effectively links input suppliers and buyers and allows for research organizations to design on-farm technology tests around recently (or soon to be) released products. All of these actions are intended to support demand-driven technology supply because adoption levels of inputs are largely determined by the production levels and marketing of the commodities they accommodate. Over 2015, WeR-ATE members bulked and marketed over 478 tons of produce through their collection centres with prices ranging between $0.38 and $0.75 per kg (data not presented). These data do not include the production and sales by individual group members and other farmers away from these collection centres, but certainly indicate that soybean production and marketing is becoming a viable concern. Climbing bean, however, has achieved less impact, in large part because farmers prefer to intercrop bush beans as an understorey with maize, a practice that avoids expensive and time-consuming trellising.

Acknowledgements

The assistance of Ms. Welissa Mulei and Mr. Wycliff Waswa in the collection and compilation of field data in this study is greatly appreciated. The N2Africa Project is funded through a grant from the Bill and Melinda Gates Foundation. This study was also conducted in conjunction with the CGIAR Humidtropics Program.

References

Chianu, J.N., Huising, J., Danso, S., Okoth, P., Chianu, J.N. and Sanginga, N. (2010) 'Financial value of nitrogen fixation in soybean in Africa: increasing benefits for smallholder farmers', *Journal of Life Sciences*, 4, pp. 50–59.

Graham, P.H. and Rosas, J.C. (1977) 'Growth and development of indeterminate bush and climbing cultivars of *Phaseolus vulgaris* L. inoculated with *Rhizobium*', *Journal of Agricultural. Science*, 88, pp. 503–508.

Kaizzi, C.K., Byalebeka, J., Semalulu, O., Alou, I., Zimwanguyizza, W., Nansamba, A., Musinguzi, P., Ebanyat, P., Hyuha, T. and Wortmann, C.S. (2012) 'Optimizing smallholder returns to fertilizer use: bean, soybean and groundnut', *Field Crops Research*, 127, pp. 109–119.

Lacy, W.B. (1996) 'Research, extension and user partnerships: models for collaboration and strategies for change', *Agriculture and Human Values*, 13, pp. 33–41.

Mpepereki, S., Javaheri, F., Davis, P. and Giller, K.E. (2000) 'Soybeans and sustainable agriculture: promiscuous soybeans in Southern Africa', *Field Crops Research*, 65, pp. 137–149.

Ojiem, J.O., Vanlauwe, B., de Ridder, N. and Giller, K.E. (2007) 'Niche-based assessment of contribution of legumes to the nitrogen economy of western Kenya smallholder farms', *Plant Soil*, 292, pp. 119–135.

Sanginga, N., Okogun, J.A., Vanlauwe, B., Diels, J. and Dashiell, K. (2001) 'Contribution of nitrogen fixation to the maintenance of soil fertility with emphasis on promiscuous soybean-based cropping systems in the moist savanna of West Africa', in G. Tian, F. Ishida and J.D.H. Keatinge (Eds.) *Sustaining Soil Fertility in West Africa*, SSSA Special Publication No. 58, Soil Science Society of America, Madison, USA, pp. 157–178.

Sinclair, T.R., Marrou, H., Soltani, A., Vadez, V. and Chandolu, K.C. (2014) 'Soybean production potential in Africa', *Global Food Security*, 3, pp. 31–40.

Tefera, H. (2011) 'Chapter 7: Breeding for promiscuous soybeans at IITA', in A. Sudaric (Ed.), *Soybean – Molecular Aspects of Breeding*, InTech Press, Rijeka, Croatia, pp. 147–162.

Woomer, P.L., Huising, J., Giller, K.E. *et al.* (2014) 'N2Africa Final Report of the First Phase 2009–2013', www.N2Africa.org, 138 pp.

Woomer, P.L., Karanja, N.K. and Okalebo, J.R. (1999) 'Opportunities for improving integrated nutrient management by smallhold farmers in the Central Highlands of Kenya', *African Crop Science Journal*, 7, pp. 441–454.

Woomer, P.L., Mukhwana, E.J. and Lynam, J.K. (2002) 'On-farm research and operational strategies in soil fertility management', in B. Vanlauwe, J. Diels, N. Sanginga and R. Merckx (Eds.) *Integrated Plant Nutrient Management in Sub-Saharan Africa*, CABI, Wallingford, UK, pp. 313–332.

Woomer, P.L., Mulei, W. and Kaleha, C. (2016) 'Chapter 6: Humidtropics innovation platform case study: WeRATE operations in West Kenya', in I. Dror, J.J. Cadilhon, M. Schut, M. Misiko and S. Maheswari (Eds.) *Innovation Platforms for Agricultural Development; Evaluating the Mature Innovation Platforms Landscape*, Routledge, UK, pp. 98–116.

Part III

Integrating nutrition, gender and equity in research for improved livelihoods

16 Balancing agri-food systems for optimal global nutrition transition

Linley Chiwona-Karltun, Leif Hambraeus and Friederike Bellin-Sesay

Introduction

All food contains a mixture of nutrients comprising macronutrients and micro-nutrients, which can be grouped into *energy-yielding* and *essential nutrients*, respectively (Whitney and Rolfes, 2002). The energy-yielding macronutrients comprise carbohydrate, fat, protein, and also alcohol although it is often classified as a drug. The essential nutrients are, in addition to water, protein or rather essential amino acids, essential fatty acids, minerals, trace elements and vitamins. Human beings need access to and to consume all these nutrients in order to be in nutritional balance. Skipping a day of consumption of one of these nutrients is not detrimental for one's nutritional status, but prolonged exclusion of some or most of the nutrients is deleterious to health.

Optimal nutrition status is secured when food intake absorption and utilization provide all essential nutrients in required amounts. Poor nutrition is caused by the lack of physical, economic, social or physiological access to the right amounts of dietary energy and nutrients. Consequently, we need to know the principles of nutrition assessment (Gibson, 2005). Diseases, water, hygiene and sanitation affect the health status as well as the care and feeding practices and dietary habits of the person. These causes – underlying, intermediate or immediate – have been elaborated upon in the UNICEF conceptual framework to assess nutrition of mothers and children (UNICEF, 1990). While the political will to internalize the framework has not been pervasively adopted, the causes and associations with high morbidity and mortality have not changed; rather we realize that much still needs to be done (Black et al., 2008). What seems to remain problematic in addressing nutrition is what were referred to already in the 1990s as 'the politics of problem definition' and 'the garbage can model' of organizational behaviour (Pelletier, 2000).

During the last decade, Scaling Up Nutrition (SUN) countries are aiming collectively to reach by 2025 the global targets agreed upon by the World Health Assembly in 2012 (Table 16.1). This was followed in 2015 by the United Nations summit for the adoption of the Sustainable Development Goals (SDG, 2015).

Table 16.1 Targets to be reached by 2025 agreed by the World Health Assembly in 2012

Target 1	40% reduction of the global number of children under 5 who are stunted
Target 2	50% reduction of anaemia in women of reproductive age
Target 3	30% reduction of low birth weight
Target 4	Increase exclusive breastfeeding 50%
Target 5	No increase in overweight childhood; reducing and maintaining childhood wasting to less than 5%.

Source: Scaling Up Nutrition (SUN) www.scalingupnutrition.org.

In this chapter, we comment on the energy and nutritional needs of humans. This is followed by the current global concerns in nutrition. In the next section, we examine the evidence around the issue of 'is there food for all' and 'what comes first: food security or food safety?' We illustrate these issues of food security and food safety with our research on two crops, potato and cassava. The chapter ends by outlining how nutritionists and those working with nutrition in an agricultural context can take action to affect food and nutrition policy.

Is there food for all? The question of food availability, food security and food safety

The ever ongoing problem is the relation between food production, population growth and food availability. Interestingly, several shifts in the focus and priorities in the global food and nutrition policy debate can be observed throughout the years.

The Malthusian 'over population concern' era

Already in the end of the 18th century, Malthus commented on the imbalance between population growth and global food production (Malthus, 1798). These concerns led to programmes and emphasis on slowing population growth through family planning during the 20th century. It was followed by one of the most contested debates, that of the 'demographic entrapment of poor countries in Africa' by Dr Maurice King. King (1993) and King and Elliott (1997) argued that a country's population would exceed the carrying capacity of its ecosystem and its 'connectedness' to other ecosystems, leading to famine, epidemics and war.

Many poor people caught in the poverty trap in Sub-Saharan Africa are acutely aware of the need to cut down on family size as land holdings dwindle and farming as a livelihood becomes precarious. At the same time, when heads of households in Niger with an average of 7–8 children were interviewed, they stated that they would like to have family planning due to land size holding decline, but that they still wanted to have 8–11 children for social security reasons (personal communication with author FBS). In low-income countries the rural poor leave old age matters in the hands of their children in the absence of a functioning social welfare system.

However, improved public health conditions have led to changes in fertility throughout the last few decades (Butler, 2004). In a global perspective, fertility rate has decreased from 6 in 1965 to 2.5 in 2015 and will probably asymptotically reach 2 by 2050 (UNFPA, 1997). Infant mortality has decreased from around 150 per 1000 to below 30 in most low-income countries and around 5 in affluent societies. At the same time, life span has increased immensely from around 40 after birth in low-income countries to 65–70 and above 80 in affluent societies. The former population pyramid has now changed to a population hexagon in relation to age starting from zero years to old age on the y-axis (Figure 16.1). The total effect of reduced fertility on total world population will consequently take some time as it is 'counteracted' by the increased life span, particularly in Sub-Saharan Africa.

Decreased numbers of children born, in combination with an increase in the number of elderly globally, has also impacted the public health perspective. Undernutrition in infancy and childhood is decreasing and rising public health costs for later adult life in non-communicable diseases, i.e. cardiovascular disease, obesity, cancer and osteoporosis are rapidly increasing (Popkin, 2009). In low-income countries, it is imperative to try to break this trend.

The agricultural focus on 'increased food production' era

Throughout the centuries, the energy flows and carrying capacities of yields in food production have changed from foraging and pastoralism to development of agricultural production from shifting cultivation via traditional farming to modern mechanized farming (Figure 16.2).

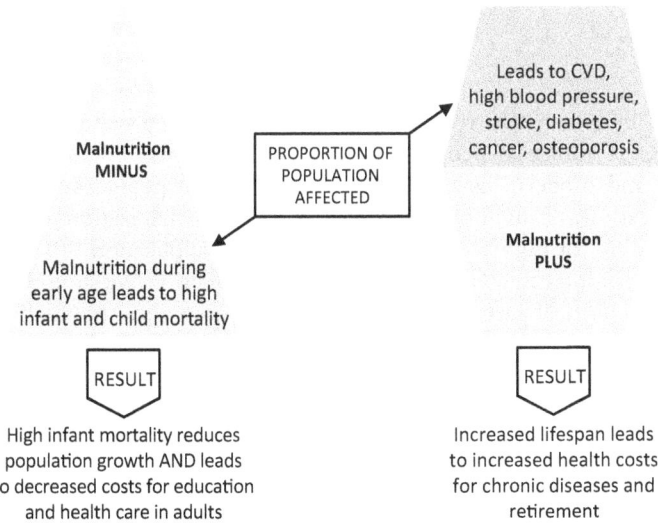

Figure 16.1 Geometric population changes and their effects on costs for health care

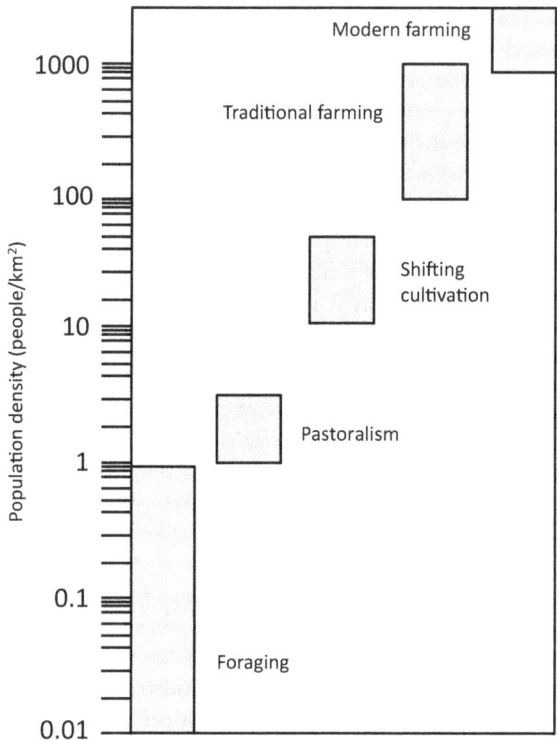

Figure 16.2 Comparison of carrying capacities of the principal modes of human food
       production

Source: Modified from Smil, 2000.

While the least productive form of agriculture, i.e. shifting cultivation could
only support about 10 people/km², traditional farming could sustain several
hundred and modern agriculture well above 1,000 people/km² arable land.
This increase was caused by the introduction of mechanization, genetic selection, breeding for high yielding varieties and the application of fertilizers and
pesticides that increased yields in agricultural production. We should not underestimate the critical parts played by the evolution between population growth
and the increase in food production (Smil, 2000).

The relation between food production and population size could also be
regarded as an illustration of the 'hen and egg problem': which came first? The
expansion of carrying capacity could not have been possible without investments in crop cultivation and human labour. However, population increase
could neither be possible without the capability to feed more individuals. There
have been uneven benefits across regions and farmers in terms of productivity gains, with farmers benefiting only where cost reductions exceeded price
reductions (Evenson and Golin, 2003).

During the 1960s and 1970s most interest was focused on the quantitative production of dietary food energy (kilocalories) and food proteins. This was also heavily stressed in international meetings in connection with the energy crisis as well as the 'protein fiasco' debate in the 1970s (MacLaren, 1974; UN, 1975; FAO and WHO, 1992)

The socio-economic paradigm shift

In the latter half of the 20th century, the debate about food availability changed its focus to include more socio-economic perspectives concerning the political means and willingness to increase food availability for the global population, especially the poor and unprivileged. Diagrams were presented to illustrate the complex situation by economists and social-anthropology representatives (UNICEF, 1990). They asked for more macroeconomic perspectives on food availability and less on nutrition and agricultural research activities, which were argued to only deal with micro-perspectives (Berg, 1993).

Interestingly, the best indicators of an optimal balance between food production and population increase seem to be public health indicators such as infant mortality, maternal mortality and life span (Figure 16.3). Socio-economic macro-oriented programmes which cannot result in positive effects on these

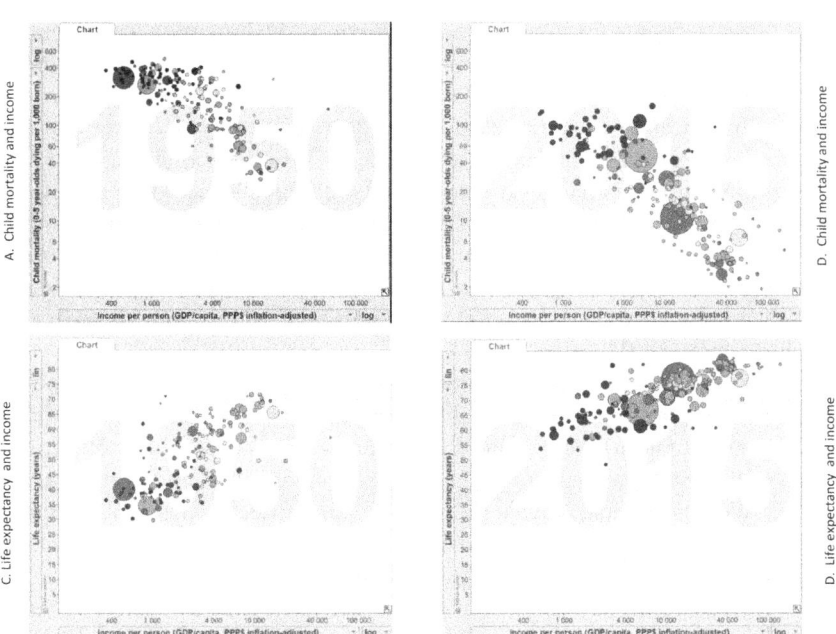

Figure 16.3 Changes in health and its relations to socio-economic development

Source: Published with permission of Gapminder.org.

public health parameters are of little value. However, better knowledge about nutrient needs in relation to changes in lifestyle is still urgently needed. This can best be achieved by providing thorough education and training in nutritional sciences and public health (Lachat et al., 2014).

The nutritional approach

In addition to the three above-mentioned concerns in the context of increasing food availability (population imbalance, increased food production and socio-economic factors), there is need for a fourth dimension to tackle the problem, a new nutritional approach. The nutritionist will no longer need to define minimal nutrient requirements to avoid undernutrition, but to define the optimal intake of nutrients taking into consideration changes in the food system, the role of retail supermarkets and the effect of lifestyle changes in physical activity and dietary habits.

Globally there is an ongoing change in dietary habits, food availability and lifestyle, referred to as the *nutrition transition* (Caballero and Popkin, 2002). In the affluent societies, the transition from hunting and gathering to modern agriculture took thousands of years and industrialization about two centuries. In low-income countries, the current nutrition transition occurs over a few decades only with increased energy intakes from oils, sugars and salty processed foods (Montiero et al., 2013)

Malnutrition is not only limited to undernutrition reflected as a low intake of required nutrients for normal bodily functions. Malnutrition also refers to overnutrition in energy intake leading to obesity and concomitant chronic health problems, i.e. obesity, cardiovascular diseases and cancer, known as non-communicable diseases. In addition, overnutrition may co-exist with vitamin and mineral deficiencies when energy density is high but nutrient density low. Lifestyle changes during the 21st century – including the introduction of mass transportation systems, increasing use of cars and motor-bikes and labour-saving devices in industry as well as in the home – have resulted in less physical activities, as has the introduction of and frequent use of IT-technology, particularly among teenagers and young adults (Popkin and Gordon-Larsen, 2004).

In short, the world has made significant progress in raising food consumption per person during the last decades. However, this progress has been accompanied by structural changes in the dietary intake from staple foods like roots, tubers and cereals, to, increasingly, simple carbohydrates, animal source foods and oils. We can nowadays divide the countries into three groups: (i) developing or low-income countries, (ii) industrial or affluent societies, and (iii) transition countries.

Energy intake, as well as protein availability, has increased in low-income and affluent societies, while protein intake has declined in transitional countries. This is partly due to higher energy density with fat and added sugars, and reduced fruit and vegetable intake in the rapidly growing middle class

(Caballero & Popkin, 2002; Pokin et al., 2012; Haggblade et al., 2014). In the case of India, protein intake is reduced due to increased energy from fats and added sugars (Pingali, 2007). Overnutrition and increased food waste also have adverse impacts on the optimal use of limited primary resources.

Severe protein-energy malnutrition in its extreme cases, *kwashiorkor* and *marasmus*, has decreased during the last decades as well as most vitamin deficiencies under 'normal conditions' in low-income countries. However, at the same time, some vitamin deficiency related diseases such as pellagra, scurvy and beriberi now reappear as a result of environmental stress situations in refugee populations (Dye, 2007). (For further reading on this issue, please consult the Lancet Nutrition Series 2008 and 2013).

A graphic presentation of the changes in the health panorama and its relation to socio-economic development is illustrated and regularly updated in Gapminder (www.gapminder.org). Indicators such as infant and maternal mortality as well as life span are illustrative of the changes in public health and the linkage between socio-economic changes and nutritional status and public health (Figure 16.3).

Priority of nutritional needs

Energy needs

Energy needs can be covered by any of the energy-yielding macronutrients, i.e. carbohydrate, fat and protein, and alcohol. It thus represents a non-specific quantitative demand on the dietary intake. The energy density of food refers to the amount of energy-yielding macronutrients per weight or volume. The energy percentage concept refers to the relative distribution of energy from the energy-yielding macronutrients in the diet.

Energy requirements are related to the basal metabolic rate (BMR) and the physical activity level (PAL) (James and Schofield, 1990). BMR is relatively stable from day to day and related to body composition, which varies with age, sex and weight. PAL is related to lifestyle, livelihood and physical activity, and is expressed as a multiple of BMR in a 24 hr perspective as a PAL factor, which varies between 1.5 in sedentary life, and up to about 3 in active athletes or labourers.

Nutrient needs

Nutrient needs refer to specific nutrients that cannot be synthesized in the body at all or in insufficient amounts, e.g. essential amino acids, trace minerals, vitamins. This represents a qualitative aspect of the dietary intake. Nutrient density refers to the amount of essential nutrients per energy unit (joule or kcal). For further reading we refer to Whitney and Rolfes (2002). In a low-income country context, food needs have seldom been prioritized in terms of nutrient quality and density.

The dietary protein dilemma – How much is needed?

Although protein is an essential nutrient, it cannot be stored in the body as many other essential nutrients, e.g. iron, fat soluble vitamins. Thus a protein intake above what is needed for protein synthesis will be metabolized and used as energy in the daily energy turnover or stored in energy pools, essentially as adipose tissue. This is of importance to remember when high protein intakes, not least of expensive animal protein, in the diets in affluent societies as well as transitioning economies, are now reported.

If energy needs are not met in the diet, protein will primarily be used as an energy source as the body gives priority to meet its energy needs. A high protein intake per se, as well as a high protein energy percent, is consequently of little use for covering protein needs if energy needs are not met.

The protein energy percent concept was introduced to express the dietary quality and protein density in the diet. However, it must be related to physiological parameters, i.e. age, gender and body weight as well as to physical activity, energy balance and protein quality (Millward and Jackson, 2003).

Interestingly, most conventional diets, vegetarian as well as mixed diets, have a protein energy percent around 10–12. In affluent societies with high intakes of animal food items it is often as high as 15–20 E%. The protein energy percent in human milk is extremely low but nevertheless enough to cover protein needs in breastfed infants when energy needs are met (Hambraeus et al., 1978).

In the case of high energy intakes, a fixed protein energy percent recommendation may lead to unnecessary overconsumption of protein. Thus, when physically active individuals consume more food to cover their energy needs, this indirectly leads to an overconsumption of protein, as the protein need is not related to the energy turnover. A high intake of protein, especially of animal protein, creates a high burden on the primary resources in food production.

The dominant global health and nutrition problem is still protein-energy malnutrition (PEM)

Already in 1930, Cecily Williams described a special malnutrition syndrome which occurred in early childhood among the poor in low-income countries, which was given the name *kwashiorkor* (Williams, 1949). When increased interest was devoted to the nutritional problems after World War II, two extreme forms of protein malnutrition were described, *kwashiorkor* and *marasmus*, and the focus was on the protein intake.

In a global perspective, it is important to emphasize that any disturbed protein synthesis results in a breakdown of the humoral and cellular defence mechanisms against infection, a nutritionally acquired immune deficiency syndrome, NAIDS (Beisel, 2001). This breakdown of the defence mechanisms is principally similar to the acquired immune deficiency syndrome (AIDS) described as a result of HIV infections. NAIDS can be corrected by active nutritional therapy and sound nutritional science knowledge. However, although HIV–AIDS breakdown is essentially an irreversible condition, it calls for nutrition therapy.

The food crop–cash crop dilemma and its impact on food security

Nutritionists must actively take into consideration *new conflicts between the 'four big Fs': Food, Feed, Fibre and Fuel.* There is always need to improve profits of production in the agricultural sector. This has led to the so-called food crop–cash crop conflict. The interest to sell food crops for cash has led to a competition between food crops and feed crops, i.e. feed for animal husbandry and fish farming as a result of increasing demand for meat and animal food items. There is, ironically, an association with increasing incomes, life expectancy and increased demand for animal-source foods as shown by Popkin (2009). In addition, today two 'new' factors have aggravated the situation: the use of agro-products for industrial production of fibre and renewable biofuels.

Food prices have a direct impact on food availability for the poor and the unprivileged in the society. Food availability consequently represents a direct potential political weapon (food power) both nationally and internationally.

Is biofuel production a threat to food security or a stimulus for the agro-economy?

The conflict between biofuels and food crops has led to the use of the concept of agro-fuel instead of biofuel in order to stress the potential harmful effects on food security and livelihoods (Borras et al., 2011; Harvey and Pilgrim, 2011). Using food crops for biofuels, which are often produced as monocultures, also leads to environmental problems. However, it should be recognized that biofuels might not necessarily be based on potential food crops (Hambraeus, 2009). Biofuels could also be produced from by-products, waste or crops on marginal lands. But this is complicated by politics, absence of policies and the lack of political will to secure land title deeds for many of the world's poor (Cotula et al., 2008; De Schutter, 2011; HLPE, 2013).

Biofuels are by definition energy sources from renewable material. The first generation of biofuels was based on the use of food crops, e.g. cereals and tubers (maize and wheat or potato and cassava) for alcohol production or oil seeds (rape seed, palm oil) for diesel production. The second ·generation is based on the biomass of non-food crops, wood and trees but also on organic wastes from crops (e.g. stems, leaves and husks), industry and society.

Agro-fuels, while controversial, may present an opportunity for rural development but this effect has so far been marginal (Borras et al., 2011; Hambraeus, 2011). However, agro-fuel producers usually seek access to quality land and water resources for agro-fuel plantations, leading to land grabbing both on a national and international level. Anseeuw et al. (2012) recently discussed the investments in the verified land grab during 2000–2012 in Africa, Asia and Latin America. Agro-fuel dominated with 58% and food and forestry constituted only 18% and 13% respectively, and precariously exacerbated food security. Africa is the dominant continent for land grabbing, representing about half.

Food security and food safety

Food security and food safety are two concepts that are easily mixed up. Food security concerns the food availability to cover nutritional needs of the population for today and tomorrow, which is essentially the main nutrition concern of low-income countries. Food safety relates to the potential risks of natural toxins, contaminants and adverse environmental effects of exogenous toxins in the food production chain (Jelliffe and Jelliffe, 1982). It is more common to discuss food safety in affluent societies, where food security rarely is a public health problem. However, food safety is also a serious problem in low-income countries where food storage, contamination and preservation cause challenges. More studies from low-income countries are emerging and they highlight the role of aflatoxins in food safety issues (Gnonlonfin et al., 2013).

Cassava and potato – Food crops for minding food security and food safety

The discussion on world food production is still concentrated on cereals, especially rice, wheat and maize. The high yield and role of potato and cassava in the fight against hunger and starvation is often forgotten, especially when it comes to the livelihoods of resource-poor households. In the 17th and 18th centuries, potato played an essential role to help the poor and underprivileged people in Europe to cover their energy and nutrient needs. Unfortunately, it also led to strong dependence on a single crop. When, in 1845, potato crops in Ireland failed it led to extreme famine, known as the potato famine, that saw millions of people migrate from Europe to North America (Salaman, 1949).

Potato and cassava both emanate from the Andean region in Latin America. Interestingly, both have been exported, or imported, to Europe and Africa during the last two to three centuries (Jones, 1959). Although this might be considered a short period in the history of humans, these exotic crops are today valuable staple food items in several communities in North America, Europe, Africa and Asia.

Almost two-thirds of the potato consumption in the world occurs in the high-income countries (Keijbets, 2008), while more than half of the cassava consumption is in the low-income countries, with Africa accounting for most of it. Cassava has been of importance for the survival of the poor and unprivileged in low-income countries, especially in times of food scarcity (Chiwona-Karltun et al., 1998). Ironically, both crops today play a less dominant role in Latin America from where they originated. However, they remain important food crops there; 20% of the global cassava consumption and only 5% of potato consumption, respectively, occur in Latin America.

A common characteristic for potato and cassava is their potential toxicity as they both contain natural toxins, which clearly illustrates that there can be a conflict between food security and food safety (Dolan et al., 2010). The solanin content in potato can be harmful, and its fruits are toxic. The potential toxicity of cassava is in the form of cyanogenic glucosides. Under certain conditions, this can lead to neurological disturbances in the form of spastic paralysis. Health problems due to cassava consumption are the result of improper processing due to food shortages or crises such as conflict and/or epidemics disturbing normal agricultural practices. Food scarcity has often resulted in impaired traditional processing that follows a very specific process, i.e. soaking, fermentation and sun drying, due to shorter modified processing time (Banea-Mayambu et al., 1997).

Regarding cassava, the engagement of scientists and their misunderstanding of how farmers utilize and perceive cassava has at times raised some unnecessary alarms around food safety and consumption of cassava rather than understanding the various roles cassava plays in food security (Chiwona-Karltun et al., 2015). As illustrated by studies from South America and Africa, communities that consume potentially toxic crops have identified ways of processing and consuming bitter and toxic cassava using a bio-cultural approach (Chiwona-Karltun et al., 2004; Dufour, 2006).

Both potato and cassava represent crops with very high energy yields per acre. Consequently, increased interest for renewable sources of energy may result in a conflict in their use. Bioenergy production from these crops may be a serious threat to cheap food availability for the poor and underprivileged in low-income countries (Nuwamanya et al., 2012). Furthermore, alternative uses for these crops, such as alcohol production by breweries, may also bring about unintended consequences in rural areas such as increased expenditure on non-food items, raising prices for food items, gender differences in expenditure for household food and consumption patterns, as well as increased domestic violence. To our knowledge, there are very few empirical studies on these emerging issues.

Concluding remarks

The problem with nutrition lies in its definition. Nutrition is basically a biological science. The challenge for nutritionists is first to have a scientifically well-based definition and understanding of the optimal diet. This should be in balance with primary resources, lifestyle, physical activity and environmental and socio-economic conditions, i.e. climate, optimal land use, water availability, greenhouse gas emission and labour costs.

In contrast to the feeding of animals, the optimal nutrition of humans is not to reach a maximal growth within the shortest time, but to primarily avoid nutrient deficiency as well as to avoid over-intake of nutrients and total energy. Human beings are a relatively slow growing species amongst the mammals. Any nutrient intake above what is required for normal growth, maturation and maintenance of health is not only a waste of primary resources, it may also be toxic during the neonatal period. This potentially could lead to health problems in the long-term, e.g. cardiovascular disease, diabetes, obesity or cancer.

As discussed in this chapter, there are no magic bullets to solve nutritional problems. Non-cereal crops, such as potato and cassava, have often not been given much attention, especially their important role for food security and nutrition, production, processing and consumption promotion. Too often as scientists or professionals we are embroiled in advancing our own interpretations of the nature of the problem and preferred solutions, and more often than not seeking to advance our own interests. It is thus no wonder that malnutrition continues to be a problem of magnitude proportions in low-income countries and requires tackling systemic institutional, environmental, professional, political and commercial interests.

References

Anseeuw, W., Boche, M., Breu, T., Giger, M., Lay, J., Messerli, P., and Nolte, K. (2012) *Transnational land deals for agriculture in the global South*, IIED, UK.

Banea-Mayambu, J. P., Tylleskär, T., Gitebo, N., Matadi, N., Gebre-Medhin, M., and Rosling, H. (1997). 'Geographical and seasonal association between linamarin and cyanide exposure from cassava and the upper motor neurone disease konzo in former Zaire', *Tropical Medicine & International Health*, 2(12), pp. 1143–1151.

Beisel, W.R. (2001) 'Nutritionally acquired immunodeficiency syndrome', in Friis, H. (ed.) *Micronutrients and HIV infection*, CRC Press, Boca Raton, pp. 23–42.

Berg, A. (1993) 'Sliding toward nutrition malpractice: time to reconsider and redeploy', *Annual Review of Nutrition*, 13(1), pp. 1–16.

Black, R.E., Allen, L.A., Bhutta, Z.A., Caulfield, L.E., De Onis, M., Ezzati, M., Mathers, C., Rivera, J., and Maternal and Child Undernutrition Study Group (2008) 'Maternal and child undernutrition: global and regional exposures and health consequences', *The Lancet*, 371(9608), pp. 243–260.

Borras Jr, S.M., Fig, D., and Suárez, S.M. (2011) 'The politics of agrofuels and mega-land and water deals: insights from the ProCana case, Mozambique', *Review of African Political Economy*, 38(128), pp. 215–234.

Butler, C.D. (2004) 'Human carrying capacity and human health', *PLoS Medicine*, 1(3), p. 192.

Caballero, B., and Popkin, B.M. (2002) *The nutrition transition: diet and disease in the developing world*, Academic Press, San Diego, CA.

Chiwona-Karltun, L., Brimer, L., Kalenga Saka, J.D., Mhone, A.R., Mkumbira, J., Johansson, L., Bokanga, M., Mahungu, N.M., and Rosling, H. (2004) 'Bitter taste in cassava roots correlates with cyanogenic glucoside levels', *Journal of the Science of Food and Agriculture*, 84(6), pp. 581–590.

Chiwona-Karltun, L., Mkumbira, J., Saka, J., Bovin, M., Mahungu, N., and Rosling, H. (1998) 'The importance of being bitter – a qualitative study on cassava cultivar preference in Malawi', *Ecology of Food and Nutrition*, 37, pp. 219–245.

Chiwona-Karltun, L., Nyirenda, D., Mwansa, C.N., Kongor, J.E., Brimer, L., Haggblade, S., and Afoakwa, E.O. (2015) 'Farmer preference, utilization, and biochemical composition of improved cassava (Manihot esculenta Crantz) varieties in southeastern Africa', *Economic Botany*, 69(1), pp. 42–56.

Cotula, L., Dyer, N., and Vermeulen, S., 2008. *Fuelling exclusion: the biofuels boom and poor people's access to land*, IIED, London, UK.

De Schutter, O. (2011). 'How not to think of land-grabbing: three critiques of large-scale investments in farmland', *The Journal of Peasant Studies*, 38(2), 249–279.

Dolan, L.C., Matulka, R.A., and Burdock, G.A. (2010) 'Naturally occurring food toxins', *Toxins*, 2(9), pp. 2289–2332.

Dufour, D.L. (2006) 'Biocultural approaches in human biology', *American Journal of Human Biology*, *18*(1), pp. 1–9.

Dye, T.D. (2007) 'Contemporary prevalence and prevention of micronutrient deficiencies in refugee settings worldwide', *Journal of Refugee Studies*, *20*(1), pp. 108–119.

Evenson, R.E., and Gollin, D. (2003) 'Assessing the impact of the green revolution, 1960 to 2000', *Science*, *300*(5620), pp. 758–762.

FAO and WHO (1992) International Conference on Nutrition – World Declaration and Plan of Action for Nutrition, Rome.

Gapminder (2016) www.gapminder.org/world *[accessed 06–06–2016]*

Gibson, R.S. (2005) *Principles of nutritional assessment*, Oxford University Press, New York.

Gnonlonfin, G.J., Hell, K., Adjovi, Y., Fandohan, P., Koudande, D.O., Mensah, G.A., Sanni, A. and Brimer, L. (2013) 'A review on aflatoxin contamination and its implications in the developing world: a sub-Saharan African perspective', *Critical Reviews in Food Science and Nutrition*, 53(4), pp. 349–365.

Haggblade, S., Gyebi, D., Kabasa, J.D., Minnar, A., Ojijo, N.K.O., and Taylor, J.R.N. (2015) 'Emerging early actions for bending the curve in Africa's nutrition transition', MAFS Working Paper No. 12, www.mafs-africa.org/uploads/files/mafs_working_paper_12_-_bending_the_curve_ver1.pdf *[accessed 06–06–2016]*

Hambraeus, L. (2009) 'Biofuels – a new challenge for nutritional science? Invited commentary', *Public Health Nutrition*, 12, pp. 2533–2335.

Hambraeus, L. (2011) 'How to balance biofuel and food production for optimal global health and nutrition – the food crop- feed crop- fuel crop trilemma', in dos Santos Bernardes, M.A. (ed), *Economic effects of biofuel production*, Intech open access publishers, pp. 57–78, www.intechopen.com *[accessed 06–06–2016]*.

Hambraeus, L., Lönnerdal, B., Forsum, E., and Gebre-Medhin, M. (1978) 'Nitrogen and protein components of human milk', *Acta Paediatrica*, *67*(5), pp. 561–565.

Harvey, M., and Pilgrim, S. (2011) 'The new competition for land: food, energy, and climate change', *Food Policy*, 36(suppl 1), S40–S51.

HLPE, 2013. Biofuels and food security. A report by the High Level Panel of Experts on Food Security and Nutrition of the Committee on World Food Security, Rome 2013 *[accessed 06–06–2016]*

James, W.P.T., and Schofield, E.C. (1990) *Human energy requirements, a manual for planners and nutritionists*, Oxford University Press, Oxford.

Jelliffe, E.F.P., and Jelliffe, D.B. (1982) *Adverse effect of foods*, Plenum Press, New York.

Keijbets, M.J.H. (2008) 'Potato processing for the consumer: developments and future challenges', *Potato Research*, 51(3–4), pp. 271–281.

King, M. (1993) 'Demographic entrapment', *Transactions of the Royal Society of Tropical Medicine and Hygiene*, 87(suppl 1), pp. 23–28.

King, M., and Elliott, C. (1997) 'To the point of farce: a Martian view of the Hardinian taboo – the silence that surrounds population control', *British Medical Journal*, *315*(7120), p. 1441.

Lachat, C., Nago, E., Roberfroid, D., Holdsworth, M., Smit, K., Kinabo, J., Pinxten, W., Kruger, A., and Kolsteren, P. (2014) 'Developing a sustainable nutrition research agenda in sub-Saharan Africa – findings from the SUNRAY project', *PLOS Medicine*, DOI: 10.1371/journal.pmed.1001593.

Malthus, T. (1798) An essay on the principle of population, Penguin Edition, London, http://www.esp.org/books/malthus/population/malthus.pdf *[accessed 06–06–2016]*

Manioc in Africa. WILLTAM 0. Jones, W.O. (1959) 'Manioc in Africa', Stanford, Stanford University Press, 1959, 315 pp.

McLaren, D. (1974) 'The great protein fiasco', *The Lancet*, 304(7872), pp. 93–96.

Millward, J.B., and Jackson, A.A. (2003) 'Protein/energy ratios of current diets in developed and developing countries compared with a safe protein/energy ratio: implications for recommended protein and amino acid intakes', *Public Health Nutrition*, 7, pp. 387–405.

Monteiro, C.A., Moubarac, J.C., Cannon, G., Ng, S.W., and Popkin, B. (2013). 'Ultra-processed products are becoming dominant in the global food system', *Obesity Reviews*, 14(S2), pp. 21–28.

Nuwamanya, E., Chiwona-Karltun, L., Kawuki, R.S., and Baguma, Y. (2012) 'Bio-ethanol production from non-food parts of cassava (Manihot esculenta Crantz)', *Ambio*, 41(3), pp. 262–270.

Pelletier, D. (2000) 'Toward a common understanding of malnutrition: assessing the contributions of the UNICEF conceptual framework', *World Bank/UNICEF Assessment of Contributions to Nutrition Policy*, Washington, DC.

Pingali, P. (2007) 'Westernization of Asian diets and the transformation of food systems: implications for research and policy', *Food Policy*, 32(3), pp. 281–298.

Popkin, B.M. (2009), 'Global changes in diet and activity patterns as drivers of the nutrition transition', *Nestle Nutrition Workshop Series, Paediatric Programme*, 63, pp. 1–10.

Popkin, B.M., Adair, L.S., and Ng, S.W. (2012), 'Global nutrition transition and the pandemic of obesity in developing countries', *Nutrition Reviews*, 70(1), pp. 3–21.

Popkin, B.M., and Gordon-Larsen, P. (2004) 'The nutrition transition: worldwide obesity dynamics and their determinants', *International Journal of Obesity*, 28, S2–S9.

Salaman, R.N. (1949) *The history and influence of the potato*, Cambridge University Press, Cambridge, Reprinted 1970.

Scaling Up Nutrition (SUN). (2016), http://www.scalingupnutrition.org *[accessed 06–06–2016]*

SDG Sustainable Development Goals (2015) http://www.un.org/sustainabledevelopment/summit *[accessed 06–06–2016]*

Smil, V. (2000) *Feeding the world: a challenge for the twenty-first century*, MIT Press, Cambridge, MA.

UN (1975) *Report of the world food conference held in Rome in November 1974*, United Nations, New York.

UNFPA (1997) UN department of economic and social affairs, Population division: 'World population ageing 1950–2050', www.un.org/esa/population/publications/worldageing1950–2050 *[accessed 06–06–2016]*

UNICEF (1990) *Strategy for improved nutrition of children and women in developing countries*, UNICEF, New York.

Williams, C. (1949) 'Kwashiorkor', *The Lancet*, 253(6556), pp. 711–712.

Whitney, E.N., and Rolfes, S.R. (2002) *Understanding nutrition*, Wadsworth Publishing Company, Belmont, CA.

17 Nutrition-sensitive landscapes

Approach and methods to assess food availability and diversification of diets

Gina Kennedy, Jessica Raneri, Celine Termote, Verena Nowak, Roseline Remans, Jeroen C.J. Groot and Shakuntala H. Thilsted

Background

Global concerns about the sustainability of food production systems as well as the quality of diets resulting from these food systems are escalating. Malnutrition, including undernutrition, overnutrition and micronutrient deficiencies together with increases in diet related non-communicable diseases (NCDs) are key developmental and political challenges (International Food Policy Research Institute, 2014). The diets of many people around the world are dominated by a single staple crop, most notably rice, maize or wheat, and lack diversity of other foods such as vegetables, fruit or animal-source foods (fish, milk, eggs and meat). Low diversity of diets is associated with lower probability of adequate intake of micronutrients from the diet (Kennedy, 2009). Current agricultural practices are moving toward intensified monocultures, which increase grain yields in the short-term, but limit dietary and biological diversity (Khoury et al., 2014). In addition, population growth, climate change and changing consumer preferences add pressure to our current food production systems.

Both health and environmental sustainability concerns have been raised about food production systems, particularly those which focus on agricultural intensification, which are seen to reduce environmental health, and contribute to the rise of NDCs (Demaio and Rockström, 2015). The Sustainable Development Goals (SDGs) recently endorsed by the member states of the United Nations make commitments toward 17 SDGs, including reducing hunger and malnutrition, and improving environmental sustainability. The nutrition-sensitive landscape (NSL) approach works within the space between the SDGs to study synergies and trade-offs related to dietary improvements and sustainable food production.

The nature of the relationships between communities, food production systems and the environment is complex and bidirectional (MA, 2005; Deckelbaum et al., 2006). The behaviour of people influences the capacity of landscapes to provide the multiple functions that are essential for human well-being, including multiple nutrition and health functions, ranging from the variety of nutrient-rich foods produced, water and air quality and soil fertility. The complexity of current global challenges requires a fresh look at how people interact with their environments in order to fulfil the goals of food and nutrition security while maintaining, restoring and securing the ecosystems upon which we are dependent.

At the heart of systems orientation is an emphasis on understanding relationships between changing factors. The NSL approach addresses the relationship between nutrition, agriculture and the environment, and aims to identify, quantify and tackle unsustainable trade-offs while generating synergies. The ecosystems component emphasizes the interaction between species and their environment and the socio-ecosystem component highlights the coupling between people and the environment.

The NSL approach does not mean that the environment can produce all nutrients required for adequate human nutrition; however, it does mean focus should be put on producing diverse sources of food within a given landscape, while also managing other ecosystem functions that are critical for environmental sustainability and human well-being.

Nutrition-sensitive landscape conceptual framework

The NSL conceptual framework combines ecosystem service functions including supporting (soil formation, primary crop production), provisioning (food, water and fuel), regulatory (air and water quality) and cultural (recreation and leisure) functions (MA, 2005) with the UNICEF conceptual framework for nutrition, in which food (dietary quantity and quality), health (disease and water and sanitation factors) and care practices (infant feeding and hygiene, maternal care) are the underlying determinants of human nutrition outcomes (Black et al., 2008). Within the NSL approach, research seeks to develop a strong knowledge-base of how nutrition and health outcomes can be improved in contexts where finite natural resources are managed to achieve multiple, and sometimes competing, objectives. The aim of this chapter is to demonstrate how better understanding a landscape's capacity to provide more diverse foods across seasons can contribute to diversified production systems for more diversified diets through three case studies in Zambia, Kenya and Vietnam.

Nutrition-sensitive landscape research methodology

The NSL methodological approach involves both qualitative and quantitative assessments that encompass aspects of food availability and access via own production, wild forage and capture, market availability, food cost and dietary intake (Figure 17.1). Seasonal food availability and access is a key theme running throughout all assessments.

In all case studies, qualitative assessments were made of (1) seasonal food availability, (2) potential barriers to achieving good nutrition, and (3) potentially

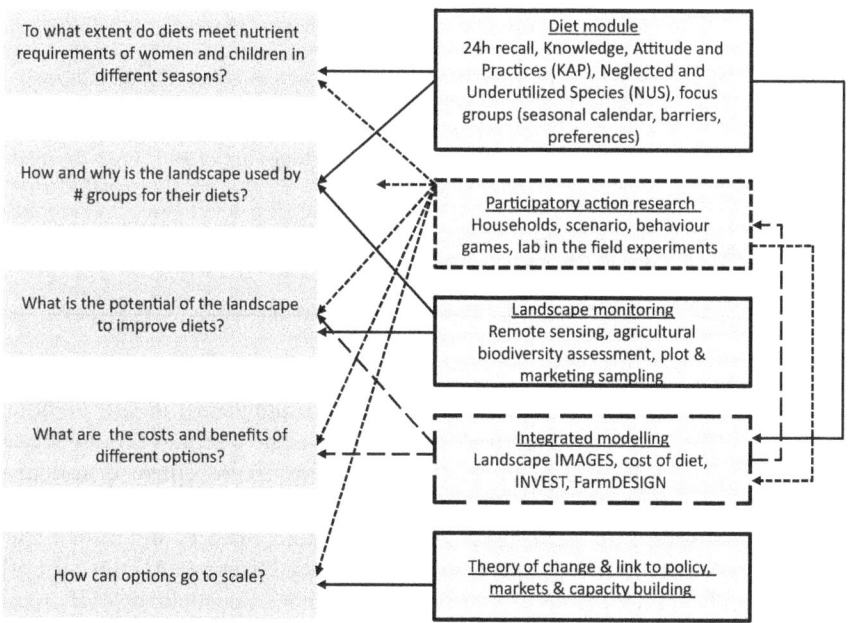

Figure 17.1 Nutrition-sensitive landscapes research questions and methodology

nutrient dense foods that have become neglected or underutilized over time. The studies were designed to specifically answer the following questions:

- How does locally available biodiversity contribute to dietary diversity and nutrition?
- How does a household's production diversity, availability and access to market and wild diversity influence dietary diversity and nutrition?
- What nutrition knowledge, attitudes and practices exist and how do they affect dietary and production diversity?
- What key household and landscape system elements can be leveraged to improve dietary diversity and quality?

Two quantitative questionnaires were developed, one focused on household production and socio-economic status, and the other on nutrition. The respondents of the first questionnaire were household heads and of the second, mothers (or primary caretakers) of children aged 12–23 months (and 6–23 months in Zambia). The nutrition questionnaire contained two components,

Table 17.1 Dietary Intake Indicators and definitions used in all case studies

Indicator name	Indicator description	Reference
Minimum Dietary Diversity of women (adults in Zambia)	Proportion of women (adults in Zambia) that consumed 5 or more food groups out of 10	FAO and FHI 360, 2016
Minimum Dietary Diversity (MDD) infants and young children	Proportion of children (12–23 months and 6–23 months in Zambia) that consumed 4 or more food groups out of 7	WHO, 2008
Average Individual Dietary Diversity Score women/ adults	Average number of food groups out of 10 consumed by women	FAO an dFHI 360, 2016
Average Individual Dietary Diversity Score infants/ young children	Average number of food groups out of 7 consumed by infants and young children	WHO, 2008

(1) a quantitative 24-hour recall of all food consumed in the last 24 hours (24HR) adapted to capture agriculture diversity in the diet, and (2) nutritional knowledge, attitudes and practices (KAP). Two non-consecutive quantitative 24HRs were conducted for the primary child caretaker and the child, over at least two seasons. Data were used to construct four primary indicators: Dietary Diversity Score (DDS) and Minimum Dietary Diversity (MDD) for both women and children (Table 17.1). In Barotse, Zambia, a qualitative 24HR recall was implemented rather than a quantitative due to resource constraints. In addition, for some case studies, plot, farm and landscape mapping, including gender disaggregated land use, were conducted. After analysis of the baseline data, communities were engaged in participatory action planning exercises, in which crops suitable for different seasons and soil types were identified and community-based diversification interventions established. The production diversity activities were accompanied by nutrition interventions, involving community nutrition groups which gathered to discuss and prepare more nutritious meals, using locally available foods.

The three selected case studies highlight findings of the seasonal food availability, dietary intake and on-farm production research activities. In the chapter by Groot et al. (this volume), the modelling tool FarmDESIGN is described, which has been modified to include nutrition as well as traditional parameters such as income, soil fertility and yield.

Case studies

Barotse, Zambia – Putting agriculture diversity on the plate in fishing communities

Zambia experiences a persistently high level of stunting (UNICEF, 2011). The 2007 Demographic and Health Survey (DHS) reported that 45% of children under five years of age in Zambia were stunted and 21% were severely stunted (Republic of Zambia, Central Statistics Office et al., 2009). The most recent DHS findings from 2013–2014 revealed a drop in the national rate of stunting to 40% (Republic of Zambia, Central Statistics Office et al., 2014). Poor nutritional status, especially of women and young children, inhibits individual growth and development and negatively impacts the overall health, productivity and economic potential of a community.

In the Barotse floodplain, food availability and dietary quality are highly seasonal, with a long hunger season. Overall, diets are low in diversity, dominated by the staple food maize. Current agricultural practices are moving towards more intensified monocultures and overfishing that can provide some short-term benefits but further limit dietary diversity and contribute to land and biodiversity habitat degradation, water nutrient loading and increased greenhouse gas emissions. In addition, changing weather conditions and unpredictable flooding are adding pressure to this vulnerable ecosystem. A key development challenge is how to improve dietary diversity sustainably throughout the year while managing natural resources.

Dietary intake surveys are reported for two time points: November 2013 (wet, lean, planting season) and February/March 2014 (wet, harvesting, maize available, but period of fish ban). Focus groups were formed in each community to complete the seasonal calendars. The food items listed in each seasonal calendar were grouped into three primary food categories, based on nutrition education materials used by the Ministry of Agriculture and Livestock: protective foods (*silelezo*), body-building foods (*ze yaha mubili*) and energy dense foods (*ze fa maata*). The protective category was divided into the sub-categories: vitamin A rich foods (*lico ze nani vitamin A*), dark leafy green vegetables (*miloho ye butala*) and other vegetables and fruit (*miloho ni litolwana zemu*). The body-building category was divided into the sub-categories: animal-source foods (*lico ze fumaniwa kwa lifolofolo*) and legumes, beans and seeds and nuts (*manawa, ndongo*).

Food consumption data were available for 1089 adults (72% women, 29% men) and 252 young children 6–23 months of age (53% girls, 47% boys). The average DDS of infants and young children 6–23 months was 2.34 (± 1.1), with no differences between girls and boys. The mean DDS for all adults was 2.88 (± 1.0). Women tended to have higher DDS values than men; however, the difference was not statistically significant. DDS for adults was affected by the survey round, being lower in February/March, and higher in November. MDD for children was at similar levels in November and February/March, while for adults 14% achieved MDD in November and 5% in February/March (Table 17.2).

Food groups consumed by adults included starchy staple foods which were consumed by all adults (100%), followed by flesh foods, consumed by about 60%, and vitamin A rich dark green leafy vegetables and other vegetables, consumed by 45% and 36% of adults, respectively (Figure 17.2). Food groups most and least commonly consumed were similar for children 6–23 months of age.

Table 17.2 Dietary Diversity Score (DDS) and Minimum Dietary Diversity (MDD) by season for children 6–23 months and adults

	Leaner season[*]	More abundant season[**]
Average Dietary Diversity Score		
Women		
Kenya	4.1 ±1.3	4.4 ±1.3
Vietnam	4.8 ±1.1	4.8 ±1.1
Zambia^	3.5 ±1	2.9 ±0.9
Children		
Kenya	4.0 ±1.0	4.1 ±1.1
Vietnam	3.7 ±1.2	3.7 ±1.1
Zambia	2.7 (±1)	2.7 (±0.9)
Minimum Dietary Diversity (%)		
Women		
Kenya	40	50
Vietnam	58	59
Zambia^	14	5
Children		
Kenya	73	77
Vietnam	57	58
Zambia	21	23

[*]Leaner season: Kenya April 2015;Vietnam Aug/Sept 2014; Zambia Nov 2013
[**] More abundant season: Kenya Sept/Oct 2014;Vietnam Nov/Dec 2014; Zambia Feb/March 2014
^Data presented for men and women, no significant difference between groups

There were significant differences in the percent of young children and adults consuming food groups by season for most groups.

Learning plots were established in the communities to allow experimentation with growing different crops to diversify production and consumption. The crops were selected in a participatory process. In several workshops, information was shared on (1) nutritional benefits of diversification and different food groups and the current food consumption patterns; (2) soil types and benefits from ecosystem services; (3) agronomical practices; and (4) drought resistances and climate change issues. Workshops included representatives from the Ministry of Agriculture and Livestock, community facilitators, farmers, Caritas Mongu and project research staff. Based on the shared information and the community's priorities, the final list of crops for the learning plots was developed.

More than 100 learning plots are being established in the communities for the following crops: cassava, orange maize, rice (supa rice purchased from different sources, New Rice for Africa, or NERICA, 1 and 4), orange sweet potatoes, sorghum and cowpeas, where cowpeas are used for intercropping with maize and cassava. Additionally, learning plots at four schools are being established with orange maize, carrots, watermelons, as well as pawpaw and citrus trees. Communities have also formed nutrition groups where they meet to learn how to cook with a variety of ingredients as well as access nutrition education materials and exchange best practices with each other.

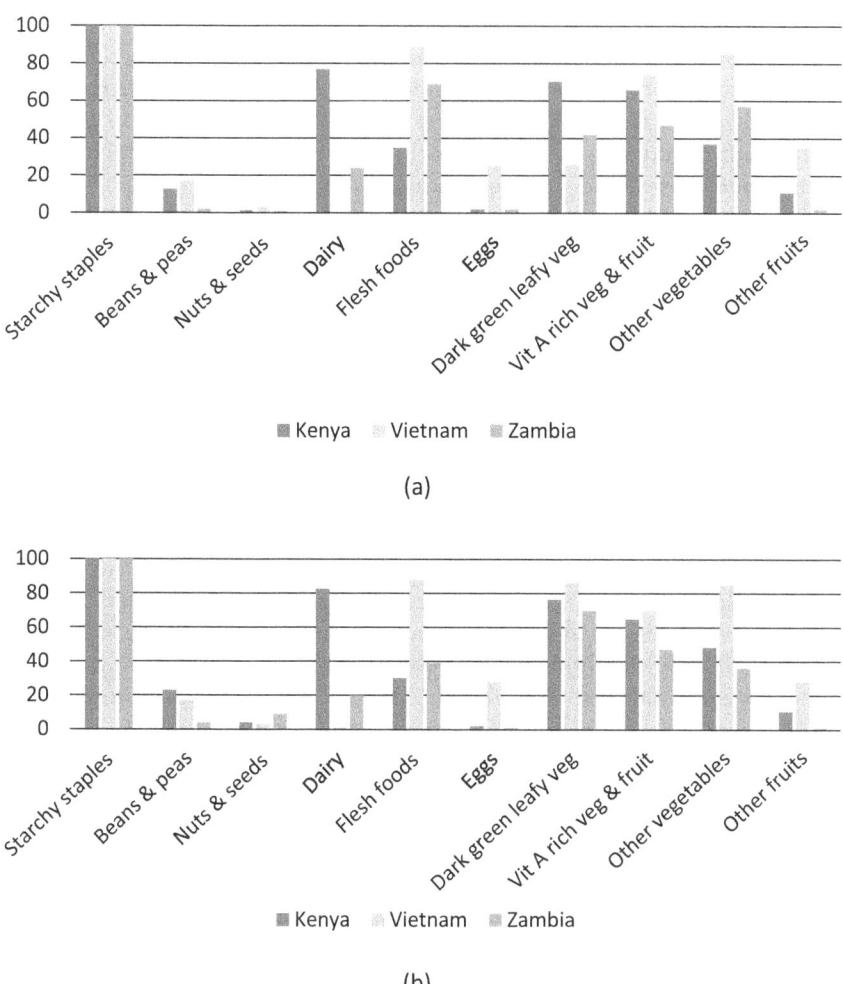

Figure 17.2 Percentage of women consuming foods from different food groups by season, (a) leaner season (b) abundant season

Vihiga, Kenya – Unlocking the potential of landscapes for dietary diversification

Vihiga County is located in western Kenya, in the Lake Victoria Basin, and mainly lies in the Upper Midland (UM) agro-ecological zone (Jaetzold et al., 2005). The extreme north-west and south-west parts of the county lie in the Humid Lower Midland (LM1) agro-ecological zone. Vihiga is well suited for sugarcane, coffee and tea as cash crops, and maize, beans and cowpeas as staple crops. The dominant ethnic group is the Luhya. According to the 2009 population and housing census, Vihiga's population is 554,622, with 45% below

15 years in age, and 24% of children under five years of age are stunted (KNBS et al., 2015).

During a first diagnostic phase, ten sub-locations were randomly selected for two rounds of household surveys, September 2014 (post-harvest, plenty season; season 1) and April 2015 (lean season; season 2). In each sub-location 40 households with children aged 12–23 months were randomly selected. Household heads were interviewed regarding farm species diversity. In total, 647 households participated in the survey over the two seasons.

The household farm species survey documented 105 different plant species, of which 67 (64%) were edible. Maize (95% of households), beans (87%), banana (61%) and cowpeas (50%) were cultivated by 50% or more households. Sixteen species were cultivated by more than 10% of the households and 14 were cultivated by just a single household. About 33% of species were fruits, 30% vegetables, 15% used as staple foods, 14% legumes/nuts and the remainder (8%) were used as spices, condiments or infusions. On average, seven species were found per farm (min. 1; max. 23) and 476 (74%) households mentioned they collected 38 different wild edible plants. The average number of wild plant species collected per household was two (min. 1; max. 6). Twelve different domesticated animal species were kept by households and 38% of households reported they collected/hunted wild animals for food, of which 15 different species were documented, with termites (*Isoptera*) being most popular followed by quail birds (*Coturnix coturnix*).

More than 80% of households experience a period during the year when they lack enough food to meet their needs. The problem is higher from January to July, with a peak in April. The hunger problem eases in July/August when households harvest for the first season. Another peak of hunger is experienced in October, before the second harvest in December. The mean DDS for children was 4.12 in the plenty and 3.99 in the lean season ($p > 0.05$) (Table 17.2). Nearly one in every four children (23%) did not meet MDD in the plenty season. This figure was slightly higher (27%) in the lean season. The mean DDS of the caregivers (based on ten food groups) was 4.4 and 4.1 for the plenty and lean seasons respectively. The difference was significant ($p = 0.001$). Sixty percent of women did not meet MDD in season 2 compared to fifty percent in season 1 ($p = 0.003$).

In general, grains, roots and tubers were consumed by almost all women and children, followed by dairy products, vitamin A rich fruits and vegetables and other fruits and vegetables by more than 80% of children in both seasons. Consumption of flesh foods (about 30%), legumes and nuts (about 20%) and eggs (almost nil) was quite low in both seasons for children. In comparison, while a similar trend was seen of around 83–77% of women consuming dairy in both seasons, less than 80% consumed vitamin A rich fruit and vegetables, and between 48% and 37% consumed other vegetables and fruits respectively between the two seasons (Figure 17.2).

In conclusion, while there is agricultural biodiversity available in Vihiga, most species are grown and consumed by few households. There is thus scope to improve the use of agricultural biodiversity to benefit dietary (and nutrition) outcomes.

For the project's phase II, five sub-locations (out of the ten participating in phase I) were targeted for intervention development. Per sub-location, six work-shops were organized to raise awareness on nutrition, present results of phase I and identify and plan agriculture for nutrition interventions. Per sub-location, 36 participants with influence in their communities (e.g., village elders and members of mothers' clubs) were selected to participate in the workshops. All sub-locations chose a combination of vegetables and legume production and poultry keeping as interventions, developed community action plans and corresponding budgets, identified the local funding mechanisms to finance the interventions, and defined ways to reach other community members with their actions.

All groups developed agricultural interventions to diversify diets. The factors determining these differences will be further analyzed and the implementation of the community plans monitored for one year together with local partners. Education materials will be developed, in collaboration with the local stakeholders and the Ministry of Health, for workshops on diets and nutrition to complement the community's agricultural activities. A final end-line survey will be organized to assess the impact of the whole approach on dietary diversity and diet quality.

Son La, Vietnam – Minority ethnic groups embrace diversity within a mono-cropped landscape

Mai Son district, in Son La Province in north-west Vietnam, covers an area of 1410 km^2, with an elevation range of 800–1500 m and a population of 147,319 (Son La Provincial Office, 2014). The total agro-forest land of the district is 93,687 ha (66% of total forest land) and the main production systems include upland rice and maize monoculture. As each minority ethnic group in the district has its unique farming and food cultures, it was decided to focus on one group in the pilot phase of the study: the Thai ethnic group. The Thai ethnic group represents 56% of the population in Mai Son and are active in both farming and marketing. Children under five years of age in Son La province suffer from higher rates of malnutrition than the national average; the rates of stunting, underweight and wasting are 35%, 22% and 12%, respectively. The proportion of young children (6–24 months) who achieved MDD was 49% (Viet Nam National Institute of Nutrition, 2014).

A sample of 416 households from 60 villages was randomly selected with data collection conducted in August/September (wet and leaner season) and November/December (dry and more abundant season) 2014. In addition to the standard NSL Dietary Intake Indicators (Table 17.2), Percent of Total Production Diversity (PTPD) was also calculated, defined as the number of species available at the landscape level from own production or wild harvest within a food group divided by the total number of species available across the landscape from own production or wild harvest (for all food groups).

Nutritional KAP of mothers was low with little recognition of the national nutrition messages. Only 58% were able to recognize the food pyramid and only 33% had heard of the 'balanced meal'.[1] While 85% of mothers reported it was important to have a diverse diet, over one-third said it was difficult to provide and only 18% reported it was important to give a diverse range of foods to prevent undernutrition. The main reason given for the difficulty of providing a diverse diet was lack of availability at the homestead.

The average DDS calculated for women and young children was 4.8 and 3.7 respectively with no significant difference between the two seasons (Table 17.2). The percentage of women and children reaching MDD was 59% and 58%, respectively. Food groups with a low consumption for both women and children were legumes, nuts and seeds, eggs, dark green leafy vegetables and vitamin A rich vegetables and fruits (Figure 17.2). However, in women and children who reached MDD, these foods groups were more frequently consumed.

Interestingly, we see a similar pattern in regards to how frequently food groups are represented in household production.[2] In total, 398 different species were available across the landscape, of which 292 were plants and 106 animals. Whilst at the landscape level there is a wide diversity of foods available, these are not evenly distributed across the different food groups. From the 398 different species produced in the landscape, only 14% were from the food groups with low consumption by both women and children (legumes, nuts and seeds, eggs, dark green leafy vegetables and vitamin A rich vegetables and fruit) (Table 17.3).

A weak positive relationship between the species richness of the non-staple food groups at landscape level, and the proportion of women and children consuming foods from these food groups, was observed. There was congruence between the non-staple food groups that represent a lower percentage of total landscape diversity, and those food groups that are consumed by a lower percent of women and children. These same food groups increased in consumption in those individuals who reach MDD. This suggests that targeting these foods for increased consumption through direct nutrition counselling and the promotion of increased production of species from these food groups in home gardens

Table 17.3 Production species richness (per food group) and subsequent food group contribution to diet

Food group	Species count	% of total production diversity	% children consumed (all seasons)	% women consumed (all seasons)	Most frequent plot cultivated
Cereals/grains	5	1	100	100	Sloping land
White roots and tubers	10	3	100	100	Home garden
Vitamin A rich vegetables and tubers	2	1	72	34	Home garden
Dark green leafy vegetables	22	6	29	12	Home garden
Other vegetables	102	29	85	42	Home garden
Vitamin A rich fruit	12	3	3	4	Home garden
Other fruit	60	17	32	31	Home garden
Flesh foods	66	19	78	73	Home garden
Fish and other aquatic	40	11	47	38	Homestead pond
Nuts and seeds	3	1	3	2	Home garden
Legumes	11	3	17	16	Home garden

could increase both the average DDS, and the percent of women and children reaching MDD.

A participatory process of consultation with the farmers was conducted to identify a key set of underutilized locally available crops from the following food groups which acted as the cornerstone of the interventions designed to improve diversity in the diets and landscape: vitamin A rich vegetables and fruit, dark green leafy vegetables and legumes, nuts and seeds. Diversity clubs implemented at the village level are used to connect both the nutrition education and agriculture capacity components. Nutrition education material was developed in consultation with the National Institute of Nutrition and other national partners to ensure that fundamental nutrition messaging was in line with national priorities, as well as adapted to ensure a nutrition-sensitive agriculture link relevant to the dietary gaps and local biodiversity available in the landscape. The interventions are currently underway in partnership with the Provincial Health Department and Commune Health Centers, and will be evaluated for their impact on the diet after a 12-month period.

Discussion and next steps

The application of the NSL approach in three countries has demonstrated how a set of tools and indicators can be used to: (1) identify community dietary gaps; (2) facilitate community innovation and identify opportunities to diversify production systems; and (3) to provide opportunities for families and individuals to directly improve their diets through direct consumption or diversified livelihoods.

All three sites reported very low consumption of eggs, legumes, nuts and seeds for both women and children, and other fruits for women across all seasons. Dark green leafy vegetables and other vitamin A rich food consumption varied greatly between the sites, and in the case of Zambia, also between seasons. A participatory process that engaged communities to identify and design solutions together with researchers interestingly resulted in solutions that were often very similar. Each site established regular meetings where nutrition and production diversity skills and information were transferred and shared, together with the development of nutrition education materials adapted from existing National guidelines. All sites selected legumes and vitamin A rich fruits and/or vegetables. However, Kenya participants were the only ones to decide to incorporate poultry keeping (primarily for egg consumption), despite all three sites demonstrating very low consumption of egg by women and children. Zambia included biofortified orange maize, cassava and rice, despite 100% of women and children consuming starchy staples. Participatory consultation and engagement with communities in which interventions will be tested is expected to be more effective in achieving buy-in from the community, as well as facilitating and connecting agriculture and nutrition government sectors with each other who engage with the implementation, compared to top down approaches. The diversity of local solutions, identified in each case study, to similar dietary gaps highlights that one-size-fits-all approaches are no longer sufficient to achieve systems improvements for nutrition.

Notes

1 Balanced meal, colouring your plate, and the food pyramid are all nutrition education messages and materials disseminated by the National Institute of Nutrition.
2 Production is defined as produced on a farm or collected/hunted in the wild.

References

Black, R.E., et al. (2008) 'Maternal and child undernutrition: global and regional exposures and health consequences', *The Lancet*, vol. 371, no. 9608, pp. 243–260.

Deckelbaum, R., et al. (2006) 'Econutrition: implementation models from the millennium village project in Africa', *Food and Nutrition Bulletin*, vol. 27, no. 4, pp. 335–342.

Demaio, A.R. and Rockström, J. (2015) 'Human and planetary health: towards a common language', *The Lancet*, vol. 386, no. 10007, pp. e36–e37.

FAO and FHI 360 (2016) Minimum Dietary Diversity for Women: a guide for measurement, FAO, Rome.

International Food Policy Research Institute (IFPRI) (2014) 'Global Nutrition Report 2014: Actions and Accountability to Accelerate the World's Progress on Nutrition', IFPRI, Washington, DC.

Jaetzold, R., et al. (2005) 'Farm Management Handbook of Kenya–Natural Conditions and Farm Management Information–West Kenya-Subpart A1', Ministry of Agriculture, Kenya in Cooperation with the German Agency for Technical Cooperation (GTZ), Nairobi, Kenya.

Kennedy, G. (2009) 'Evaluation of dietary diversity scores for assessment of micronutrient intake and food security in developing countries', PhD Dissertation, GVO printing, Ede, Netherlands.

Kenya National Bureau of Statistics (KNBS), et al. (2015) 'Kenya Demographic and Health Survey 2014, Key Indicators', KNBS, Ministry of Health, National AIDS Control Council, Kenya Medical Research Institute, National Council for Population and Development, Nairobi, Kenya.

Khoury, C.K., et al. (2014) 'Increasing homogeneity in global food supplies and the implications for food security,' *Proceedings of the National Academy of Sciences*, vol. 111, no.11, pp. 4001–4006.

Millennium Ecosystem Assessment (MA). (2005) 'Ecosystems and Human Well-Being: Biodiversity Synthesis', World Resources Institute, Washington, DC.

Republic of Zambia, Central Statistical Office, et al. (2009) 'Zambia Demographic and Health Survey 2007', CSO and Macro International Inc., Rockville, MD.

Republic of Zambia, Central Statistical Office, et al. (2014) 'Zambia Demographic and Health Survey 2013–14', Republic of Zambia Central Statistical Office, Republic of Zambia Ministry of Health, and IFC International, Rockville, Maryland.

Son La Provincial Office (2014) Data provided on request from Son La Provincial Office, Son La, Vietnam.

UNICEF (2011) 'Zambia: Nutrition', Available from: <http://www.unicef.org/zambia/5109_8461.html>. [Accessed: 21 June 2015].

Viet Nam National Institute of Nutrition, et al. (2014) 'Nutrition Surveillance Profiles 2013', Ha Noi, Viet Nam.

World Health Organization (WHO) (2008) Indicators for assessing infant and young child feeding practices: conclusions of a consensus meeting held 6-8 November 2007 in Washington D.C., USA, world Health Organization (WHO), Washington D.C.

18 Integrated systems research in nutrition–sensitive landscapes

A theoretical methodological framework

Jeroen C.J. Groot, Gina Kennedy, Roseline Remans, Natalia Estrada-Carmona, Jessica Raneri, Fabrice DeClerck, Stéphanie Alvarez, Nester Mashingaidze, Carl Timler, Minke Stadler, Trinidad del Río Mena, Lummina Horlings, Inge Brouwer, Steven M. Cole and Katrien Descheemaeker

Introduction

South Asia and sub-Saharan Africa are two regions of the world with the highest concentration of nutritionally vulnerable populations that depend to a large extent on agriculture as an important source of livelihood (Gillespie et al., 2015). The vast majority of farmers in these regions have small landholdings due to land fragmentation (Jayne et al., 2014; Valbuena et al., 2015) and are often constrained in their access to resources and agricultural inputs (Herrero et al., 2010), especially women (e.g., Cole et al., 2015). As a consequence, productivity levels are low, and because income sources are also limited, dependence on surrounding landscapes and ecosystem services is high in terms of safeguarding supplies of clean water, human and animal foods, construction materials and fuel wood. People shape their physical landscapes (Ellis, 2015), influenced by cultures, values and livelihood opportunities (Horlings, 2015). People's utilization of their physical landscapes is shaped by various conditions such as soil properties, topography, climate and flooding patterns. People's dependence on their physical landscapes is strong and expected to increase due to climate change, resulting in gradual but persistent changes including adjustments in frequency, timing and severity of anomalies such as droughts and floods (Naylor et al., 2007; Gornall et al., 2010).

Agricultural research has focused primarily on increasing crop productivity at the field level and on improving the provisioning of livestock products and fish at the animal level. Similarly, agriculture interventions aimed at improving nutrition mainly rely on income as a pathway to achieving more nutritious diets, or increasing production that leads to greater food availability and diversity (World Bank, 2007; Herforth and Ahmed, 2015). Both approaches have so far yielded inconclusive evidence about the impact on nutrition outcomes (Masset, 2011). Crop-oriented research has concentrated on cultivation of monocultures, which can be very productive if managed uniformly and efficiently, but which also depend on high levels of external inputs (e.g., fertilizers and pesticides). The strong focus on cereal crops resulted in the promotion of monotonous, and therefore poor in nutrients, cereal-based human (and animal) diets (Defries et al., 2015). The research focus at animal level has led to intensified animal production systems, using improved breeds that are highly dependent on uniform conditions, antibiotics-based animal health care and high quality feeds such as cereals cultivated on arable lands (NRC, 1999; Burkart, 2007). While most technological innovations are not accessible for smallholders (Kiers et al., 2008; Herrero et al., 2010), especially resource-poor women, their implementation elsewhere has frequently led to loss and degradation of soils and other natural resources (Matson et al., 1997; IAASTD, 2009) and compromised human health (Richter et al., 2015). Moreover, important place-based cultural habits and local adaptations such as maintaining seed systems and locally-adapted agrobiodiversity are at risk of being lost and gender and social inclusion issues, while many, are rarely addressed (Villamor et al., 2014).

To foster more equitable, ecologically-supportive and economically-viable interactions with people and their landscapes, we espouse the use of a systems-oriented, place-based and multi-disciplinary approach to analyze the integrated pathways leading to improvements in well-being, income and nutrition livelihood outcomes and in ecosystem services, especially for marginalized and vulnerable populations. A place-based approach considers a setting as a node in a network, established in practices, constituted in and through relations and interactions (Massey, 2005). Practices are embedded in and structured by cultural beliefs (Hebinck, 2016). A place-based approach analyzes the multiple relations that are expressed between the land and the economy, nature and society, rural and urban, as well as at the unique intersection of social, economic, cultural and political relations mapped over multiple localities and result in the distinctiveness of places (Woods, 2013).

We hypothesize that better and more sustainable livelihood, nutrition and ecosystem services provision gains can be obtained by ecological intensification of agriculture and improved management and utilization of the (agro-) biodiversity and natural resources (cf. Powell et al., 2015; Wood et al., 2015). Ecological intensification "proposes landscape approaches that make smart use of the natural functionalities that ecosystems offer" (Tittonell, 2014, p. 53; see also Doré et al. [2011] for a review). The espoused approach aims to support the development of people and their landscapes to better enable them to safeguard their biophysical landscape and ecosystem services.

To investigate and support the envisioned development pathways in these social-ecological systems through research, a diverse portfolio of concepts and methods needs to be mobilized. Integrated systems research is one of the pillars in the nutrition-sensitive landscapes (NSL) approach that is being implemented in case-study sites in Zambia and Kenya from which some results will be presented later in this chapter. For each site we describe and explain current systems and systematically explore windows of opportunity for sustainable redesign and innovation in landscape and farm systems for improved nutrition using whole-farm and landscape models. Central to the methodology is an inclusive, gender-sensitive, participatory approach in all phases of the experiential learning and innovation cycle. In this chapter we only discuss the strengths and challenges of systems approaches in agricultural research based on interactions with smallholder communities from the three diverse settings.

Integrated systems research

Integrated systems research (ISR) in agriculture embraces approaches that: (1) aim at place-based system intensification and diversification beyond increases in single crop productivity; (2) pursue system intensification by minimizing trade-offs and exploiting synergies and complementarities between system components particularly tree-crop-livestock-soil-water interactions; (3) underpin system-level improvements in productivity and natural resource integrity with larger-scale enabling policies, and institutions that promote market development and information flows; (4) frame the integrated research approach within the multi-scale and multi-dimensional context of farming systems, thus allowing nuanced approaches to the scaling out of best-fit technologies; and (5) strengthen the science–policy interface that informs governments and international bodies for enabling changes on the ground to rural people.

ISR fosters connectivity with markets and value chains and collaboration among farmers and development partners. The activities and components in agricultural systems interact, and by explicitly considering these interactions ISR can help to quantify and foresee how proposed changes affect the overall performance of the system for different productive, socio-economic and environmental performance indicators. As such, ISR allows putting newly developed innovations and technologies in a larger perspective. By doing this, the focus shifts from smaller to larger scales and from one (e.g., yield) to multiple criteria, for example, by evaluating what the effect of a new crop variety has on biophysical aspects of the farm and landscape (e.g., productivity, mitigation of pollution), but also on socio-economic aspects of the household and community (e.g., intra-household resource allocation, gender equity).

ISR addresses the heterogeneities in landscapes and populations that are encountered when deploying innovations to larger target groups and when scaling out (Tittonell et al., 2006). It acknowledges that the requirements for innovation and adaptation are dependent on the local biophysical conditions, the resource endowment and the socio-economic and institutional context of the household (Franke et al., 2014; Van Wijk, 2014). The biophysical heterogeneity can be assessed and analyzed through remote sensing, geographical information systems and spatial sampling of vegetation, land-use and soil characteristics. Spatial analysis and household typologies are available to analyze the heterogeneities in endowment and other socio-economic features and to use them to support scaling out of technologies (Vanlauwe et al., 2014; Cortez-Arriola et al., 2015). Moreover, ISR can be used to analyze the dynamics of systems over time (Falconnier et al., 2015), thus informing a stepwise approach to sustainable intensification of agricultural production (Giller et al., 2008).

Since ISR focuses on multiple performance dimensions (or goals) of systems at the same time, it allows quantifying trade-offs and synergies among indicators in a relatively straightforward and intuitive way (Tittonell et al., 2007; Groot et al., 2009). It provides insight into implications of adoption and behavioural changes at larger scales beyond the plot level (Martin et al., 2013), and thus is highly suitable for evaluation of development outcomes and can support identification of appropriate policy instruments, i.e., to choose between different incentive schemes and extension efforts (e.g., Parra-López et al., 2009).

The tools used in ISR allow the construction of 'what if' scenarios and exploration of windows of opportunity for future development and system dynamics (Van Ittersum et al., 1998). These explorations can be performed under different scenarios of changes in external conditions such as policies, markets, and biophysical and climatic conditions (Groot and Rossing, 2011). This allows a quantitative assessment of adaptability and resilience to, for instance, climate change, policy regimes and market volatility.

Despite its strengths, the use of ISR in research and development projects has been limited and debated. This may be because ISR follows a design-oriented integrative research approach that builds on already collected scientific insights and data, rather than conducting reductionist analysis, i.e., analysis-oriented research. This type of research starts from a question arising from curiosity and not necessarily linked to any application. In analysis-oriented research the aim of obtaining knowledge about the functioning of biological systems is supported by methods that study existing structures in (eco-)systems to reveal their functions and hence their purpose. This leads to increased understanding of system functioning and is translated into research outputs. The design-oriented research process synthesizes existing knowledge on functions that should be mobilized to achieve the purposes, and to elaborate one or more land-use configurations that will support these functions. The result is integrated knowledge and takes the form of inventions or decisions. Implementing these may trigger new questions that feed into the analysis-oriented research cycle. Design-oriented ISR requires a different skill set and research team composition. In this era of big data, information and knowledge, ISR could be part of a movement toward an agriculture that is less-resource but more knowledge-intense; a positive step forward for all involved. Capacity development to build a strong evidence base will be key.

Farm and ecosystem indicators

The NSL approach investigates how improved land-use and management of resources in farms and landscapes can improve the livelihoods of rural people, their nutritional status and the ecosystem services that landscapes provide.

These analyses respond to constraints and objectives of local people and other stakeholders involved. However, a large range of indicators can be mobilized to analyze farming systems and landscapes. These indicators can be quantified experimentally or by using models such as whole-farm and landscape models such as FarmDESIGN (Groot et al., 2012), FarmSIM (Van Wijk et al., 2009) and Landscape IMAGES (Groot et al., 2007, 2010).

Agronomic analysis focuses on crop and animal productivity levels in terms of dry or fresh matter yields and the content of nutrients, and the variability therein, which can lead to formulation of risk indicators (Mandryk et al., 2014). Quantification of inputs and outputs of farms and of production levels and the allocation of products (incl. animal excreta) allows analysis of nutrient flows and budgets, from which nutrient losses can be estimated (Groot et al., 2012). To obtain more detail regarding losses and emissions, models can be used to calculate the losses of nutrients through different loss pathways (e.g., Tittonell et al., 2009; Shah et al., 2013). Besides the nutrient dynamics, the organic matter flows on farms and within landscapes are also quantified to assess impacts on soil organic matter changes (e.g., van der Burgt et al., 2006) that strongly affect soil properties such as fertility, water-holding capacity and abundance of soil biota.

Socio-economic indicators of farm models include the balances between required and available labour (Groot et al., 2012), and gendered labour-use profiles throughout the year to analyze peaks and constraints in labour allocation. This should be monitored and analyzed per person, so that the distribution of tasks between female and male household members can be assessed. Economic indicators of the farm enterprise calculate crop and animal margins, operating profit and the returns to labour and the distribution of these benefits among the household members, and should be differentiated by sex and age. Risk indicators can also be calculated to help assess the vulnerability of livelihoods and policies and insurance schemes.

Environmental indicators quantify the impacts of management and use of farms and landscapes. In addition to the nutrient losses, greenhouse gas emissions can also be determined on the basis of model calculations (Seebauer, 2014). Such losses can be aggregated to the landscape level. However, in addition, at the landscape level other processes and services that exceed the farm level are relevant, such as biodiversity and habitat quality and connectivity (Groot et al., 2007, 2010). Modelling approaches at this level can also be used to quantify erosion, hydrology and the dynamics of pests, diseases and bio-control agents in a landscape (Bianchi et al., 2013). Landscape features and functions can be related to cultural identity, habits and history that are important for the use and appreciation by inhabitants and other users or visitors of landscapes (Groot et al., 2010).

Nutrition indicators

The aim of the NSL approach is to link both dimensions of human health and nutrition and environmental health to the production of food at the farm and landscape level. An explicit connection is made with the food provisioning from the household farm, from the surrounding landscape and through interactions with markets. The landscape is often either underutilized or overexploited due to complex causes such as lack of regulation or information about the system and its thresholds. Looking beyond the potential of a single household's production system to provide sufficient nutrients to that household allows for a more comprehensive, holistic and realistic perspective and understanding of how communities, within a landscape, access and consume foods. This helps to identify opportunities to improve the availability of diverse and nutrient dense foods at the landscape approach that might be otherwise missed with household level only interactions (for example, household typologies that specialize in production of vitamin A rich vegetables and others on animal source foods). The NSL methodology is described in more detail in the chapter by Kennedy et al. (this volume). Moreover, an integrated approach allows analyzing to what extent the provisioning of sufficient amount of healthy food can be combined with improving income generation of farmers and strengthening ecosystem services. To this aim, the conventional approach of evaluating productivity at the field/crop and animal levels in terms of dry matter or at best caloric value used in agricultural ISR has to be extended. For this purpose, various indicators that quantify the nutrition diversity and sufficiency are available.

Two categories of indicators are used: production- and consumption-oriented metrics. The production-oriented indicators are production diversity (diversity of food groups and nutritional functional diversity) and nutritional system yield; these indicators only consider produced nutrients and do not or only partly account for losses of nutrients during processing and cooking (Remans et al., 2011; Defries et al., 2015). The consumption-oriented indicators include the modelling estimations of dietary diversity scores (diversity of food groups produced and nutritional functional diversity), food patterns and nutrient adequacy. Although the two types of nutrition indicators are linked, they are not necessarily tightly correlated due to effects of storage, processing, packaging and preparing of food materials as well as the purchasing of foods from markets.

Nutritional system yield (NSY) is an adjustment of the 'nutritional yield' metric proposed by Defries et al. (2015) and uses system productivity in terms of balanced nutrient supply (for human consumption based on nutrient requirements) rather than food item yield expressed in dry matter amount. NSY quantifies the number of consumer units that can obtain their complete daily dietary reference intake (DRI), particularly the recommended dietary allowance that accounts for the quantities required by 97–98% of healthy people (Otten et al., 2006) of different micro and macro nutrients for a year per unit of area of a production system. The production system can be a field, a farm or a broader landscape where one or more crops are cultivated, animals are kept and/or 'wild' foods grow. The consumer unit can be a reference adult female or male. The inverse of NSY is the area required of a production system to feed a consumer unit with energy or individual nutrients during a year.

Nutritional functional diversity (NFD; Remans et al., 2011) quantifies the fraction of diversity in nutrients that is produced relative to the potential diversity that is present in a landscape or region. The potential diversity is captured in a dendrogram wherein available food items are clustered on the basis of their nutritional traits. The NFD metric then assesses which part of the dendrogram, i.e., fraction of potential diversity, is produced in the production system or consumed by individuals or households. The metric can be up-scaled to national and global levels (Remans et al., 2014). Although the metric is usually applied to the production situation, it is generic and can also be used to determine the diversity of supplied and consumed nutrients.

Dietary diversity scores are proxy indicators that provide qualitative measures of food consumption reflecting micronutrient adequacy of the diet (Kennedy et al., 2010; Kennedy et al., this volume). Individual dietary diversity indicators often focus particularly on women and young children as these groups are among the most vulnerable to malnutrition. The minimum dietary diversity score for women (MDD-W; FANTA, 2014) is the most recent indicator. This indicator classifies foods into ten food groups of which at least five should be consumed by women to increase the likelihood of meeting their micronutrient needs compared to women consuming foods from fewer food groups (FANTA, 2014).

A direct indicator for food consumption is the assessment of nutritional adequacy, which quantifies the deviations between the consumption of nutrients and the daily DRI, as used in linear programming models for diet composition (e.g., Maillot et al., 2010; Frega et al., 2012). Recommendations for improvements of human nutrition based on the nutrition adequacy assessments should allow for factors such as cultural food consumption patterns, acceptable foods (available, affordable and regularly consumed), realistic food portion sizes and the impact of recommendations on other nutrient intake (such as through displacement of nutrients) and the environment (Ferguson et al., 2006). Food patterns can be quantified by calculating the amounts and proportions of the different foods and food groups that are consumed, as an indicator of the suitability of the proposed nutrition interventions for the food habits of the target group.

Inclusive learning cycles

To address and include the culture, values and priorities of people (Norton, 2005) in nutrition-sensitive landscapes during the investigation and support of the envisioned development pathways, an inclusive approach to learning cycles is used. Participatory approaches foster co-learning among stakeholders in each step of the problem-solving cycle, aiming to ensure that the identified constraints are of high priority, that the tested options are relevant and that the developed solutions match the farmer context.

InDEED is an iterative cycle derived from the classical steps of the problem-solving or experiential learning cycle (cf. Kolb, 1984), which includes:

(1) DESCRIBE the landscape or land-use system to be managed, define the problem(s), formulate the goals and select the indicators; (2) EXPLAIN the current performance of the system in terms of the indicators; (3) EXPLORE the window of opportunity for change and development by generating alternative farm and landscape configurations using a basket of innovative practices and technologies; and (4) DESIGN and implement the most desirable alternative, supported by suitable policies and monitoring, and followed by adaptive management and new DEED learning cycles (Giller et al., 2008; Tittonell, 2008). In each of these phases, iteration or feedback to previous phases may take place when more knowledge becomes available (Groot and Rossing, 2011). The inclusive 'In' aspects from the InDEED approach facilitate:

- Interacting with stakeholders using participatory processes.
- Incorporating the community culture and values while simultaneously addressing economy, ecology and equity (McDonough and Braungart, 2002), and applying a gender transformative approach (Cole et al., 2014; Kantor et al., 2015).
- Inspecting all possible consequences and externalities to avoid deterioration of economic benefits, equity, culture, health, natural resources and ecosystem services (Robèrt et al., 2002).
- Investing in positive outcomes by identifying truly effective and new options, avoiding lock-in and tuning of sub-optimal and inefficient solutions (McDonough and Braungart, 2002; Kirk et al., 2007).
- Informing learning by positive experiences, applying principles of appreciative inquiry (AI) and dialogue in the context of multi-stakeholder processes. AI is used for the development of potential social systems, but also as a way to facilitate cooperation and joint learning (Whitney et al., 2010). The approach is not centred around problems but on positive strengths and possibilities for change, and starts with discovering the very best in the shared experiences (Cooperrider and Whitney, 1999).

Illustration: Participatory analysis, exploration and design in action

In this section we illustrate the approach with case study examples. The ISR applied to nutrition-sensitive landscapes uses a broad range of methods to arrive at inclusive development pathways. People are actively engaged at different stages of the process through an array of participatory research methodologies using a gender-sensitive lens. For the case study examples we conducted: (1) participatory interviews to understand productive systems; (2) focus group discussions (FGDs) to understand food availability (e.g., seasonal calendars), decision making around crop and land allocation, critical ecosystem services and landscape knowledge; (3) participatory mapping to assess the heterogeneity of the elements in the landscape, the use of the landscape by people for multiple purposes (ecosystem services), the location of farmer fields and the values that are attached to spatial elements; and finally (4) field work and transect walks to assess gradients in the biophysical conditions in the landscape, presence of vegetation and animals, practices that are used to cultivate and harvest, and relation with the diet of community members.

The information collected by the different participatory methods helped us to triangulate the information, validate and complement the perceptions of women and men across communities. This resulted in cartographic and descriptive information of the inherent heterogeneity and the presence of highly valued places in the landscape, which helped to create awareness and initiate open discussions on what is working well and on possible improvements in practices to enhance livelihoods and increase availability of more diverse foods. In short, it helped to identify windows of opportunities, constraints and challenges associated to each development pathway with the people who could directly benefit from the research process.

Figure 18.1 illustrates examples of NSL methodology work packages in one of the case study sites in the Barotse Floodplain (part of the Zambezi River Basin) in Western Zambia. Results obtained with and for women and men in the study communities include descriptions of valuable places, use of terrestrial

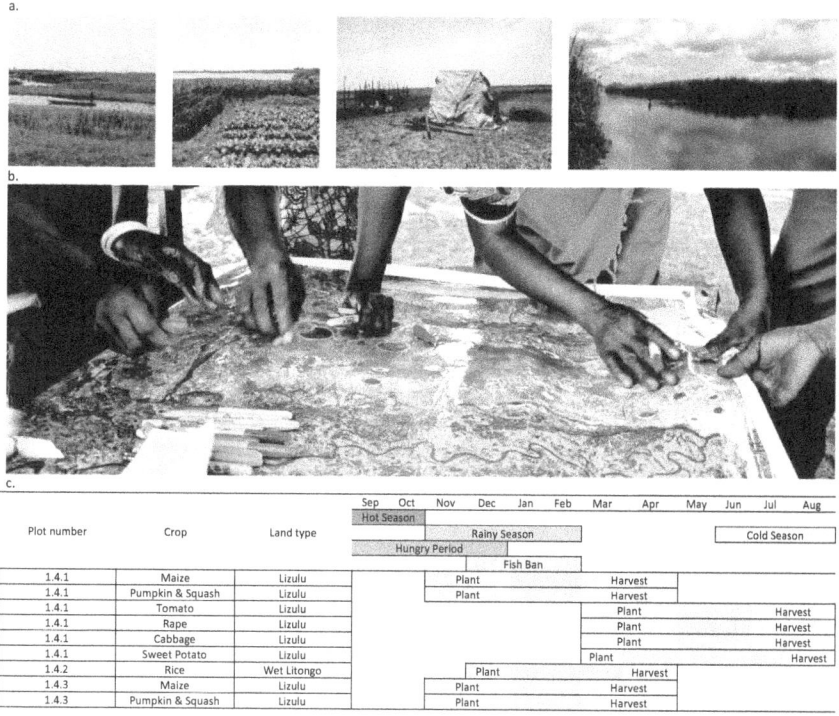

Plot number	Crop	Land type	Sep	Oct	Nov	Dec	Jan	Feb	Mar	Apr	May	Jun	Jul	Aug
			Hot Season		Rainy Season						Cold Season			
			Hungry Period											
						Fish Ban								
1.4.1	Maize	Lizulu				Plant			Harvest					
1.4.1	Pumpkin & Squash	Lizulu			Plant				Harvest					
1.4.1	Tomato	Lizulu								Plant				Harvest
1.4.1	Rape	Lizulu								Plant				Harvest
1.4.1	Cabbage	Lizulu								Plant				Harvest
1.4.1	Sweet Potato	Lizulu							Plant					Harvest
1.4.2	Rice	Wet Litongo					Plant			Harvest				
1.4.3	Maize	Lizulu				Plant			Harvest					
1.4.3	Pumpkin & Squash	Lizulu				Plant			Harvest					

Figure 18.1 Results of transect walks and participatory mapping in the Barotse floodplain, Western Zambia

and aquatic resources and of place and time-determined food consumption and farming practices. Results also indicate which nutritious crops should be incorporated in the farming systems to reduce the extent and the intensity of the hunger season. Those results were used to design place-specific interventions such as learning plots in which farmers are generating knowledge, sharing and beginning to experiment with sustainable practices.

The DESCRIBE phase of farming systems analysis entails a characterization of farms, describing the use of landscape elements for farming, choices of types of crops and animals associated with their management throughout the year, and the allocation of resources such as residues, manures and inputs like seeds and fertilizers. Farmers together with researchers construct gendered seasonal calendars of cultivation, production and labour allocation, showing which and when food is available in relation to where their fields are located (see Figure 18.1c). These characteristics can be used to EXPLAIN the performance of the system by quantifying farm productivity, nutrient cycling and gender-specific labour profiles and associated economic analysis of costs and benefits. As an example, nitrogen cycles for two representative farms in the study site in Vihiga, Kenya, are shown in Figure 18.2. The nitrogen cycles combine farm nitrogen flows, and the flows from the farm and from the surrounding landscape or market to the household for human nutrition. Table 18.1 demonstrates the difference in the farm configurations and the values of selected nutrition indicators between the two farms.

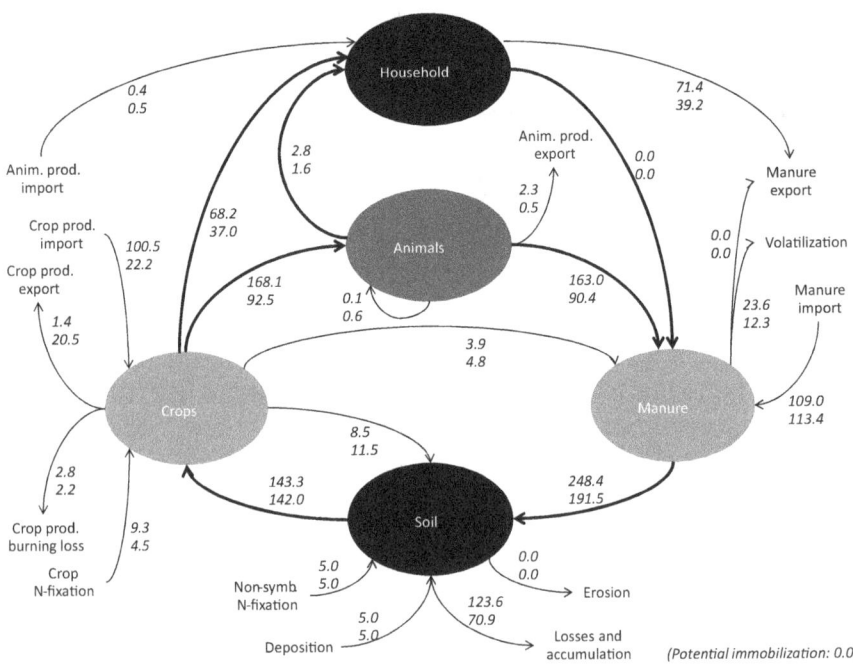

Figure 18.2 Nitrogen cycles of two representative farms in Vihiga, Kenya

Table 18.1 Farm characteristics and nutrition indicators of two representative farm households in the villages of Masana and Mambai in Vihiga district, Western Kenya

	MASANA	MAMBAI
Farm area (ha)	0.36	0.4
Crop area (ha)		
Tea	–	0.1
Maize and common bean	0.18	0.1
Napier grass	0.08	0.07
Banana	0.0376	0.0272
Eucalyptus	0.06	0.07
Kale	–	0.03
Avocado	0.0012	0.0012
Papaya	–	0.0016
Mango	0.0012	–
Animal numbers		
Cow	1	1
Heifer	1	–
Calf	1	1
Goat	3	2
Poultry	9	15
Nutrition indicators		
MDD-W[1]	7	7
NFD (produced)[2]	0.164	0.190
NFD (consumed)[2]	0.262	0.318
NSY (consumer units per ha)[3]		
Dietary energy	11	8
Carbohydrates	33	20
Dietary fiber	17	10
Protein	42	38
Phosphorus (P)	39	35
Potassium (K)	47	48
Magnesium (Mg)	19	13
Manganese (Mn)	97	58
Calcium (Ca)	8	10
Sodium (Na)	3	5
Iron (Fe)	343	325
Zinc (Zn)	36	37
Vitamin A	3	8
Vitamin C	6	13
Tiamin	22	13
Riboflavin	22	18
Folate	17	10
Niacin	14	8
Vitamin B-6	19	15
Vitamin B-12	11	15
Copper (Cu)	17	20

[1] MDD-W: Minimum Dietary Diversity for women. Expressed as number of food groups included in
the diet

[2] NFD: Nutritional Functional Diversity. Expressed as fraction of the total diversity that is present on
the farm, in the landscape and on the market

[3] NSY: Nutritional System Yield. Expressed as number of consumer units that can be fed sufficiently
with the nutrient

The nutrition analyses and participatory research in the EXPLAIN phase resulted in quantification of nutrition indicators and the identification of the possible constraints in the diet. Particularly, FGDs were used to generate lists of desirable new crops and animals that could be cultivated in the farms that have the potential to fill the nutrient and food group gaps identified in the diet, which were also under-produced across the landscape. In model-based explorations, informed by the FGDs and on-site data collection, consequences of adaptation are analyzed both in the existing systems with the currently used crops, animals and resources but a different allocation, and after adoption of innovative practices and technologies.

To further inform the exchanges among farmers, researchers and other stakeholders, we developed 'discussion support tools' that foster identification of a rich diversity in development options and perspectives in the EXPLORE phase, rather than single-scenario decision support tools. With these tools the possible trade-offs and synergies involved in such decisions are analyzed with exploratory whole-farm or landscape modelling tools (Groot et al., 2010, 2012). These combine the indicator calculations and use Pareto-based multi-objective optimization algorithms. The generated options and the windows of opportunities are visualized. The outcomes are translated into narratives that were fed back to farmers in the communities through interactive and participatory methods including focus group discussions with farmers and other community members for selection and fine-tuning for implementation in the DESIGN phase. After implementation an adaptive management approach can be adopted for further fine-tuning to the context of the local environment and for dynamic adjustment of landscape management to environmental changes and natural processes (Groot and Rossing, 2011). If an efficient exploratory framework for generating, evaluating and selecting natural resources management (NRM) alternatives is in place, the adaptive management of the system can be supported by repeated design cycles.

Conclusions

The NSL approach is a systems-oriented, place-based and multi-disciplinary methodology that is beginning to enable integrated assessments of nutrition security, agricultural production, market interactions, gendered processes and natural resources management. The goal of such assessments is to identify with local people inclusive, testable and scalable development pathways to help ensure they are healthier and food secure and live in more resilient landscapes that provide a wide range of ecosystem services vital for improving their well-being.

References

Bianchi, F.J.J.A., Schellhorn, N.A. and Cunningham, S.A. (2013) 'Habitat functionality for the ecosystem service of pest control: reproduction and feeding sites of pests and natural enemies', *Agricultural and Forest Entomology*, 15(1), pp. 2–23.

Burkart, M.R. (2007) 'Diffuse pollution from intensive agriculture: sustainability, challenges, and opportunities', *Water Science and Technology*, 55, pp. 17–23.

Cole, S.M., Puskur, R., Rajaratnam, S. and Zulu, F. (2015) 'Exploring the intricate relationship between poverty, gender inequality and rural masculinity: a case study from an aquatic agricultural system in Zambia', *Culture, Society & Masculinities*, 7, pp. 154–170.

Cole, S.M., van Koppen, B., Puskur, R., Estrada, N., DeClerck, F., Baidu-Forson, J.J., Remans, R., Mapedza, E., Longley, C., Muyaule, C. and Zulu, F. (2014) *Collaborative effort to operationalize the gender transformative approach in the Barotse Floodplain*, Penang, Malaysia: CGIAR Research Program on Aquatic Agricultural Systems, Program Brief: AAS-2014-38.

Cooperrider, D.L. and Whitney, D. (1999) *Collaborating for change: appreciative Inquiry*, Berret Koehler, San Francisco, CA.

Cortez-Arriola, J., Rossing, W.A.H., Scholberg, J.M.S., Groot, J.C.J. and Tittonell, P. (2015) 'Leverages for on-farm innovation from farm typologies? An illustration for family-based dairy farms in north-west', *Agricultural Systems*, 135, pp. 66–76.

DeFries, R., Fanzo, J., Remans, R., Palm, C., Wood, S. and Anderman, T.L. (2015) 'Metrics for land-scarce agriculture', *Science*, 349(6245), pp. 238–240.

Doré, T., Makowski, D., Malézieux, E., Munier-Jolain, N., Tchamitchian, M. and Tittonell, P. (2011) 'Facing up to the paradigm of ecological intensification in agronomy: revisiting methods, concepts and knowledge', *European Journal of Agronomy*, 34(4), pp. 197–210.

Ellis, E.C. (2015) 'Ecology in an anthropogenic biosphere', *Ecological Monographs*, 85(3), pp. 287–331.

Falconnier, G.N., Descheemaeker, K., Van Mourik, T.A., Sanogo, O.M. and Giller, K.E. (2015) 'Understanding farm trajectories and development pathways: two decades of change in southern Mali', *Agricultural Systems*, 139, pp. 210–222.

FANTA (2014) *Introducing the Minimum Dietary Diversity – Women (MDD-W) Global Dietary Diversity Indicator for Women*, Washington DC, 2 pp. URL: http://www.fao.org/fileadmin/templates/nutrition_assessment/Dietary_Diversity/Minimum_dietary_diversity_-_women__MDD-W__Sept_2014.pdf. Accessed: 14 November 2015.

Ferguson, E.L., Darmon, N., Fahmida, U., Fitriyanti, S., Harper, T.B. and Premachandra, I.M. (2006) 'Design of optimal food-based complementary feeding recommendations and identification of key "problem nutrients" using goal programming', *The Journal of Nutrition*, 136(9), pp. 2399–2404.

Franke, A.C., van den Brand, G.J. and Giller, K.E. (2014) 'Which farmers benefit most from sustainable intensification? An ex-ante impact assessment of expanding grain legume production in Malawi', *European Journal of Agronomy*, 58, pp. 28–38.

Frega, R., Lanfranco, J.G., De Greve, S., Bernardini, S., Geniez, P., Grede, N., Bloem, M. and de Pee, S. (2012) 'What linear programming contributes: world food programme experience with the "cost of the diet" tool', *Food and Nutrition Bulletin*, 33(3 Suppl), pp. 228–235.

Giller, K.E., Leeuwis, C., Andersson, J.A., Andriesse, W., Brouwer, A., Frost, P., Hebinck, P., Heitkönig, I., Van Ittersum, M.K. and Koning, N. (2008) 'Competing claims on natural resources: what role for science', *Ecology and Society*, 13, p. 34.

Gillespie, S., van den Bold, M., Hodge, J. and Herforth, A. (2015) 'Leveraging agriculture for nutrition in South Asia and East Africa: examining the enabling environment through stakeholder perceptions', *Food Security*, 7(3), pp. 463–477.

Gornall, J., Betts, R., Burke, E., Clark, R., Camp, J., Willett, K. and Wiltshire, A. (2010) 'Implications of climate change for agricultural productivity in the early twenty-first century', *Philosophical Transactions of the Royal Society B – Biological Sciences*, 365, pp. 2973–2989.

Groot, J.C.J., Rossing, W.A.H., Jellema, A., Stobbelaar, D. J., Renting, H., and Van Ittersum, M.K. (2007) 'Exploring multi-scale trade-offs between nature conservation, agricultural profits and landscape quality—A methodology to support discussions on land-use perspectives', *Agriculture, Ecosystems & Environment*, 120(1), pp. 58–69.

Groot, J.C.J., Jellema, A. and Rossing, W.A.H. (2010) 'Designing a hedgerow network in a multifunctional agricultural landscape: Balancing trade-offs among ecological quality, landscape character and implementation costs', *European Journal of Agronomy*, 32(1), pp. 112–119.

Groot, J.C.J., Oomen, G.J.M. and Rossing, W.A.H. (2012) 'Multi-objective optimization and design of farming systems', *Agricultural Systems*, 110, pp. 63–77.

Groot, J.C.J. and Rossing, W.A.H. (2011) 'Model-aided learning for adaptive management of natural resources: an evolutionary design perspective', *Methods in Ecology and Evolution*, 2(6), pp. 643–650.

Groot, J.C.J., Rossing, W.A.H., Tichit, M., Turpin, N., Jellema, A., Baudry, J., Verburg, P.H., Doyen, L. and van de Ven, G.W.J. (2009) 'On the contribution of modelling to multifunctional agriculture: learning from comparisons', *Journal of Environmental Management*, 90(Suppl 2), pp. S147–S160.

Hebinck, P. (2016) 'Local maize practices and the cultures of seed in Luoland, west Kenya', in Dessein, J., Battaglini, E. and Horlings, L. (Eds) *Cultural sustainability and regional development; Theories and practices of territorialisation*, Routledge, London and New York, Routledge studies in culture and sustainable development, pp. 206–218.

Herforth, A. and Ahmed, S. (2015) 'The food environment, its effects on dietary consumption, and potential for measurement within agriculture-nutrition interventions', *Food Security*, 7(3), pp. 505–520.

Herrero, M., Thornton, P.K., Notenbaert, A.M., Wood, S., Msangi, S., Freeman, H., Bossio, D., Dixon, J., Peters, M., van de Steeg, J., Lynam, J., Parthasarathy Rao, P., MacMillan, S., Gerard, B., McDermott, J., Seré, C. and Rosegrant, M. (2010) 'Smart investments in sustainable food production: revisiting mixed crop-livestock systems', *Science*, 327(5967), pp. 822–825.

Horlings, L.G. (2015) 'Values in place: a value-oriented approach toward sustainable place-shaping', *Regional Studies, Regional Science*, 2(1), pp. 256–273.

International Assessment of Agricultural Science and Technology for Development (2009) Global Report, Island, Washington, DC.

Jayne, T.S., Chamberlin, J. and Headey, D.D. (2014) 'Land pressures, the evolution of farming systems, and development strategies in Africa: A synthesis', *Food Policy*, 48, pp. 1–17.

Kantor, P., Morgan, M. and Choudhury, A. (2015) 'Amplifying outcomes by addressing inequality: the role of gender-transformative approaches in agricultural research for development', *Gender, Technology and Development*, 19(3), pp. 292–319.

Kennedy, G., Ballard, T. and Dop, M.C. (2010) *Guidelines for measuring household and individual dietary diversity*, Nutrition and Consumer Protection Division, Food and Agriculture Organization of the United Nations, Rome, 53 pp.

Kiers, E.T., Leakey, R.R.B., Izac, A.-M., Heinemann, J.A., Rosenthal, E., Nathan, D. and Jiggins, J. (2008) 'Agriculture at a crossroads', *Science*, 320, pp. 320–321.

Kirk, E., Reeves, A. and Blackstock, K. (2007) 'Path dependency and the implementation of environmental regulation', *Environment and Planning C: Government and Policy*, 25(2), pp. 250–268.

Kolb, D. (1984) *Experiential learning: experience as the source of learning and development*, Prentice-Hall, Upper Saddle River, NJ.

Maillot, M., Vieux, F., Amiot, M.J. and Darmon, N. (2010) 'Individual diet modeling translates nutrient recommendations into realistic and individual-specific food choices', *American Journal of Clinical Nutrition*, 91(2), pp. 421–430.

Mandryk, M., Reidsma, P., Kanellopoulos, A., Groot, J.C.J. and van Ittersum, M.K. (2014) 'The role of farmers' objectives in current farm practices and adaptation preferences: a case study in Flevoland, the Netherlands', *Regional Environmental Change*, 14(4), pp. 1463–1478.

Martin, G., Martin-Clouaire, R. and Duru, M. (2013) 'Farming system design to feed the changing world: a review', *Agronomy for Sustainable Development*, 33(1), pp. 131–149.

Masset, E. (2011) 'A review of hunger indices and methods to monitor country commitment to fighting hunger', *Food Policy*, 36(Suppl. 1), pp. S102–S108.

Massey, D. (2005) *For Space*, Sage, Los Angeles, London, New Delhi, Singapore.

Matson, P.A., Parton, W.J., Power, A.G. and Swift, M.J. (1997) 'Agricultural intensification and ecosystem properties', *Science*, 277, pp. 504–509.

McDonough, W. and Braungart, M. (2002) *Cradle to cradle: remaking the way we make things*, North Point Press, New York, 193 pp.

Naylor, R.L., Battisti, D.S., Vimont, D.J., Falcon, W.P. and Burke, M.B. (2007) 'Assessing risks of climate variability and climate change for Indonesian rice agriculture', *Proceedings of the National Academy of Sciences USA*, 104, pp. 7752–7757.

National Research Council (1999) *The use of drugs in food animals: benefits and risks*, National Academies Press, Washington, DC.

Norton, B.G., (2005), *Sustainability: a philosophy of adaptive ecosystem management*, The University of Chicago Press, Chicago, IL, 608 pp.

Otten, J.J., Hellwig, J.P. and Meyers, L.D. (2006) *Dietary reference intakes (DRI) – the essential guide to nutrient requirements*, Institute of Medicine, of the National Academies, The National Academies Press, Washington, DC, 1327 pp.

Parra-López, C., Groot, J.C.J., Carmona-Torres, C. and Rossing, W.A.H. (2009) 'An integrated approach for ex-ante evaluation of public policies for sustainable agriculture at landscape level', *Land Use Policy*, 26(4), pp. 1020–1030.

Powell, B., Thilsted, S.H., Ickowitz, A., Termote, C., Sunderland, T. and Herforth, A. (2015) 'Improving diets with wild and cultivated biodiversity from across the landscape', *Food Security*, 7(3), pp. 535–554.

Remans, R., Flynn, D.F.B., DeClerck, F., Diru, W., Fanzo, J., Gaynor, K., Lambrecht, I., Mudiope, J., Mutuo, P.K., Nkhoma, P., Siriri, D., Sullivan, C. and Palm, C.A. (2011) 'Assessing nutritional diversity of cropping systems in African villages', *PloS One*, 6(6), p. e21235.

Remans, R., Wood, S.A., Saha, N., Anderman, T.L. and DeFries, R.S. (2014) 'Measuring nutritional diversity of national food supplies', *Global Food Security*, 3(3–4), pp. 174–182.

Richter, C.H., Custer, B., Steele, J.A., Wilcox, B.A. and Xu, J. (2015), 'Intensified food production and correlated risks to human health in the greater Mekong subregion: a systematic review', *Environmental Health*, 14(1), p. 43.

Robèrt, K.-H., Schmidt-Bleek, B., Aloisi de Larderel, J., Basile, G., Jansen, J.L., Kuehr, R., Price-Thomas, P., Suzuki, M., Hawken, P. and Wackernagel, M. (2002) 'Strategic sustainable development – selection, design and synergies of applied tools', *Journal of Cleaner Production*, 10(3), pp. 197–214.

Seebauer, M. (2014) 'Whole farm quantification of GHG emissions within smallholder farms in developing countries', *Environmental Research Letters*, 9(3), p. 035006.

Shah, G.A., Groot, J.C.J., Shah, G.M. and Lantinga, E.A. (2013) 'Simulation of long-term carbon and nitrogen dynamics in grassland-based dairy farming systems to evaluate mitigation strategies for nutrient losses', *PloS One*, 8(6), p. e67279.

Tittonell, P. (2008) *Msimu wa Kupanda – Targeting resources within diverse, heterogeneous and dynamic farming systems of East Africa*, PhD Thesis, Wageningen University. http://edepot.wur.nl/121949, Accessed: 14 November 2015.

Tittonell, P. (2014) 'Ecological intensification of agriculture – sustainable by nature', *Current Opinion in Environmental Sustainability*, 8, pp. 53–61.

Tittonell, P., Leffelaar, P., Vanlauwe, B., Vanwijk, M. and Giller, K. (2006) 'Exploring diversity of crop and soil management within smallholder African farms: a dynamic model for simulation of N balances and use efficiencies at field scale', *Agricultural Systems*, 91(1–2), pp. 71–101.

Tittonell, P., Rufino, M.C., Janssen, B.H. and Giller, K.E. (2009) 'Carbon and nutrient losses during manure storage under traditional and improved practices in smallholder crop-livestock systems – evidence from Kenya', *Plant and Soil*, 328(1–2), pp. 253–269.

Tittonell, P., van Wijk, M.T., Rufino, M.C., Vrugt, J.A. and Giller, K.E. (2007) 'Analysing trade-offs in resource and labour allocation by smallholder farmers using inverse modelling techniques: a case-study from Kakamega district, western Kenya', *Agricultural Systems*, 95(1–3), pp. 76–95.

Valbuena, D., Groot, J.C.J., Mukalama, J., Gérard, B. and Tittonell, P. (2015) 'Improving rural livelihoods as a "moving target": trajectories of change in smallholder farming systems of Western Kenya', *Regional Environmental Change*, 15(7), pp. 1395–1407.

van der Burgt, G.J.H.M., Oomen, G.J.M., Habets, A.S.J. and Rossing, W.A.H. (2006) 'The NDICEA model, a tool to improve nitrogen use efficiency in cropping systems', *Nutrient Cycling in Agroecosystems*, 74(3), pp. 275–294.

Van Ittersum, M.K., Rabbinge, R. and Van Latesteijn, H.C. (1998) 'Exploratory land use studies and their role in strategic policy making', *Agricultural Systems*, 58(3), pp. 309–330.

Vanlauwe, B., Coyne, D., Gockowski, J., Hauser, S., Huising, J., Masso, C., Nziguheba, G., Schut, M. and Van Asten, P. (2014) 'Sustainable intensification and the African smallholder farmer', *Current Opinion in Environmental Sustainability*, 8, pp. 15–22.

van Wijk, M.T. (2014) 'From global economic modelling to household level analyses of food security and sustainability: How big is the gap and can we bridge it?' *Food Policy*, 49, pp. 378–388.

van Wijk, M.T., Tittonell, P., Rufino, M.C., Herrero, M., Pacini, C., De Ridder, N. and Giller, K.E. (2009) 'Identifying key entry-points for strategic management of smallholder farming systems in sub-Saharan Africa using the dynamic farm-scale simulation model NUANCES-FARMSIM', *Agricultural Systems*, 102(1–3), pp. 89–101.

Villamor, G.B., van Noordwijk, M., Djanibekov, U., Chiong-Javier, M.E. and Catacutan, D. (2014) 'Gender differences in land-use decisions: shaping multifunctional landscapes?' *Current Opinion in Environmental Sustainability*, 6, pp. 128–133.

Whitney, D., Trosten-Bloom, A. and Rader, K. (2010) 'Leading positive performance: a conversation about appreciative leadership', *Performance Improvement*, 49(3), pp. 5–10.

Wood, S.A., Karp, D.S., DeClerck, F., Kremen, C., Naeem, S. and Palm, C.A. (2015) 'Functional traits in agriculture: agrobiodiversity and ecosystem services', *Trends in Ecology & Evolution*, 30(9), pp. 531–539.

Woods, M. (2013) 'Regions engaging globalization: a typology of regional responses in rural Europe', *Journal of Rural and Community Development*, 8(3), pp. 113–126.

World Bank (2007) *From agriculture to nutrition: pathways, synergies, and outcomes*, World Bank, Washington, DC.

19 Gender and systems research

Leveraging change

Cynthia McDougall

Introduction

There may be nothing new under the sun[1] – but there are novel and potentially potent ways of perceiving, approaching, researching, and engaging in the work of research for development. This chapter is a reflection of that. While neither gender nor systems research are new, applying a systems perspective to understanding the role of gender in agricultural development research offers much needed new insights into how this research may contribute to lasting and significant increases in productivity, food security, and livelihoods.

The chapter tackles the above by exploring the significance of gender to and in agricultural systems research for and in development. In doing so, it also connects to one of the overarching questions of the International Conference on Integrated Systems Research:[2] what is the value-added of systems research to achieving global sustainable development outcomes, including poverty reduction, increased food, and nutrition security? The chapter begins by exploring two key concepts: complex systems and gender. The former is explored through a conceptual framework for understanding the nature of complexity in systems (Snowden's *Cynefin* framework, http//:cognitive-edge.com). Next, the chapter explores the roles and significance of gender in and to systems research. In doing so, it starts by noting the more familiar roles of gender in agricultural research and research for and in development (such as enhanced research relevance), then focuses on the more novel role: the significance that emerges when gender is considered through the lens of systems thinking (namely, gender as a leverage point). The chapter illustrates this with examples from CGIAR[3] research on a range of issues. The chapter closes by presenting three key points that emerge from this exploration: the first is a broader insight regarding the relevance of transformability to systems research; the second is a response to the overarching conference question of what is the value of systems research; and the third synthesizes a response to the key question of this chapter, namely what is the significance of gender in systems research.

Key concepts

Gender

Gender is widely understood to refer to socially constructed differences in roles, behaviours, and identities between men and women. As such, gender is distinct in meaning from biological sex, and is temporally, culturally, and contextually specific. While often (mis)interpreted and (mis)applied as if it were synonymous with women or women's empowerment, gender embodies the 'space in between', i.e., the connections and power relations between women and men. Moreover, the concept recognizes that gendered identities, roles, and relations of all men and women are cross-cut by a host of other factors of significance, including but not limited to wealth, ethnicity or caste, and class (Leach et al., 2016). Gender – and cross-cutting socio-economic identities or 'intersectionalities'[4] – shape the lives of poor farmers and fishers and their families (as they do the lives of men and women more generally). This influence spans from farmers' and fishers' aspirations, needs, experiences, opportunities, safety and risk, access to and control over assets and resources, voice in decision-making, productivity, to their broader health, economic security, and wellbeing (Shields, 2008; Carr and Thompson, 2014).

While gender and systems research, to date, have not necessarily been readily integrated – perhaps in part because of divergent epistemological and methodological foundations (Kawarazuka et al., 2016) – gender has increasingly been recognized as central to agricultural research (see Meinzen-Dick et al., 2011). While here we bundle this under the term gender analysis, the overall CGIAR framing has evolved towards a minimum distinction between gender-integrated research and gender strategic research.[5] Strides have continued to be made in this area of enhanced attention to gender in research. However, global gender inequities remain pervasive (Meinzen-Dick et al., 2011; WorldBank, 2012). While these are widely recognized as social justice issues, gender imbalances in education, access to resources, and other areas have also been increasingly recognized as instrumental issues, in that they limit global development outcomes (Meinzen-Dick et al., 2011; Laven et al., 2012; Leach et al., 2016).

Complex systems

Systems research involves the holistic analysis of components within a defined agro-ecological space, the interactions, trade-offs, and synergies among the components within the system, and the management and improvement of the system aimed at livelihoods enhancement for farmers and communities and agro-ecological sustainability. Systems research has roots in multiple fields, including the field of systems thinking, which refers to a way of viewing an issue such that the parts – or the 'thing' being studied – is understood as part of a larger set of elements or components that interact in multiple ways to shape outcomes (Meadows and Wright, 2008; Aronson, ND). Systems thinkers such as Dave Snowden have highlighted that systems and the challenges embedded within them can embody different levels of complexity, and thus reflect a need for research and development interventions of differing natures. Snowden's Cynefin Framework[6] (see http//:cognitive-edge.com; Snowden and Boone, 2007) illustrates this, indicating a more ordered side manifest in simple and complicated systems and a less ordered side manifest in complex and chaotic systems. As illustrated in Figure 19.1, a simple system is ordered, linear, and predictable. Following this thinking, engaging in a simple system can be seen as akin to preparing *garri* (cassava) or *dhal* (lentils): straightforward, replicable instructions are likely to reproduce the same results each time. Complicated systems and challenges within them can be illustrated with the example of building a tractor – more complicated, more parts, more challenging, yet still ordered. Complex systems – on the less ordered left side of the framework – are

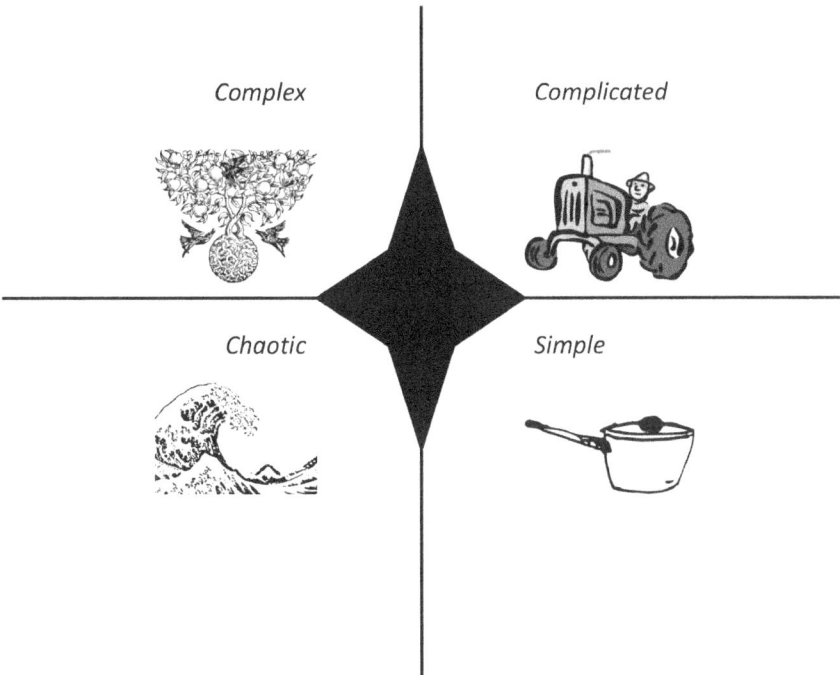

Figure 19.1 Analogies for system types: preparing *garri* or *dhal* (simple system); building a tractor (complicated system); a rainforest (complex system); disaster situation such as a tsunami event (chaotic system)

Source: Snowden and Boone, 2007 (http://cognitive-edge.com).

dynamic and less predictable, they involve multiple feedback loops, and no single point of control (i.e., they are self-organizing). As Snowden and Boone (2007: 74) describe:

> In a complicated context, at least one right answer exists. In a complex context, however, right answers can't be ferreted out. It's like the difference between, say, a Ferrari and the Brazilian rainforest. Ferraris are complicated machines, but an expert mechanic can take one apart and reassemble it without changing a thing. The car is static, and the whole is the sum of its parts. The rainforest, on the other hand, is in constant flux – a species becomes extinct, weather patterns change, an agricultural project reroutes a water source – and the whole is far more than the sum of its parts. This is the realm of 'unknown unknowns', and it is the domain to which much of contemporary business has shifted.

Chaotic contexts – as illustrated by disaster situations – increase again in uncertainties and unpredictabilities relative to other contexts. In chaotic contexts, it is not possible to readily determine cause and effect relationships because they are in constant flux and manageable patterns do not exist – only turmoil (Snowden and Boone, 2007).

This is relevant to the question of systems research – and gender – because closer consideration suggests a fit between different types of research areas and different quadrants, or at least halves, of the framework. For example, research into: growing bigger fish or increasing sweet potato production on station (simple-complicated systems); increasing productivity in integrated agricultural/aquacultural landscapes, and landscape-scale natural resource governance (complex systems); and agriculture and development in contexts deeply affected by significant climate change events or contexts of conflict (complex-chaotic systems). This suggests that systems research – that engages with a breadth of actors, across scales, and with the multiple uncertainties and 'messiness' that accompanies those – operates in the domain of complex systems.[7] This is of importance here because, accordingly, insights from the field of complex systems thinking also apply to the field of systems research. In particular, in this chapter we connect with the notion of *leverage points* in complex systems. Drawing on Meadows (2009: 41), leverage points are 'places within a complex system (a corporation, an economy, a living body, a city, an ecosystem) where a small shift in one thing can produce big changes in everything'. After it highlights the roles of gender in research, the following section of this chapter picks up on this idea of leverage points. Specifically, it draws on the concept of leverage points to signal the particular significance of gender in systems research and the value-added of systems research to development.

Significance of gender in systems research

Before exploring systems thinking-specific roles of gender, we note one overarching potential role of gender in research: anchoring research in gender analysis can also draw attention to other aspects of social equity in relation to research and development. Specifically, because of the inextricable intersectionality between gender and other dimensions of socio-cultural and economic diversity – and the inherent equity focus of gender – engaging with gender effectively can open the door to also addressing these other equity issues in an integrated way. This role is not a certainty, however, and the risk of an over-emphasis on gender as a binary category is well-noted in the literature (Carr and Thompson, 2014).

Turning to the focus of this chapter – the role of gender (and social equity) in relation to systems research – the significance of gender emerges as not one role, but as multiple roles or layers of roles. These can be seen as corresponding with movement across the conceptual framework above, from research engaging with more simple to more complex systems. Here we have clustered these into four roles: while the first three are likely quite familiar to most readers, the fourth is both more specific to systems research and more novel.

Three fundamental roles of gender analysis in research

Here we consider: relevance and accuracy; knowledge base and engagement; and targeting of research. These focus on the instrumental values of gender, which complement the ethical or social justice aspect.

i) Relevance and accuracy

Drawing on learning from decades of agricultural research in simple to more complicated and complex contexts, such as in crop variety development through to forestry research, we see the significance of gender analysis in terms of the value addition to research quality (Meinzen-Dick et al., 2011). Specifically, recognizing and nuancing research questions, data collection, and analysis in terms of gender means that we are not only asking (and answering) 'what technologies or innovations work?', but also investigating for whom, when, in what conditions and why. This contributes to the relevance and accuracy of research. Gender analysis, for example, leads to more relevant and accurate research through enabling findings that are reflective not only of perspectives of dominant groups, i.e., those that are often taken to represent 'communities', such as natural resource committee members, who may be predominantly male, or so-called male head of households. Rather, gender analysis enables a range of actors' views to be recognized, surfacing commonalities, as well as differences and even tensions. Both of these contribute to more effective research outcomes, as illustrated in counterpoint, for example in Wilde and Vainio-Mattila's (1995) recounting of an analysis of 'local' preferences for tree species that resulted in 3,000 hardwood seedlings being delivered to a community. The analysis had been based on the input of a select group of men (only). The seedlings were planted – and subsequently all died. Further study indicated that in fact in this area women were the keepers of the tree nursery, not men. It also indicated that had women been included, it would have emerged that the women's needs were for fast-growing softwoods for fuelwood and fodder – not hardwoods. The gender-imbalanced analysis had led to a loss of seedlings, no benefit for the community, and a waste of development investment.

ii) Knowledge base and engagement

The second role of gender analysis in systems research is that it adds value to the knowledge base of the research. This refers to gender analysis enabling the building of knowledge through inclusive research that draws on the breadth of human knowledge and capital, rather than only on a narrow (gender, socio-economically, or otherwise limited) range of knowledge (Vernooy, 2006). This is relevant to commodity research that involves farmers directly as well as to research in complex systems with multiple actors, such as natural resources management, governance, and climate change research. This breadth is signifi-cant in that the diverse knowledge and perspectives, such as an integration of local, indigenous, and scientific knowledge, enables a broader range of potential insights, innovations, and development pathways.

Relevant research that involves the knowledge of diverse women and men can also increase the gender-balanced interest of women and men farmers and fishers in engaging in and co-owning research. As this relates to action and participatory action research, gender-inclusive research can also contribute to the extent to which capacity is built amongst diverse male and female actors in relation to innovation and development.

This role of gender vis-à-vis knowledge and capacity has broader signifi-cance. In Röling's words (2002: 25), 'humans have become a major force of nature' and 'our main human predicament' is an anthropogenic phenomenon. As such, agricultural and development research are primarily tackling a host of anthropogenic challenges. The implication of this in terms of ways forward is that having 'all hands on deck' – meaning active and extensive broad social and gender-balanced knowledge, capacity, and engagement – is a far stronger position than having just the hands of a select, albeit dominant, few involved.

iii) Targeting of research

In addition to the above roles, social and gender analysis can help determine which actors are the most vulnerable and why. In other words, while the first above role is about strengthening accuracy and relevance of research in relation to a specific technical issue or challenge (such as a crop or raising tree seedlings), this role of gender is about being more effective in identifying and addressing the needs of a particular vulnerable social group.

Together, these three roles are important because of the ways in which they contribute to enhanced research quality and utility. In particular, they have implications for the effectiveness of research in contributing to intended devel-opment outcomes. Innovations that respond accurately to both women's and men's needs and perspectives, that draw on a range of relevant knowledge and engage diverse women and men, and are targeted to those who most need them, are more likely to be effective in contributing to development outcomes than those which are not. And yet, while this is clearly significant, it begs the question: is this the full potential of gender analysis? And, more broadly, is this the full potential of systems research? Or, does gender present an opportunity to aim for more fundamental contributions in and by systems research?

A systems research–specific role: Leveraging change

From a systems perspective, and drawing on social learning theory, the above roles and influences embody 'single loop' types of influence and improvements (Figure 19.2). This single loop influence refers to the research contributing to 'doing better' within the given system or way of doing things (Argyris, 1977; Jiggins et al., 2007; Loeber et al., 2007), such as increasing productivity through genetic enhancements.

While learning about and strengthening various parts of the system – such as enhancing genetics or farming practices to increasing productivity – the question emerges: will this (on its own) have the kind of influence that is needed for long-term, widespread, inclusive, and equitable socio-ecological sustainability and achieving our collective sustainable development goals? Or do we also need to engage with the system itself on a more fundamental level? This

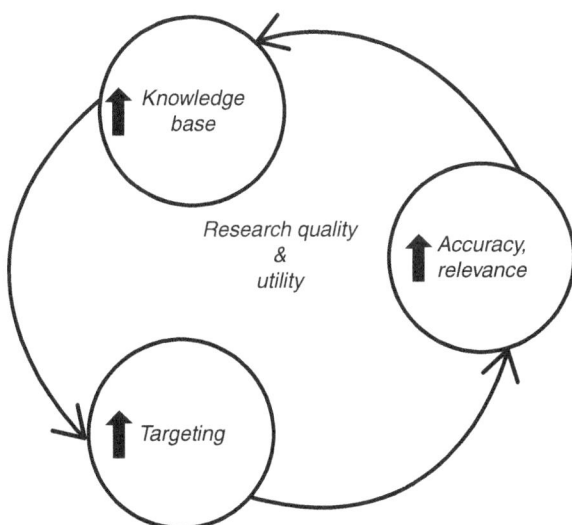

Figure 19.2 Three roles of gender analysis in research that reflect *single loop* influence: relevance and accuracy; knowledge base and engagement; and targeting of research. These roles are about gender analysis contributing to research 'doing better' within the given system or way of doing things

denotes 'double loop learning' (Figure 19.3). Double loop influence refers not to doing things better within the given system, but rather to understanding, questioning, and engaging with the assumptions and variables of the system (Argyris, 1977; Jiggins et al., 2007; Loeber et al., 2007). This can be colloquially understood as a difference between doing better while 'following the rules' versus doing better through 'changing the rules' (https://organizationallearning9.wordpress.com/single-and-double-loop-learning/). As systems researchers, we have a systemic vantage point. We are thus uniquely positioned to consider the system (be it food, agro-ecological, multi-stakeholder landscape, or other) as a whole and thus to identify potential shifts in it that would better leverage movement towards development goals.

So what is the role of gender in relation to this double loop learning? From a systems perspective, we propose that the fourth and most compelling significance of gender is its potential to enable system research to operate at this double loop level. In other words, the novel role of gender proposed here is that it may be an entry point to leverage change in the system in a way that enables greater sustainability, greater achievement of development goals: i.e., gender may represent a leverage point in development systems. Linking to the explanation of leverage points in systems thinking above, this refers to the idea that there may be issues and places to engage in a system that may shape the configuration – and thus outcomes – of the system. Gender and global food security offers one such example. As highlighted by the FAO (2011), if women had the same access to agriculture as men, with current resources on-farm yields could increase by 20–30% – and this could reduce global hunger by 12–17%.

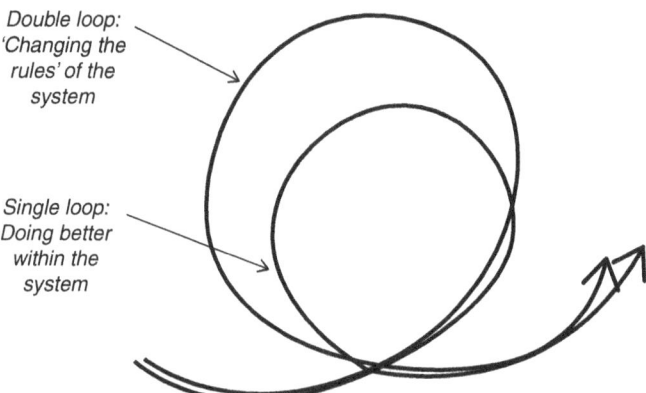

Double loop:
'Changing the
rules' of the
system

Single loop:
Doing better
within the
system

Figure 19.3 Single and double loop learning and influence. While single loop learning focuses on improvements within the given system or way of doing things (such as technical improvements), double loop learning focuses on questioning and engaging with the underlying assumptions about how the system works and its goals (in other words, 'changing the rules' of the system)

Source: Adapted from McDougall, 2015.

In this example, women + equal access to resources does not just equal women + equal access to resources. It equals substantial increases in food production and reductions in hunger. It is a 1 + 1= 3 relationship, in which gender equality leverages positive change in elusive productivity and food security goals. This particular 'if–then' example of the role of gender in larger changes in the system highlights a different role from the above gender analysis as improving the quality of technical research or even targeting of vulnerable areas. Within systems research, this more novel and profound role is not only about improving research outcomes in the given system; rather it is about research contributing to changes in the system itself that better enable us to collectively contribute to the desired outcomes.

And yet, while the above example illustrates that shifting gender access to resources contributes to significant changes, when we link back to systems thinking and leverage points, this is unlikely to be the whole story or lesson for research. Specifically, Meadows (2009) underscores that leverage points are not about numbers: numbers appear in last place on the list of effective points to intervene in a system. Applying this to gender in agricultural systems tells us that leveraging significant change in a system is not (only) about changing the number of women who are given equal access to land, new technologies, or credit or other important resources – at least not in the first instance. Rather, as Woodhill (2008) underscores, leveraging change in development relies on changing 'the rules of the game' (see also Leeuwis and Pyburn, 2002). These 'rules' refer to institutions – meaning informal institutions such as beliefs, norms, behaviours, and practices, along with formal institutions, such as policies – that shape development.

So while gender-equal access to resources and other forms of visible gender equity such as gender balance in trainings are critical, it must also be asked: are the most potent 'leverage points' at this visible gaps level? Or, are there more fundamental points with which to engage, i.e., at the level of beliefs, norms, behaviours, and practices that underlie visible gender gaps? Growing experiences and studies point to the latter. Evidence from Bangladesh shows the centrality of values and norms, illustrating that gender norms limit women's ability to utilize technologies, and thus ability to engage in, benefit from, and contribute to income and food security (Kantor et al., 2015). A recent study from Zambia (Cole et al., 2015) found that masculine identities in rural Zambia – reflecting gender norms around male drinking and extra-marital relations – contribute to household poverty. A review of the literature on micro-credit – widely hailed as significant to enabling development outcomes – indicates that access (alone) to micro-credit is insufficient to generate its intended outcomes. In fact, due to underlying gender norms and values, micro-credit schemes may even contribute to perverse outcomes such as jealously and gender-based violence (Cole and Muyaule, 2015). As Wilde (in Smith, 2015, n.p.) highlights, gender-equitable control (not only access) is crucial: 'when women earn an income that they control, you will see more investment in food, and education needs within the household'. Reflecting back to Meadows

(2009), it is highly significant that control is not a 'number', but rather a reflection of gender values and norms.

The above signals the importance of gender values, norms, attitudes, and behaviours as potential leverage points. And yet, recognition of this is one thing; the possibility for research to engage with them towards constructive systemic change is another. Does systems research have the potential to engage in this way? There is growing evidence from within the CGIAR that this may be possible. A multi-site, multi-year study on community forest governance indicated that research can use participatory action research to engage with and contribute to constructively shifting social and gender attitudes – even in contexts in which these are deeply entrenched, such as Nepal (McDougall and Banjade, 2015). Moreover, the study indicated that by doing so it changed the configuration of the community forestry systems, in the sense of shifting who is perceived as legitimate actors in the system, who makes decisions and how they are made, what is prioritized, and who acts as if they have 'ownership' of community forests. For example, the study illustrated that based on changes towards recognition of the poor as legitimate members with needs that should be addressed equitably, community forest user groups shifted from protection orientation towards a mixed protection and use orientation that allowed income generation while still conserving forest resources (McDougall and Banjade, 2015). Changes in recognition of the value of women as members led to both greater involvement of women, more gender-equitable leadership roles, and increased opportunities for women, especially poor women (McDougall and Banjade, 2015). Moreover, these systemic shifts triggered quantitative positive impacts not only on more gender- and socially-equitable engagement in decision-making and benefit sharing, but also triggered overall increases in small loans, and income generation activities, as well as the implementation of forest protection activities (McDougall and Banjade, 2015). Focusing on aquaculture, Kantor et al. (2015) present a study in Bangladesh that indicates the need for engagement with gender at this fundamental level through what they refer to as a gender transformative approach.[8] Similarly, while still in progress, a study in Zambia has begun to demonstrate that a gender transformative research approach can effectively shift underlying gender norms in relation to women's control over (and benefits from) micro-credit and savings (Cole and Muyaule, 2015). Broader evidence for women's control over financial assets suggests that these types of changes may have significant impacts in relation to children's education and nutrition (Meinzen-Dick et al., 2012).

Final thoughts

Before we return to the original questions posed – what is the significance of gender in systems research, and what is the value-added of systems research? – we synthesize a lesson for systems research that emerges from the above exploration. This lesson is about transformability of systems as the focal area of research. Both author experience and scan of systems research-related literature for this chapter reveals a preponderance of emphasis in systems research on resilience. And yet, as fundamental as is resilience, the above exploration signals that resilience in systems is not the whole story in terms of achieving development outcomes. Rather, it illuminates that the systems concept of *transformation* or transformability should be equally fundamental to our discourse and designs as systems researchers.

What does this mean in terms of how research could usefully be engaging with transformation? In line with the above, there is growing evidence that systems research – and other forms of agricultural research for and in development – can seek to enable locally-driven, locally-valued, and accepted shifts in those gender and social attitudes, norms, and behaviours that underpin unequal access to and control over resources. This includes in relation to gender-equitable engagement in decision-making and markets at multiple scales. This is in line with the call for development to engage with institutions – including norms – that shape development. Research that engages with the transformation of systems through transformation of informal institutions such as gender norms thus emerges on the development research agenda as having the potential to contribute to development outcomes in a way that research focused (only) on 'doing things better' within the existing system configuration (single loop research) may not.

At the outset of this chapter, we suggested we would also link to the question of the value-added of systems research. This exploration has highlighted a particular role of systems research: systems research may offer the perspective needed to identify leverage points in development. These can be engaged with to shift the system towards more sustainable and equitable configurations and trajectories. This would make a formidable complement to critical single loop research that enables 'doing better' within the same configuration, such as enhancing productivity.

Finally, we return to the central question of the chapter: what is the significance of gender in systems research? The above exploration indicates that gender is not only of value to research in general, but is specifically significant to the above value-added of systems research. In particular, gender presents itself as a potentially potent leverage point for development. While there is growing evidence of this potential through isolated cases, the more profound opportunities plausibly rest in a shared focus of research centres, programmes, and partners. In this scenario, many – as opposed to a few – work on the same leverage points, in particular, gender. If this were the case, and a plethora of actors and initiatives were engaging in this fundamental way with gender and social equity in different spheres, locations, and at multiple scales, the momentum for constructive transformation would be considerable. Considerable enough, perhaps, to bring our collective sustainable development goals within reach at last.

Acknowledgements

The author would like to acknowledge WorldFish and the CGIAR Research Program on Aquatic Agricultural Systems for having funded participation in the Conference on which this book is based. Additionally, sincere thanks go to Dr Patrick Dugan for his generous insights and encouragement leading up to the keynote presentation on which this chapter is based—inspiring and invaluable as always.

Notes

1 Proverb – Latin from the Hebrew.
2 The International Conference on Integrated Systems Research. March 3–6, 2015. Hosted by the CGIAR Collaborative Research Programs of Humidtropics, Drylands, and Aquatic Agriculture Systems. Ibadan, Nigeria.
3 The CGIAR, which hosted the conference at which this keynote was presented, is a not-for-profit global agricultural research partnership for a food secure future.
4 Following Shields (2008), 'intersectionality' refers to the mutually constitutive relations among social identities.
5 Gender integrated research 'integrates consideration of gender into technical research which is the principal topic of study, for example, plant breeding, aquaculture, postharvest technology development, systems intensification' (CGIAR, 2015). Strategic gender research refers to studies in which gender is 'the primary topic in a social analysis designed to understand what the implications of gender are for agriculture. E.g. how men and women allocate labour resources in intra-household decision-making about farm production' (CGIAR, 2015).
6 *Cynefin* (pronounced ku-*nev*-in) comes from the Welsh, referring to 'the multiple factors in our environment and our experience that influence us in ways we can never understand' (Snowden and Boone, 2007: 1).
7 Or perhaps more aptly, systems research operates across multiple levels of complexity: it offers an opening to link with other types of research such as commodities research. It is a 'connecting up' type of research.
8 A gender transformative approach, embodying a range of strategies, is defined as an approach that

> . . . can be applied within research to examine, question and, most fundamentally, enable changes in inequitable gender norms, attitudes, behaviours and practices and the related imbalances of power (IGWG, 2010). Through encouraging critical awareness among men and women of social inequality and practices, GTAs [gender transformative approaches] help people challenge and re-shape distribution of and control over resources, allocation of duties between men and women, and access to and influence in decision making (Caro, 2009). They also enable men and boys to question the effects of harmful masculinity, not only on women, but also on men themselves.
> (Meng, 2015 in McDougall et al., 2015; see also McDougall, 2015)

References

Argyris, C. (1977) 'Double loop learning in organizations', *Harvard Business Review*, *55*(5), pp. 115–125.

Aronson, D. (n.d.) *Overview of systems thinking*, The Thinking Page, http://www.thinking.net/Systems_Thinking/OverviewSTarticle.pdf, Accessed September 30, 2015.

Caro, D. (2009) *A manual for integrating gender into reproductive health and HIV programs: from commitment to action* (2nd Edition), Population Reference Bureau, Washington, DC.

Carr, E.R. and Thompson, M.C. (2014), 'Gender and climate change adaptation in agrarian settings: current thinking, new directions, and research frontiers', *Geography Compass*, *8*(3), pp. 182–197.

Cole, S. and Muyaule, C. (2015) *Successes and challenges implementing a gender transformative savings and lending group pilot project in Western Zambia*, Presentation to SG2015: Savings Group Conference – The Power of Savings Groups, November 10–12, 2015, Lusaka, Zambia.

Cole, S.M., Puskur, R., Rajaratnam, S. and Zulu, F. (2015)' Exploring the intricate relationship between poverty, gender inequality and rural masculinity: a case study from an aquatic agricultural system in Zambia', *Culture, Society & Masculinities*, 7(2), pp. 154–170.

CGIAR (2015) Definitions of Gender Research for CRP Gender Budgets Prepared by the CGIAR Gender and Agriculture Research Network. Updated June 2015 (draft Aug 2014) https://library.cgiar.org/bitstream/handle/10947/4057/DEFINTION%20 OF%20GENDER%20RESEARCH%20FOR%20BUDGETING%20v.june%202015. pdf?sequence=1, Accessed July 10, 2015.

Food and Agriculture Organization of the United Nations (FAO) (2011) *The state of food and agriculture 2010–2011, women in agriculture: closing the gender gap for development*, FAO, Rome.

Interagency Gender Working Group (IGWG) (2010), www.igwg.org.

Jiggins, J., Röling, N. and van Slobbe, E. (2007) 'Social learning in situations of competing claims on water use', in Wals, A.E.J. (ed) *Social learning towards a sustainable world*, Koninklijke Van Gorkum, Assen, pp. 419–434.

Kantor, P., Morgan, M. and Choudhury, A. (2015) 'Amplifying outcomes by addressing inequality: the role of gender-transformative approaches in agricultural research for development', *Gender, Technology and Development*, 19(3), pp. 292–319.

Kawarazuka, N., Locke, C., McDougall, C, Kantor, P. and Morgan, M. (2016) *Bringing gender together with resilience analysis in small scale fisheries research: Challenges and opportunities*, UEA Working Paper 53, Ambio, Norwich, UK.

Laven, A., Pyburn, R. and Snelder, R. (2012) 'Introduction', in Laven, A. and Pyburn, R. (eds) *Challenging chains to change: gender equity in agricultural value chain development*, Royal Tropical Institute (KIT), Amsterdam, pp. 1–12.

Leach, M., Kehtam, L. and Prabhakaran, P. (2016) 'Sustainable development: a gendered pathways approach', in Leach, M. (ed) *Gender equality and sustainable development*, Earthscan, London, pp. 1–33.

Leeuwis, C. and Pyburn, R. (2002) 'Social learning in rural resource management', in Leeuwis, C. and Pyburn, R. (eds) *Wheelbarrows full of frogs: social learning in rural resource management*, Koninklijke Van Gorkum, Assen, pp. 11–22.

Loeber, A., van Mierlo, B., Grin, J. and Leeuwis, C. (2007) 'The practical value of theory: conceptualizing learning in the pursuit of a sustainable development', in Wals, A.E.J. (ed) *Social learning towards a sustainable world*, Koninklijke Van Gorkum, Assen, pp. 83–98.

McDougall, C. (2015) *Leveraging change: how norms matter for development*, The Fish Tank, http://blog.worldfishcenter.org/2015/11/leveraging-change-how-gender-norms-matter-for-development/.

McDougall, C. and Banjade, M.R. (2015) 'Social capital, conflict, and adaptive collaborative governance: exploring the dialectic', *Ecology and Society*, 20(1), http://dx.doi.org/10.5751/ ES-07071-200144.

McDougall, C.L. (2015) Adaptive Collaborative Governance of Nepal's Community Forests: Shifting Power, Strengthening Livelihoods, Wageningen University, Wageningen, The Netherlands, 322 pp.

McDougall, C., Cole, S., Rajaratnam, S., Brown, J., Choudhury, A., Kato-Wallace, J., Manlosa, A., Meng, K., Muyaule, C., Schwartz, A. and Teioli, H. (2015) 'Implementing a gender transformative research approach: early lessons', in Douthwaite, B., Apgar, J.M., Schwarz, A., McDougall, C., Attwood, S., Senaratna Selamuttu, S. and Clayton, T. (eds) *Research in development: lessons learned from AAS*, AAS Program Report 2015–16, WorldFish, IWMI, and Bioversity, Penang, Malaysia.

Meadows, D. (2009) 'Leverage points: places to intervene in a system', *Solutions*, 1(1), pp. 41–49.

Meadows, D.H. and Wright, D. (2008) *Thinking in systems: a primer*, Chelsea Green Publishing, White River Junction, VT.

Meinzen-Dick, R., Behrman, J., Menon, P. and Quisumbing, A. (2012) 'Gender: a key dimension linking agricultural programs to improved nutrition and health', *Reshaping Agriculture for Nutrition and Health*, pp. 135–44.

Meinzen-Dick, R., Quisumbing, A., Behrman, J., Biermayr-Jenzano, P., Wilde, V., Noordeloos, M., Ragasa, C. and Beintema, N.M. (2011), *Engendering agricultural research, development and extension* (Vol. 176), International Food Policy Research Institute, Washington, DC.

Meng, K. (2015) *Gender transformative approaches briefing paper: aquatic agricultural systems (AAS) programme in Tonle Sap Hub, Cambodia*, AAS Briefing Paper, WorldFish, Penang, Malaysia.

Röling, N. (2002) 'Beyond the aggregation of individual preferences: Moving from multiple to distributed cognition in resource dilemmas', in Leeuwis, C. and Pyburn, R. (eds) *Wheelbarrows full of frogs: social learning in rural resource management*, Koninklijke Van Gorlum, Assen, pp. 25–48.

Shields, S.A. (2008) 'Gender: an intersectional perspective', *Sex Roles* 59, pp. 301–311.

Smith, G. (27 January 2015) *Engendering bigger impact: making the other half count*, CIAT Blog, http://www.ciatnews.cgiar.org/?s=engendering+bigger+impact, Accessed January 30, 2015.

Snowden, D. (n.d.) *Cynefin Framework*, http://cognitive-edge.com/, Accessed October 5, 2015.

Snowden, D.J. and Boone, M.E. (2007) 'A leader's framework for decision making', *Harvard Business Review*, 85(11), pp. 68–76.

Vernooy, R. (2006) *Social and gender analysis in natural resource development: learning studies and lessons from Asia*, Sage, New Delhi.

Wilde, V. and Vainio-Mattila, A. (1995) *Gender analysis and forestry: section A—how forestry can benefit from gender analysis*, FAO, Rome.

Woodhill, A.J. (2008) 'Shaping behaviour: how institutions evolve', *The Broker*, 2008(10), pp. 4–8.

World Bank (2012) *World development report 2012*, Worldbank, Washington, DC.

20 Gender norms and agricultural innovation

Insights from Uganda

Anne Rietveld

Introduction

The main objective of agricultural research for development (R4D) is the production of innovations that increase agricultural productivity and profitability for smallholder farmers. Adoption of such innovations, however, has generally been low amongst female farmers, particularly in Africa (Doss, 2001). This is not surprising as literature overwhelmingly points out that women are in general less educated than men, and have less access to productive assets such as land, labour and cash (Peterman et al., 2009; Quisumbing and Pandolfelli, 2010; Ochieng et al., 2014). Many agricultural development interventions have not been able to overcome such existing gender disparities (Cornwall, 2003; Quisumbing et al., 2015) either because of a bias towards male farmers or because the focus was on technical solutions to *'technical farming problems'* ignoring the social, economic or political implications of alternative technologies (Fairhead and Leach, 2005, p. 88). The social context in which agriculture is embedded dictates to a large extent the actions and roles for men and women and their space to manoeuvre (Knight and Ensminger, 1998). In order to understand men's and women's abilities to adopt and benefit from agricultural innovations, we have to understand the underlying gender norms that shape divisions of labour, ownership and management of farms and natural resources in farming systems (Meinzen-Dick et al., 2012) as well as the larger social environment (Fairhead and Leach, 2005; Okali, 2011b).

GENNOVATE

This chapter is a first result of a large-scale global study based on comparative case-study analysis that attempts to contribute to such understanding. One of the key questions is: how do gender norms shape poor men's and women's abilities to adopt and benefit from agricultural and natural resources management (NRM) innovations? The study is called GENNOVATE and is an initiative of the CGIAR gender research network (Badstue et al., 2015). The individual case-studies provide the basis for rigorous analysis of gender norms and how these shape roles of men and women at local level. This chapter is merely based on the results of two GENNOVATE case-studies conducted in Uganda following the GENNOVATE standardized, qualitative methodology (Petesh, 2014).

The subject of each GENNOVATE case-study is per definition a social group living in a specific locality referred to as village or community. The case-studies generate a wealth of data on the gender aspects of many different topics ranging from education, social cohesion and household bargaining to factors that hinder innovation, aspirations for the community and falling into poverty. The whole methodology consists of seven data-collection instruments including (1) literature review, (2) key-informants interviews, (3) semi-structured interviews and (4) single-sex focus group discussions (FGDs). Each case-study is led by one principal investigator (PI) and data is collected by a field team of four members, two females and two males.

Aim

The aim of this chapter is to shed light on the relation between gender norms and the adoption of one particular agricultural innovation for each of the two sites in Uganda. Each agricultural innovation is placed in a social context to provide insight into some of the consequences of the innovation's adoption. We assess which innovations came up or were introduced in selected communities in Uganda over the past decade and were reported as most important by men and women. Second, the focus is on one innovation per site to go in-depth as to how, why and for whom these innovations made the most impact and what these impacts were. The analysis also looks at the social groups for which the specific innovations were not accessible or interesting and why. Lastly, the paper discusses to what extent men's and women's abilities to adopt and benefit from this innovation were shaped by gender norms.

Methodology

This chapter is based on data collected for the GENNOVATE study and therefore data collection followed the GENNOVATE standardized, qualitative methodology (Petesh, 2014). For this chapter data from the focus group discussions (FGDs) was used. For each case-study, three different FGD guidelines were followed and each targeted another social group. Since every tool was conducted with men and women separately, a total of six FGDs were done per case-study. A detailed description of each FGD follows below and is based on the GENNOVATE methodological guide (Petesh, 2014):

1) FGD: Ladder of Life

   The objective is to explore perceptions and experiences on normative framework shaping gender roles; enabling and constraining factors for innovation and their gender dimensions; the culture of inequality in the village, including factors shaping physical mobility, socio-economic mobility and poverty trends – and their gender dimensions and intimate partner violence. The participants should be between 30 and 55 years old and be part of the poor socio-economic group of their community.

2) FGD: Capacities for Innovations

The objective is to explore experiences with and perceptions on community trends in prosperity; enabling and constraining factors for agricultural and natural resources management (NRM) innovation; the local opportunity structure for agriculture and entrepreneurship; social cohesion and social capital – and their gender dimensions and gender norms surrounding household bargaining over livelihoods and assets. The participants should be between 25 and 55 years old and be part of the middle socio-economic group of their community.

3) FGD: Aspirations of Youth

The objective is to explore gender norms, practices and aspirations surrounding education; gender norms and practices surrounding livelihoods and capacities for innovation; women's physical mobility and gender norms shaping access to economic opportunities and family formation norms. The participants should be between 16 and 24 years old.

Case-study sites

The case-studies were conducted in (1) Kisweeka in Kiboga district in the Central region, Uganda, and (2) in Kabaare in Isingiro district in the Western region, Uganda. The two case-studies were both conducted in localities where Bioversity International had an ongoing project within the CGIAR Research Program (CRP) on Roots, Tubers and Bananas (RTB). One of the sites (Kiboga) is also a site for the CRP on Integrated Systems for the Humid tropics (Humidtropics). The rationale for this site selection was that results could be fed back into the ongoing projects and programs to improve gender responsiveness.

Data collection for the Kiboga site took place as one of the first case-studies under GENNOVATE (April 2014). Later on, some changes were made in the method guide and, as a result, data from Kiboga does not 100% conform to the data of the Isingiro case-study conducted after June 2014.

Site 1 – Kisweeka parish, Kiboga district, Central Uganda (Kiboga)

Kisweeka is a parish in Central Uganda situated 3 km away from Lwamata, a small trading center along the national Kampala-Hoima road and 10 km distance from the district town Kiboga. People from at least five ethnicities live in the community. The Baganda and the Banyakore are the majority, each making up about a third of the total population. The average household consists of a man and a woman with their children. Few households are polygamous, and about 10% of all households are headed by women. Agriculture is the main occupation in the community.

The main agricultural good produced in the community is coffee, followed by maize and bananas (both cooking and juice types). All these crops are mainly in the domain of men although women can also have (small) banana plantations of their own. Beans are an important crop for women. Often these crops are all intercropped, although maize can also be found in single stands. Most jobs available in the community are related to agriculture, with farming (on own farm and/or as casual labourer) scoring highest for both men and women. Trading in agricultural goods is important for men. Few people in the community have legal land titles, most are tenants. Men usually rent much larger plots than women. Average cropping area is 5 acres. There are nursery and primary schools in the community but no secondary schools. Saving/ credit and religious groups are important and most people in the community are members of such groups.

Site 2 — Kabaare parish, Isingiro district, Western Uganda (Isingiro)

Kabaare parish is located in Isingiro district in Western Uganda. Isingiro district borders Tanzania and is also near the Rwandan border. The nearest city is Mbarara, a trading hub for cooking-banana and an intersection of national roads. The most common ethnicity is the Banyankole (75%) who speak Ryankole, but six other ethnicities live in the community. About 25% of households are headed by a woman.

The majority of households are smallholders with on average 2 acres of land. Approximately 30% are large landowners with land sizes up to 200 acres and 10% of the population is landless. Few women own land and those that do own small plots of around 0.75 acre. Share-cropping and renting of land is common among women. Main orientation in the area is agricultural and primarily focused on the cultivation of cooking-banana. Collective marketing of especially cooking-banana is common in the parish. Other important crops include maize, beans, sweet potato and millet. Intercropping of banana and beans is not common; cooking-banana is mainly mono-cropped. Cattle keeping is important in the area; people keep cattle, mainly local breeds that free-graze in large groups and move over large distances, and zero-grazed cattle, mainly exotic dairy breeds.

There is a primary school in the parish and almost all boys and girls are enrolled. There is no secondary school in the parish itself. Enrolment for secondary school is 50% of boys and 35% of girls.

Results

Important innovations in agriculture and natural resources management

In every FGD, participants were asked to list the main innovations in agriculture and NRM that came up or were introduced in the past ten years. An innovation was defined as any new knowledge; a technology, tool or social organizational form, which is utilized in an economic or social process (OECD, 1999). Then they jointly assessed each innovation listed on its importance for the men and women in the community. The results of the latter assessment are compiled in a table for each site. As a result of the small adaptations made in the methodology guide after June 2014 (see Methodology), Tables 20.1 and 20.2 present data in a different way with the Isingiro data being more specific as compared to Kiboga.

Kiboga innovations

The innovation most mentioned by both men and women in the different FGDs as important was 'the use of herbicides'. This innovation was not introduced nor promoted by external organizations and farmers did not receive any training on the use of herbicides. Bioversity International had a project in this community (ended 2014) in which a system was promoted that integrated improved banana management: zero-grazing of goats and cultivation of fodder in banana-based systems. Different elements of this project were mentioned in the FGDs (Table 20.1).

Access to improved seeds was only mentioned as important to men even when brought up by women. Partly because the crops mentioned (maize, coffee, vegetables) are all cash crops in the domain of men. But women also mentioned access to cash and credit as a constraint. Men usually have more ready access to cash and are also more mobile to purchase input outside of the village.

Table 20.1 Most important innovations of the past 10 years in Kisweeka, Kiboga district, Uganda. Data from six FGDs (3M, 3F): 'what are the 3 most important innovations?'

Men	Women
Keeping goats under zero-grazing method	Herbicides 3X*
Rearing goats to provide manure to improve soil fertility	Livestock rearing piggery and poultry
Herbicides 2X*	Zero-grazing – goats, cattle and pigs
Clonal coffee growing	Planting method of bananas
Improved beans	Improved banana plantlets
Improved maize	New types of maize seed
Good management of banana	New cash crops like tomatoes, vegetables and fruits
Savings and Credit Cooperation Organizations (SACCO) was started to provide credit to farmers	Cultivation both for food and sale

* Referring to the number of FGDs in which this innovation was mentioned as among '3 most important innovations'

It seems that women do not access improved seeds or planting material for the crops that they cultivate. The only exception was for the 'improved banana management' mentioned by both men and women which meant use of newly introduced (local) banana varieties.

The majority of innovations mentioned are directly related to crop cultivation or livestock, so-called technical or 'hard' innovations, but two innovations refer to socio-organizational or 'soft' innovations. These are (a) 'SACCO was started to provide credit to farmers' and (b) 'cultivation both for food and sale (for women)'. SACCO is the acronym for Savings and Credit Cooperative Organizations; it fills the institutional gap left by the absence of banks and commercial credit providers. The SACCOs in Kiboga handle relatively small amounts; savings per person per week vary between 2,000 and 10,000 UGX per week (1 USD = 3,300 UGX). Loans provided by the SACCO in Kiboga need to be repaid within two months with an interest rate of 20%. SACCOs are important to both men and women as they enable investment in farm production and other businesses or help with large expenditures such as school fees for children. They also provide security in case of sickness, death or other casualties: *'Members save and get loans in case they have problems'* (Male FGD C). 'Cultivation for both food and sale (for women)' means that women can now sell some of the crops they produce instead of only producing for household consumption. This refers to a shift in gender norms that affects the options for women and their position within the household; now they can sell part of their produce and earn cash which they can use to *'support their children and families'* (Female FGD C). It is mentioned, however, that this depends on the husband as *'sometimes the men sell off your crops and you do not earn anything, or they take some of the money from you'* (Female FGD C).

Isingiro innovations

Improved banana management and livestock or zero-grazing of cows are consistently mentioned as number 1 and 2 innovations for men by both men and women (Table 20.2). This is not surprising as commercial cooking-banana production and cattle rearing are the two main pillars of men's farming in this area. When it comes to the women much less consensus exists; three of the innovations mentioned by men about women are still about banana management but focusing on specific activities normally performed by women: hand-weeding, mulching and manure application on banana. Women mention banana as important in two FGDs when talking about women but refer more in general to the new management practices.

Both men and women mention beans and maize, but men talk about good management of these crops whereas women mention new varieties. Women also mention livestock – cattle, goats and chickens – but men do not talk about livestock in reference to women at all. From the discussions, it was clear that cattle are always controlled by men, but women appreciate the milk they provide to the household. Chicken can be managed by women; rearing them in 'houses' for increased productivity since they are easy to manage, provide some cash income from sales of eggs and contribute to household food. Men mention

Table 20.2 Most important innovations of the past 10 years in Kabaare, Isingiro district, Uganda. Data from six FGDs (3M, 3F): 'what are the 2 most important innovations of those listed for men and for women?'

Men about men	Men about women
Good management practices in banana 3X*	Hand weeding in banana plantation
Control Banana Xanthomonas Wilt (BXW) 2X*	Manure application on banana
Rearing livestock (especially cattle and goats)	Good management of beans and maize
	Growing vegetables
	Growing of orange-flesh sweet potatoes
	Mulching the banana plantations

Women about men	Women about women
New banana management practices 3X*	New banana management practices 2X*
Zero-grazing cows 2X*	Rearing chicken in houses
Livestock: cattle and goats	Livestock: cattle and goats
	New bean varieties
	New maize varieties

* Referring to the number of FGDs in which this innovation was mentioned as among '2 most important innovations'

that vegetable growing and the cultivation of orange-flesh sweet potato are important to women, but women do not mention these.

Focus innovations

Kiboga – 'The use of herbicides'

The agricultural innovation most consistently mentioned by women and men in Kiboga was 'the use of herbicides' which was especially mentioned in relation to maize cultivation. It would be easy to report on this innovation referring to its benefits and impact for 'farmers': '*The use of herbicides has made a large difference in maize cultivation as farmers are now able to increase the size of their maize areas considerably as labour availability for weeding is not a limiting factor anymore*' and '*consequently farmers have been able to increase income from maize cultivation*'. Although this would be correct reporting, it does not depict any of the conditions, potential disparities or actual contributions to households' livelihoods underlying the use of herbicides in maize cultivation.

In Kiboga men and women have in general their own plots to cultivate. Men though are the owners of the land and they often allocate relatively small and unfertile land to their wives. Men rely on their wives to provide labour on their plots before taking care of their own plots: '*The men demand that you work on their farms first before you work on yours and you must go to their farms in the morning when you are still fresh so the only time that you have for your own crops is in the evening when you are already tired*' (Female FGD C). Women are especially responsible for weeding and as such their labour availability was limiting production before the use of herbicides was common practice. As a consequence, the use of herbicides has had some benefits for women because the demand on their labour for weeding has decreased. For men, the main benefit is the increased volumes of maize they produce: '*now men grow acres and acres of maize and have increased yields*' (Female FGD D). This, however, does not always benefit the rest of the household. Women complained that recently men tend to leave responsibilities that used to be in their domain, such as paying school fees for children, up to women: '*A good number of men are not taking care of their homes, so you have to pay school fees, pay hospital bills and feed the children*' (Female FGD D). The shifting of responsibilities within the household goes hand in hand with increased opportunities for women to earn an income from the sale of agricultural products: '*We were not allowed to work 10 years ago. Today, we can farm separate plots and earn an income*' (Female FGD D). Looking at the community as a whole, no considerable change in wealth has taken place over the past decade, and economic dynamism was low.

Although women appreciate the fact they can now earn money, this opportunity is still contested, and women face many constraints. Many of these constraints are related to access to resources and deeply entrenched prejudice against 'working women'. Men and women mention the reluctance of men to allow their wives to move around freely: '*Most men in Kiboga keep their wives at home and prevent them from attending the agricultural training, fearing that when they go to these groups they will get bad advice from their fellow women'* (Male FGD D); *'If you walk around in this community and you are a young woman, they say that you are a prostitute'* (Female FGD E). They also do not want women to have their own personal money: *'She will "grow wings"'* (Female FGD E); *'Women become stubborn when they get money; they can never respect their husbands again'* and *'If you give her a chance to look for money, she will one day disappear with other men'* (Male FGD E). The responses of male youth on the topic of 'working women' are more stringent and disapproving than those of adult men in general. Similarly, female youths are more outspoken about the limitations they face, and they seem to be more restricted in, for instance, 'moving around in the village' than elder women.

Some women mention they would like to grow maize but generally men do not want this. Also, they cannot get information or funds to buy improved seeds or herbicides. If they want to sell maize or any other produce, they often have to act via their husbands as *'A woman cannot go along with her produce on top of a lorry!'* (Female FGD D). Opinions are divided whether it is good or not to sell produce through your husband. Some women mention advantages such as men being better informed about prices and therefore less likely to be cheated by traders. Others argue that some men will only give you part of the revenue and keep the rest for themselves. Men mostly preferred to be engaged in the sales efforts of their wives and some even considered this as a condition for allowing their wife to sell produce: *'He may allow her because he will be able to know how much she earns and accordingly he will plan for its use/spending'* (Male FGD E).

Herbicides have mainly been adopted by men to support maize cultivation. The majority of men who own land seem to use herbicides; this is especially true for the 'better off' share of the population as herbicides require cash. Although women are interested in using herbicides, they face some gender-specific constraints which range from limited access to and control over resources such as land, (improved) seeds and labour to a reduced capacity to make decisions in general compared to men. Women's agency, defined as 'the ability to define one's goals and act upon them' (Kabeer, 1999, p. 438), is limited compared to men's as they need their husband's permission for many things ranging from day-to-day activities such as going to the market, to larger decisions such as, for example, choosing which crop to cultivate and whether to use inputs and how to sell. Large differences, though, exist between individual households. Some women merely inform their husband about certain farming-related decisions. This accounts especially for older women with adult children or women whose husbands are not staying with them on the same farm. Other women really need approval before they can do something.

Isingiro – 'Improved banana management'

The improved or new management practices for banana include: use of manure, de-suckering or removing corms, trench digging, spacing, hand-weeding and mulching. Banana Xanthomonas Wilt (BXW) control measures, including de-budding, are sometimes mentioned separately but are often included in the 'improved banana management' discourse. All these practices were introduced by National Agricultural Advisory Services (NAADS), the governmental extension program that promotes commercial farming since 2004. Some of these practices require a lot of physical strength, and this is given as a reason why they are almost exclusively performed by men. This is, for instance, the case with de-suckering, by completely removing the corms of new suckers through uprooting, and the digging of trenches for water retention in the plantation. Improved banana management also imposes changes in the way weeding is done. Instead of using hoes as before, weeding is now primarily done by hand in order not to damage the banana roots. This has negatively impacted women who are responsible for weeding. Hand-weeding is not only time-consuming, but it also involves a lot of bending which causes back pain. Children are also required to assist with weeding when weed pressure is high and consequently miss out from school. Herbicides are not used because it is believed in the community they harm the soil. The same is believed for chemical fertilizer, and therefore only cow manure is used for fertilization. With the increased focus on commercial banana production, the variety of cultivars grown is decreasing in favour of market-preferred varieties: *'some varieties such as Mbururu are disappearing due to their small sizes; they are being replaced with bigger and high yielding banana varieties such as Mbwazirume'* (Male FGD E).

There is a wide consensus in Isingiro that 'improved banana management' has had a large impact on livelihoods in the community. Banana is the most important source of cash for the majority of male farmers and for the household as a whole. Therefore, women also mention these banana innovations as paramount because banana is the main source of household food and income and pays, for instance, children's school fees. The community leaders praise the community: *'People in this parish have constructed good houses compared to other parishes. In addition, the parents here try to educate their children'* and *'Many people have become rich because of the cooking-banana'*.

Both men and women make clear that the banana plantations are controlled by men: *'Most plantations are owned by men and the bananas are main source of income, so the men's interest is there'* (female FGD C). Having a banana plantation as a man determines his status in the community. A 'good farmer' or even 'good husband' owns per definition a banana plantation. The only category of women that regularly owns banana plantations in the community are widows. They can sell independently and sometimes even perform the required labour normally done by men, such a de-suckering if they cannot afford to hire labour.

With men focused on banana, women have more opportunity to grow other annual crops: *'Ten years ago, the seasonal crops that were grown were the only source of income and they were controlled by men but now men are on bananas and women can get some income from the seasonal crops'* (Female FGD C). However, getting access to land for women is an issue because land is predominantly owned by men. Some men will give their wife a plot to cultivate, but other women have to rent land. Finding land for cropping is becoming less feasible with land formerly used for grazing or annual crops now being turned into banana plantations which makes finding available land difficult. In addition, land rent prices have increased.

Commercial cooking-banana cultivation using the new management practices has caused revolutionary changes in many people's livelihoods, especially for those with large land holdings. This has increased wealth differences among the households. Fewer households live under the poverty-line, as defined by the community itself, and a new class of 'super-rich' has been created. These rich farmers have access to all the resources they need, own vehicles and trucks and keep expanding their plantations, and they often employ dozens of farm labourers. This is in stark contrast with the landless or smallholders owning less than one acre who *'don't have anything of their own'* (Male FGD C). Their only option is to *'work for the rich'* to earn in cash or kind. Being knowledge-able about improved banana management practices will help them to get more work *'because the rich also want to use those who are good'* (Female FGD C). In some cases, a poor landless worker can work one or more years for a rich farmer, after which the rich farmer will buy a piece of land. This is mentioned as one of the few ways landless poor can move out of poverty.

Commercial cooking-banana cultivation using improved banana manage-ment practices has boosted the local economy and the large majority of Kabaare Parish (male/female, poor/rich, young/old) benefit to some extent. The valu-able cooking-banana plantations, however, are firmly controlled by men and gender division of labour for banana practices is strict; the only exception to the rule seems to be for widows. With men's increased focus on cooking-banana, some space was created for women to earn money through cultivation of annual crops but women's access to inputs, primarily land, is limited. Also women, especially those married to men owning middle-large plantations, spend a lot of time working in their husband's banana plantation. Work which, in the case of weeding, could probably be substituted by using herbicides. In addition, married women have less ability than their husbands to mobilize labour from other household members, such as children, as they cannot 'dictate' and they also generally lack the resources to hire external labour.

Discussion

Since the focus of this chapter is on one specific innovation per community, only (a small) part of the livelihoods of women and men in the communities are dis-cussed here. This chapter does not claim to give a complete overview or analysis of all gender norms that interfere with agricultural innovation. Nonetheless, some clear patterns came up surrounding these two innovations in relation to gender norms and the roles of women and men in agriculture in general.

Gender domains

In both case-studies divisions of labour and domains are on the one hand much gendered, while on the other hand many linkages exist, and borders between the domains are superfluous, heterogeneous and negotiable, for instance the degree to which crops should be considered male or female. Cooking-banana sales and income are strictly controlled by men in Isingiro, but at the same time women spend most of their time working on this crop and emphasize that derived revenues benefit the household at large. Calling cooking-banana a men's crop is not useful under these conditions, as it ignores the central role this crop plays for both women and men and the investment both make in terms of finance or labour inputs. It is rather, as Meinzen-Dick et al. (2012, p. 5) argue, that boundaries between men and women's crops are less rigid then they initially appear. The same accounts for farm plots: even as women often have their own plot to cultivate, the land is still owned by the husband or rented from another man. Or the revenue from the produce is managed, at least partially, by the husband and he is consulted and gives permission on which crops to plant or what inputs to use. For both cases, there is no complete gender separation of plots, as is observed elsewhere in the region (Lambrecht et al., 2016).

Women's lack of agency

Women's agency is often constrained in these communities when it comes to economic participation. First of all, 'earning an income' or 'having money' is perceived as part of men's identities and not those of women. Men should be the main breadwinner in a household and the main decision-maker on income. Other scenarios are believed to lead to conflicts in the household as a woman would not accept the authority of a man when she brings in more money than he. Secondly, the stigma that lies on women who move around a lot, inside but especially outside the community, effectively keeps women from employing in activities outside of the homestead and the community. According to the dominant norm, much of the functioning of a household is based on male authority. Men should have control over their wives and children and they decide, for instance, on labour allocation. They also control other productive assets such as land, transport means and large livestock. In return, the norm prescribes that men should provide children's school fees (often the largest cost for households), (staple) food for household consumption, the financial means to pay for health care, and other household necessities such as salt, soap, clothes and kitchen utensils. In many cases, however, women and men deviate from the norm, or negotiate alterations amongst themselves. Especially in Kiboga, traditional gender norms, concerning for instance mobility and women earning money, seem to be changing. Issues surrounding these norms were the centre of heated debates in the women's and men's FGDs. Both women and men provided arguments in favour and against these changing norms.

Heterogeneity

Gender norms are not static. Not only can gender norms change over time, but their urgency can also vary for different kinds of social actors, women and men. For instance, widows have more space to manoeuvre around certain norms, such as related to the gendered division of labour or land ownership, than married women. Yet, married women can negotiate (more) access to land or increased mobility at times. This seems to be more likely among mature couples with older children. Young married women with small children seem to be more restricted in their gendered roles. Not only do they spend more time on caring, but there was also a tendency among male participants of the FGDs to picture them as promiscuous and not loyal and as such in need of being controlled. Apart from age, number and ages of children and marital status, there are other factors that interact with gender norms and determine the heterogeneous social behaviour of women and men. An important one is wealth. Poor women can, for instance, be allowed by their husband to work on other people's farm, out of necessity.

Gender equity and benefits from agricultural innovations

The two main innovations discussed in this chapter, 'the use of herbicides' and 'improved managed for cooking-banana' were both more in the domain of men. In both cases, men controlled both the management of the crops concerned (maize and cooking-banana) and the (extra) income accruing from these innovations. Nonetheless, these innovations were also mentioned in the female FGDs; they were listed as important by the women and as very beneficial to them and their households at large. It is implied that even if benefits accrue from innovations that can be perceived as unequal from a gender perspective (because women do not face the same conditions as men to use these innovations), their impacts can still contribute to the wellbeing of both sexes. The impact of such innovations on gender equity can be neutral, empowering or disempowering. Or both of the latter at the same time, as was for instance the case in Isingiro. Men's increased focus on cooking-banana gave room to women to engage in the cultivation of annual crops for consumption and sales, a development perceived as empowering in the dominant development discourse (Okali, 2011a). At the same time, the labour demanded from women in their husband's banana plantation has increased and become more physically demanding. Also, land availability for cultivation of annual crops by women reduced because of the large demand for land for establishing new cooking-banana plantations. When studying or aiming at improving gender equity, it is important to take all these different, often contradicting, developments into account.

Conclusions

The existence of distinct enterprises managed by either a woman or man, without any involvement of the other spouse, is rather the exception than the rule in Central and Western Uganda. In general, women and men will have their own specific roles in the farm system and household. The two main innovations elaborated on in this chapter showed that adoption of an innovation by men will affect other household members, notably the wife, too. As was shown for both innovations, these effects can be multifold, with positive and negative elements. A holistic analysis of women's and men's roles in agriculture, as was conducted in these case-studies, can shed light on certain aspects of adoption that usually remain hidden. This can be very specific, for instance, related to norms that underlie certain gender-specific constraints to accessing information in a locality. Yet, it can also enable identifying different, often contradicting, trends and developments that constrain, promote or enable agricultural innovation on a higher level. Knowledge of both the specific and the higher level trends will enable actors working in agricultural R4D, development or policy to design programs and policies that benefit both women and men and reduce gender inequities.

References

Badstue, L., Kantor, P., Prain, G., Ashby, J. and Petesh, P. (2015) GENNOVATE Concept Note, https://gender.cgiar.org/wp-content/uploads/2015/09/CN-Global-Study-GenderNormsAgency_modified-Jan-2015.pdf.

Cornwall, A. (2003) 'Whose Voices? Whose Choices? Reflections on Gender and Participatory Development', *World Development 31*(8), pp. 1325–1342.

Doss, C.R. (2001) 'Designing Agricultural Technology for African Women Farmers: Lessons from 25 Years of Experience', *World Development 29*(12), pp. 2075–2092.

Fairhead, J. and Leach, M. (2005) 'The Centrality of the Social in African Farming,' *Institute of Development Studies Bulletin 36*(2), pp. 86–90.

Kabeer, N. (1999) 'Resources, Agency, Achievements: Reflections on the Measurement of Women's Empowerment', *Development and Change 30*, pp. 435–64.

Knight J. and Ensminger J. (1998) Conflict over Changing Social Norms; Bargaining, Ideology, and enforcement. In: The New Institutionalism in Sociology. Mary C. Brighton & Victor Nee, (eds.), pp. 105–126. Stanford, CA; Stanford University Press.

Lambrecht, I., Vanlauwe, B. and Maertens, M. (2016) 'Agricultural Extension in Eastern Democratic Republic of Congo: Does Gender Matter? *European Review of Agricultural Economics*, pp. 1–33.

Meinzen-Dick, R., van Koppen, B., Behrman, J., Karelina, Z., Akamandisa, V., Hope, L. and Wielgosz, B. (2012) 'Putting gender on the map; Methods for Mapping Gendered Farm Management Systems in Sub-Saharan Africa', *IFPRI Discussion Paper 01153.*

Ochieng, J., Ouma, E. and Birachi, E. (2014) 'Gender Participation and Decision Making in Crop Management in Great Lakes Region of Central Africa', *Gender, Technology and Development 18*(3), pp. 341–362.

OECD (Organization for Economic Cooperation and Development) (1999) *Managing National Innovation Systems*, OECD, Paris, France.

Okali, C. (2011a) 'Achieving Transformative Change for Rural Women's Empowerment', Expert paper prepared for the Expert Group Meeting, 'Enabling rural women's economic empowerment: Institutions, opportunities and participation', UN Women, Accra, Ghana.

Okali, C. (2011b) Searching for new pathways towards achieving gender equity beyond Boserup and 'women's role in economic development', *ESA Working Paper* No. 11–09, FAO, http://www.fao.org/3/a-am314e.pdf.

Peterman, A., Behrman, J. and Quisumbing, A., (2009) 'A review of empirical evidence on gender differences in non-land agricultural inputs, technology, and services in developing countries', *ESA Working Paper No.* 11–11, FAO, 2011.

Petesh, P., (2014) 'Methodology guide for global study; innovation and development through transformation of gender norms in agriculture and natural resource management', unpublished.

Quisumbing, A.R. and Pandolfelli, L. (2010) 'Promising Approaches to Address the Needs of Poor Female Farmers. Resources, Constraints, and Interventions', *World Development 38*(4), pp. 581–592.

Quisumbing, A.R., Rubin, D., Manfre, C., Waithanji, E., Van den Bold, M., Olney, D., Johnson, N. and Meinzen-Dick, R. (2015) 'Gender, assets, and market-oriented agriculture: learning from high-value crop and livestock projects in Africa and Asia', *Agricultural Human Values 32*, pp. 705–725.

21 Gender transformative approaches in agricultural innovation

The case of the Papa Andina Initiative in Peru

Silvia Sarapura Escobar, Helen Hambly Odame and Amare Tegbaru

Introduction

This chapter examines a case study of agricultural research for development that moves from gender mainstreaming or integration towards an approach that seeks change and innovation within agricultural systems through gender transformative approaches which are derived from critical social science theories. Given the Andean context and the nature of the program, the case suggests that gender transformative outcomes occur when a gender neutral program design is abandoned in favour of gender responsive processes achieved through participatory and applied methodologies that foster collective work, communication and individual/group learning among diverse groups of stakeholders. All of these processes interpret, analyze and influence changes in gender norms, perceptions and relations entrenched within social systems, in this case the Central Andes of Peru. It is within this context that an international research organization, the International Potato Center (CIP) works with various partners in the public, non-profit and private sectors. While this paper is focused on a specific initiative from agricultural research, it offers potential lessons for the wider agricultural and agri-food sectors and development organizations across the Latin American region. The CGIAR began to pay attention to gender issues in the 1970s. Various landmark initiatives and studies are apparent, most recently with a treatise from a group of male scientists for 'integrating the social and technical in an inclusive, critical farming systems approach that aims to foster transformation in the gendered power relations that constrain the potential of poor and marginalized women and men from joining in and benefiting from agricultural innovation processes' (Tegbaru et al., 2015, p. 137). The authors use the metaphor of a 'landscape approach' to understand systems' environmental and technical complexities without disciplinary neglect of social relations and roles (defined as human roles, rules, norms, resources, activities and power). Based on a transformative theory of change more widely discussed by various gender specialists (Meinzen-Dick et al., 2010; Kantor, 2013; Njuki et al., 2016) and discussed below, these gender analysts situate resource-poor women's and men's empowerment as ensuring integrity and equity across a range of land-scapes and levels within systems (e.g. household, community, value chains, State, and so forth) so that social inclusive processes enhance opportunities for agri-cultural innovation across the system.

The purpose of this chapter is to explore such ideas with a specific case. Here we have a closer look at the Papa Andina Initiative[1] and its legacy of actor platforms in the native potato value chains of Peru to illustrate how agricultural researchers might move towards gender transformative approaches (GTA) in agricultural systems and innovation processes. The discussion uncovers three dimensions of practical and theoretical value: (1) how institutional aspects iden-tified within Papa Andina influenced transformation and change in agricultural innovation from a gender perspective; (2) how actors, interactions, method-ologies and tools from the initiative could have been directed towards gender transformative change in agricultural innovation systems; and (3) how individ-uals' sense of self impacts capacity for gender transformative change. The chap-ter ends with considerations for the future. In the next section, we address why after numerous gender activities the CGIAR must still re-examine its strategies.

CGIAR and gender transformative approaches

Since the 1970s, a steadily increasing effort has called agricultural systems to account for gender equality, socially inclusive participation and empowerment.[2] Several major gender initiatives are evident in the CGIAR's 45-year-old history (Figure 21.1). These depart basically from the emphasis on women's issues such as access to resources and division of labour to a more systematic approach to gender research.

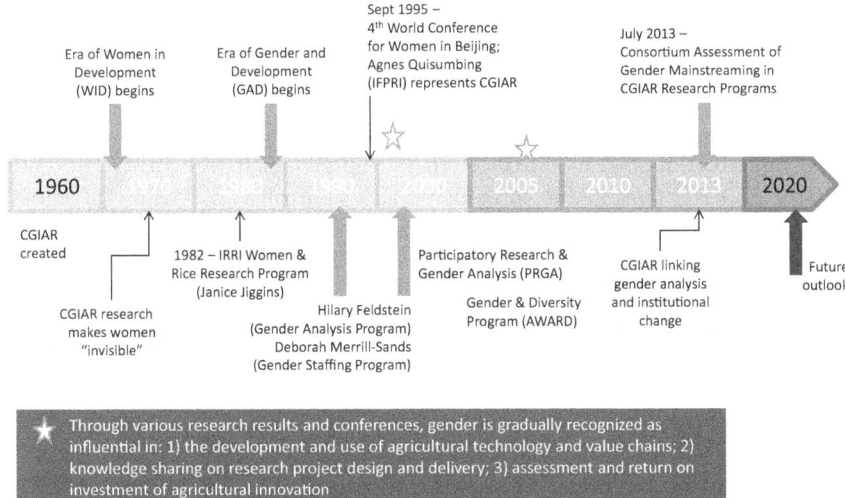

Figure 21.1 CGIAR's gender pathway

Specifically, by the beginning of Papa Andina in 1996, two system-wide efforts had already attempted to integrate gender into CGIAR work. The first, the Gender Analysis Program (Sims Feldstein, 1995; Merrill-Sands et al., 1999), focused on the development of analytical tools and capacity building processes and was later merged with participatory plant breeding activities to become the Participatory Research and Gender Analysis Program (Ashby and Sperling, 1995; Prain et al., 2000). Several empirical studies focusing on the involvement of women in technology decision-making processes emerged over the decades (Sperling et al., 2001; Quisumbing, 2003; Quisumbing and McCafferty, 2006; Paris and Rola-Rubzen, 2009). The second effort targeted the organizational level with an initial research focus on gender and staffing within National Agricultural Research Systems (NARS) (Brush et al., 1995; Merrill-Sands et al., 1999) and later, in 1999, attempting to institutionalize gender equity within the CGIAR as well as NARS partners with the Gender and Diversity Program. The latter program initially spearheaded women's leadership and networking and later moved towards advancing diversity awareness and action within the CGIAR (Joshy et al., 2000). In general, the work of the CGIAR from the 1990s to 2010 espoused a managerial path of targeting women in the design, development and adoption of agricultural technologies as well as the quantification of their participation in project activities (Tegbaru et al., 2010). The managerial approach was characterized by critics as 'add women and stir' because the established power relations within science policies, programs and projects otherwise remained the same (Harding, 1991). Little effort was made to understand women in relation to men, to challenge hegemonic power relations or elitism in order to open up and reconceptualize research processes from a more socially diverse gender standpoint.

By 2010, the leadership of the CGIAR, the Consortium Board, commissioned a study by the International Center for Research on Women which identified several shortcomings in the managerial approach and gender mainstreaming. While substantial progress was noted in projects that integrated gender into technology development and evaluation, consistency in building competencies, sustaining gender research and achieving impact in gender-related results was limited (Kauck et al., 2010). Actually, the gender transformative potential of the CGIAR's work was weak or non-existent on two counts: (1) achieving equal opportunities within and from agricultural Research and Development (AR&D), and (2) accomplishing gender equity in terms of system impact because power-based gender relations were rarely challenged or changed. Specifically, efforts to collect and analyze sex-disaggregated data for gender roles and resource-access models did not, for example, articulate with institutional processes in terms of gender policy, investment strategies or impact. Theories of change that bridged technical excellence with social inclusion were weak.

Although recognition and resources allocated to gender are expected to improve in the most recent reform process for collaborative CGIAR Research Programs (CRPs), agricultural research institutions, individual scientists and resource managers are still struggling with the relevance of gender methodologies.[3] This is due, in part, to limited exposure to underlying feminist concepts, states Kauck et al. (2010). This is not a situation unique to the CGIAR, as Njuki et al. (2016) suggest in their recent collection of analyses on global food security projects from a range of non-governmental organizations and public sector agencies.

The challenge, therefore, is for the CGIAR to adopt and integrate a more critical gender transformative approach into the CRPs that commits to the assessment of two key dimensions: (1) how gender norms act as social constraints on innovation, and (2) innovations resulting from using gender transformative approaches to challenge and change gender norms, perceptions and relations. In this chapter, we will look through a GTA lens at the efforts of Papa Andina and ongoing systems activities involving actor platforms and value chains for native potato in Peru. Based on Sarapura (2013), however, we would like to add a third, crucial element to the above-mentioned key dimensions: the interpretation and influence of socialized gender norms not just at the level of organizations and institutions, but also at the level of the individual who is engaged in challenging and changing her/himself. This we refer to as work at the level of one's sense of self.[4] Feminist researchers view the concept of selfhood (autonomy and self-identity) as distinct from social norms. We therefore argue that there is an emotional as well as an intellectual commitment to gender transformative innovation that implicates 'hearts as well as minds' because of the ethical dilemmas that it stirs up (Sarapura Escobar and Puskur, 2014). In this case, we emphasize that the sense of one's self is fundamental to personal actions, particularly as they relate to participation within actor platforms. A person's own sense that they are empowered to act on their human agency becomes, strategically and *en masse*, the potential for recognizing and influencing social transformation.[5]

The Papa Andina Initiative

Papa Andina was a regional initiative coordinated by the International Potato Center (CIP) from 1996 to 2010. It fostered pro-poor innovation in market chains to improve food security and market access and to reduce poverty of peasant producers in the Central Andes (Horton et al., 2011) and can be characterized as 'agricultural research for development' (AR4D). In Papa Andina a framework for understanding and managing institutional and organizational change processes in agricultural systems was developed (Devaux et al., 2013). CIP led the coordination team but acted as a second-tier innovation broker (Devaux et al., 2010). Strategic and operational partners in three countries (Peru, Bolivia and Ecuador) were the main part of the alliance. In Peru, the Project for Innovation and Competitiveness of the Potato (Proyecto INCOPA) was a coalition of key private and public organizations (Devaux et al., 2010; Devaux et al., 2011).[6] The coalitions were so embedded within national systems that they eclipsed the coordination by CIP and identity with Papa Andina (Bebbington and Rotondo, 2010). The coalitions effectively involved small-scale farmers or peasant producers – women and men – in the development of tech-nology and added-value market chains with researchers, agricultural service providers, policy-makers and market agents (Horton et al., 2011). Overall, the initiative led to innovations such as the Participatory Market Chain, Multi-stakeholder Platforms, and Horizontal Evaluation, which invigorated the native potato industry.[7]

Over time, Papa Andina resulted in diverse facilitating and decentralized knowledge broker networks that continued CIP's long-standing and continu-ing experience with participatory approaches for on-farm research and diverse public-private sector partnerships (Thiele et al., 2011). In the case of Peru, INCOPA platforms structured along market chains linked up peasant produc-ers with traders, processors, supermarkets, researchers, chefs and others to foster the creation of new products with a greater possibility of added value and pro-poor innovation (Devaux et al., 2009). Other platforms structured around geographically delimited supply areas addressed market coordination problems in ensuring volumes and meeting quality and timeline constraints associated with a supply chain made up of many dispersed and peasant producers. These platforms were very successful and continue to the present day addressing coor-dination problems in the subsidiary markets for support services and comple-mentary inputs, bringing NGOs and others in to provide technical support or access credit (Thiele et al., 2011; Sarapura, 2013).

Papa Andina employed a strategy that built on the assets of peasant produc-ers, their traditional values (i.e. Andean cosmo-vision) and local knowledge of biodiversity rather than on the transfer of external solutions. Working with partners on developing new technologies, managing knowledge and support-ing local and national groups to facilitate innovation processes eventually meant that Papa Andina was engaged in the realm of policies and institutions at local, regional and national levels and across the countries where it worked. They supported the formation of structures to include lines of accountability to mul-tiple stakeholders and different types of interaction, fostering greater mutual communication and understanding. Many of these approaches are consistent with gender transformative approaches. Nevertheless, two key elements were absent in the Papa Andina programming. Firstly, Papa Andina did not include any specific gender component when it was designed and implemented (Hor-ton et al., 2011). Later, it was recognized that gender responsive innovation processes should have been included (Devaux et al., 2011). For example, in the early years, R&D centred on improving production technology in the con-ventional way. After frustrating results due to marketing problems, Papa Andina began to search for new ways to engage small farmers in market chains. Towards the end of the program, Papa Andina became more responsive to resource-poor women acknowledging gender dynamics within some coalitions and gender gaps within innovation processes, particularly in terms of gendered access to and control over resources and assets for native potato value chains (Sarapura, 2013). Currently as of 2016, CIP and the CRP on Roots, Tubers and Bananas are aware of the need to integrate gender in R&D while exploring opportuni-ties for gender transformative innovation.[8]

In the following section of this paper, we will have a closer look at how Papa Andina achieved some degree of gender transformation by overcoming gender norms which acted as social constraints on innovation. Gender impact of agricultural innovation is less well understood at the level of individuals whose own sense of self is often challenged by their participation.

Tangible contributions from Papa Andina towards achieving gender transformative change in innovation processes and systems

Papa Andina made use of a strategy that developed the livelihoods of the poor and excluded women and men from peasant communities in the Andes through innovation processes. The participatory and applied methodologies and tools fostered institutional development and change through collective work, communication and mutual learning from a diverse group of stakeholders which helped to probe why and how peasant producers were required to be part of the coalitions. Without a specific gender strategy, they were able to shed light on female producers' knowledge and biodiversity, a role that male farmers widely acknowledged. To some extent, therefore, Papa Andina supported peasant female producers to be part of new processes of institutional, organizational, commercial, technological and social innovation and these outcomes were only first fully documented by Sarapura (2013). The relationship among different stakeholders created new social relations, attitudes and opportunities for peasant women and men such as recognizing the intrinsic values of nurturing and protection within germplasm management (women's reproductive role in agricultural R&D) and supporting the marginalized in Peruvian society. This became evident in unprecedented representation and visibility in national culinary fairs, cultural events and also in political discourse that identifies aspects of gender differences and inequalities at the community level and in the wider social organization of gender in the highlands (Sarapura, 2013).

Sarapura (2013) documents how women leaders supported other women (single and more vulnerable) to be part of the Papa Andina and to maintain this role within their communities. Mobility and participation in the decision-making processes allowed women and men to gain self-confidence and respect inside the Papa Andina project, their communities and own households; they became knowledgeable persons. While some sets of common values, beliefs, language and practices were agreed upon by the stakeholders (Miron et al., 2004), there were other less obvious or 'hidden' beliefs that Papa Andina brought to light. One finding reported by Sarapura (2013) was that male leaders initially expressed resentment and tension when women and poor men were chosen to represent the group in any event. Male hegemonic ideas were expressed in the group. For example, men leaders argued that women or poor men were not prepared to do these tasks because of their low level of literacy and incongruous clothing. Such expressions of power influenced gender relations and roles, negatively affecting the implementation and development of activities.

Organizational and collective skills were *the* participatory and applied meth-odologies and tools that fostered institutional development and change through collective work, communication and mutual learning from a diverse group of stakeholders, which helped to probe why and how peasant producers were required to be part of the coalitions promoted and monitored in Papa Andina. CIP facilitated the institutional capacity to identify, accommodate and facilitate innovation processes endorsing the work of strategic partners like FOVIDA, *Fondo de Vida* (an NGO in Peru), to build collective and individual capacities that strengthened social inclusion and participation in the agricultural innova-tion system. FOVIDA facilitated local training including group management; internal savings and lending; basic business skills; and the ability to access, adapt, share and apply new knowledge and technologies to manage resources. Women as well as men joined networking events, profit-sharing arrangements and other collaborations between private- and public-sector partnerships (Sarapura, 2013). The FOVIDA collaboration also supported quantitative, qualitative and process-related analyses of the innovation system when carrying out horizontal evaluations. FOVIDA's progressive approach to gender equality in its training opened up an opportunity. Unfortunately, as the subsequent discussion will allude to, Papa Andina did not generate gender analyses of the innovation sys-tems or connect to institutional changes.

Using Papa Andina to envision a possible approach for gender transformative agricultural innovation

Innovation capacity entails more than technological interventions or options or the expertise and information within research organizations that are required to produce those (Klerkx et al., 2013). The capacity for innovation also includes the processes through which research-based knowledge and context-specific knowledge are combined for the development of solutions that actually work in a specific context. This suggests that learning capacity (individual and organi-zational) is closely linked to innovation (Alegre and Chiva, 2008). The learning capacity grew within the coalitions of Papa Andina (Thiele et al., 2011). Less evident was the engagement from key stakeholders from public and private organizations such as financing entities, academia and research institutes which would offer more critical perspectives (or counterfactuals) against which the status quo of Papa Andina could be considered. Specifically, these critical view-points would likely come from academia and social advocacy groups. The lack of connection with these groups of stakeholders may also have limited the pro-duction of interdisciplinary research for policy and institutional change at the system level (i.e. social research that might inform technology adoption, adult education, legal rights awareness, etc.). This implicates a steady integrated flow of gender research that would also make use of sex-disaggregated data gener-ated from R&D. In the future, such an approach could allow for the emergence, coexistence and evaluation of diverse ways of creating, accumulating and utiliz-ing social and legal knowledge for agricultural innovation (Biggs, 2007).

In this respect the leak of gender data and analysis reduces the 'institutional leverage' of CGIAR centres which constitutes a strong tool to influence its own and other global partnerships for research, development, organizational culture and policy dialogues. It can bridge system efforts that ensure fundamental human rights (which are not basic needs but strategic and legally enforceable rights) for all people and communities. The benefit of working with different groups of people and contexts places the CGIAR in a global leadership position to foster institutional and organizational practices and performances for transformative change in agricultural innovation systems. Accountability for this leadership role is possible, including using tools such as 'gender audits' to track, on an ongoing basis, technical and institutional gender outcomes and impact. With the example of FOVIDA in mind, involvement of contributing partners is important. If Papa Andina had commissioned such ongoing research or gender audits, it might have captured the achievements discussed in the first scenario as well as identified future action agenda within and across the coalitions.

Conducting gender assessments with local advisory teams, and ideally before the design and implementation of the program, would have provided the management of Papa Andina with an opportunity to include and conduct more effective data collection and critical gender analysis. Such assessments would likely identify other relevant stakeholders from diverse sectors (e.g. health, social services) to support further social equity and gender responsiveness within Papa Andina. Sharing the vision, knowledge and ideas of social inclusion and gender equality within innovation processes implicate deeply rooted cultural meanings, attitudes and behavioural changes. By opening gender issues up from all sectors and sides for discussion and action, the load is shared. In this respect, social change rests not only on the shoulders of agricultural scientists but on all partners, organizations and communities. Such an approach would have ensured that Papa Andina, by leading and demonstrating changes in the way applied research created space for reflection and analysis of value chains, also connected the technical with social concepts such as empowerment. It is this lack of critical reflection on unjust structures as well as a lack of effort to finding alternatives to dominant power relationships and social norms that sustain social and gender inequality, poverty and exclusion (Kantor, 2013).

While Papa Andina contributed positively to capacity development, there was a missed opportunity for communication strategies that could enable peasant women and men with little or no education to engage in multi-stakeholder partnerships. For most of the Papa Andina producers, it was the first time they had to deal with external stakeholders (Sarapura, 2013). They were not prepared to talk in public, analyze and grasp even basic technical and institutional information. Their individual sense of self was tremendously important to whether they spoke out or remained silent, and how peasant women and men communicated with stakeholders from outside their communities. Again, contributions from local or national universities, social advocacy groups and other relevant civil rights stakeholders could have been engaged to consider how to successfully deal with such situations. Also, the inclusion of potential stakeholders to support resource-poor women and men over the long-term, such as social workers, teachers or church leaders, would have benefited Papa Andina's long-term impact.

The inclusion of 'one's sense of self' within the gender discourse is overdue. It will require institutions around the world to look inside their organizations and support changes in organization culture to create more critical partnerships and programs (United Nations, 2013). Without reaching to the individual level of agricultural innovation, translating gender into research, action and practice are solely managerial responsibilities. Papa Andina informs us that the capacities of the implementing staff and key partners to address gender and social transformative issues cannot be neglected, or else important impact pathways are missed. Challenging expressions of elitism and gender norms of dominance and machismo, developing and monitoring R&D ethics protocols or taking action to ensure the property rights of peasant producers from commercialized research, are examples of strategies that were potentially within the grasp of Papa Andina but not realized in any systematic manner. Papa Andina was unable to open up normative ethical or gender practices to be challenged because to do so would shake the status quo of the program, but without doing so inhibited learning and implementation of gender transformative processes.

Conclusion

In this chapter, an agricultural innovation initiative from the CGIAR, the Papa Andina Initiative, is used to illustrate the relevance of social inclusion within agricultural innovation and what must be done to achieve transformative gender relations and empowerment of resource-poor women and men. The CGIAR managerial approach to integrate gender in its research for development has proven resistant to change if recent Independent Evaluation Arrangement (IEA) analysis is considered. Further analysis expected in 2016 will look at system-level changes that might be possible if more strategically R&D processes and technologies dealing with institutional change to foster social transformation or systemic change are being applied. Analysis of the Papa Andina case study of agricultural innovation also identifies the influence of social norms and why gender matters to agricultural systems. Papa Andina contributed greatly to pro-poor inclusion by expanding visibility, representation and participation of peasant producers from the Andes. By taking a multi-actor approach, they supported women and poor people to be part of institutional, organizational and technological innovations as they developed or strengthened individual and collective capacities to relate to other stakeholders and add value to their production of native potato. Nevertheless, values, norms, culture, practices and beliefs associated with gender roles and relations were never fully analyzed or addressed in Papa Andina. It is never too late to consider how gender transformative approaches can create positive changes in the way poor and vulnerable groups, especially women, invest in, produce and market their products;

CIP and the CRPs have opportunities in this regard. As Sarapura Escobar and Puskur (2014) have summarized, this requires 'transformation in hearts and minds' which needs men and women in management and staff at different levels (scientific and non-scientific) to engage in practical and critical reflexive dialogue to internalize gender in their lives and work performance. The future of gender and socially inclusive agricultural systems is still largely unknown but over the past decades it has become clear – without attention to gender, especially transformative approaches, major development goals and the effectiveness of agricultural R&D will be hampered, not only in terms of project and program management, but for moving forwards on strategic goals of human rights and sustainable development.

Notes

1   For brevity, hereafter referred to as Papa Andina.
2   By definition gender is both dynamic and contextual. Gender analysis examines socially constructed gender roles and relations (e.g. both and between men and women and intersecting with identities of race, ethnicity, age, class, sexuality, etc.) as well as the ways that technology and institutions exercise power and control over the distribution of resources and benefits within societies. Empirical analysis that examines one case over time may be limited contextually, but, as we argue here, be a useful way to identify opportunities for gender transformative innovation.
3   This statement is premature since the Independent Evaluation Arrangement (IEA) will conduct an assessment of gender in 2016 (see: http://iea.cgiar.org/news/iea-2016-work-plan). Yet, this would be within the line of expectations given recent IEA comparative results for five CRPs (Independant Evaluation Arrangement, 2015).
4   Silvia Sarapura's methodology was heavily based on interpersonal communication and included in-depth interviews, participatory research techniques and video documentaries with peasant producers of native potato in the Junín and Huancavelica regions of Peru as well as various key informants participating in multi-stakeholder knowledge mobilization and added-value market chains supported by Papa Andina (Sarapura 2012, 2013).
5   Conceptually, this implicates a systems approach that links the macro (enabling environment), meso (e.g. peasant communities or Papa Andina areas of work) with the micro (household) level but includes also the individual (self) level of agricultural innovation systems. Windows of opportunities and constraints are identified in the innovation system which we see as embedded in a macro level (the enabling environment and the culture), the meso level (organizations and institutions – local performances), which can be expected to have impacts on the micro and/or household levels, and the individual level (the personal level). Changes in the household in favour of gender justice have a direct impact on the meso and macro levels since individuals influence organizations and the overall institutional and regulatory environment. Changes at the macro level have impacts on organizations and institutions at all levels and these have an influence on the household dynamics at the personal or individual level. And finally, changes in the individual level help to identify major constraints and opportunities that individual female producers face and their implications for the household, the meso and macro levels. By accounting for transformative change across scales, level of interactions and actors' diversity, the nuanced understandings of system needs and actions are identified. Sustainability of action is also considered in relation not only to the biophysical, natural and ecological system, services and technical changes, but also to the opportunity and choice of decision-making processes and the aspects of equity within these systems (Kemp and Parto, 2005). We have

to consider that people and their systems need to be socially structured in processes of change in which innovation is a necessary element.

6 Cordoba et al. (2014) provides a review of the discussion of the technical-institutional experience of Papa Andina in Bolivia.

7 For a review of the technical and institutional achievements of Papa Andina see Devaux et al. (2011). And for the external evaluation see Bebbington and Rotondo (2010).

8 See http://cipotato.org/gender/ and http://www.rtb.cgiar.org/gender/

References

Alegre, J. and Chiva, R. (2008) 'Assessing the impact of organizational learning capability on product innovation performance: an empirical test', *Technovation* 28(6), pp. 315–326.

Ashby, J. and Sperling, L. (1995) 'Institutionalizing participatory client driven research and technology development in agriculture', *Development and Change* 26(4), pp. 753–770.

Bebbington, A. and Rotondo, E. (2010) *Informe de la evaluacion externa de la fase 3 de Papa Andina*, Papa Andina, CIP, Lima, Perú.

Biggs, S. (2007) 'Building on the positive: an actor innovation systems approach to finding and promoting pro-poor natural resources institutional and technical innovations', *International Journal of Agricultural Resources, Governance and Ecology (IJARGE)*, 6(2), pp. 144–164.

Brush, E.G., Merrill-Sands, D., Gapasin, D.P. and Mabesa, V.L. (1995) 'Women Scientists and Managers in Agricultural Research in the Philippines', *ISNAR Research Report No. 7*. The Hague: International Service for Agricultural Research.

Córdoba, D., Jansen, K. and González, C. (2014) 'The malleability of participation: the politics of agricultural research under neoliberalism in Bolivia', *Development and Change*, 45, pp. 1284–1309.

Devaux, A., Horton D., Velasco, C. Thiele, G., Lopez, G., Bernet, T., Reinoso, I., and M. Ordinola (2009) 'Collective action for market chain innovation in the Andes', *Food Policy* 34, pp. 31–38.

Devaux, A., Andrade-Piedra, J., Horton, D., Ordinola, M., Thiele, G., Thomann, A., and Velasco, C. (2010) Brokering innovation for sustainable development: The Papa Andina Case. ILAC Working Paper 12. Institutional Learning and Change Initiative, Rome.

Devaux, A., Andrade Piedra, J., Horton, D., Ordinola, M., Thiele, G., Thomann, A. and Velasco, C. (2011) 'Brokering Innovation for Sustainable Development: The Papa Andina Case', in Devaux, A., Ordinola, M. and Horton, D. (eds) *Innovation for Development: The Papa Andina Experience*, International Potato Centre, Lima, Peru, pp. 76–110.

Harding, S. (1991) 'Feminism Confronts the Sciences: Reform and Transformation', in *Whose Science? Whose Knowledge? Thinking from Women's Lives. Cap. 2*, Cornell University Press, Ithaca, NY, pp. 105–163.

Horton, D., E. Rotondo, R. Paz, G. López, R. Oros, C. Vesco, F. Rodriguez, G. Hareau, and Thiele, G. (2011) The participatory market chain approach in the Andean Change Alliance: Implementation and results in 4 cases. International Potato Center (CIP), Lima, Peru. Social Sciences Working Paper No. 2011-1.

Independent Evaluation Arrangement (2015) *IEA Workplan and Budget for 2016 and Report on 2015 Activities*, CGIAR, Italy.

Joshi, J., Wilde, V. and Masters, A. (2000) 'Gender and diversity in the CGIAR: a new baseline', *Working Paper 25*, Gender and Diversity Program, CGIAR.

Kantor, P. (2013) 'Transforming gender relations: a key to lasting positive agricultural development outcomes', CGIAR Research Program on Aquatic Agricultural Systems, *Brief: AAS-2013–12*, Penang, Malaysia.

Kauck, D., Paruzzolo, S. and Schulte, J. (2010) *CGIAR Gender Scoping Study*, International Center for Research on Women, Washington, DC.

Kemp, R., Parto, S. and Gibson, R. (2005) 'Governance for sustainable development: moving from theory to practice', *International Journal of Sustainable Development* 8(1–2), pp. 13–30.

Klerkx, L. and Aarts, N. (2013) 'The interaction of multiple champions in orchestrating innovation networks: conflicts and complementarities', *Technovation* 33(6–7), pp. 193–210.

Meinzen-Dick, R., Quisumbing, A., Behrman, J., Biermayr-Jenzano, P., Wilde, V., Noordeloos, M., Ragasa, C. and Beintema, N. (2010) 'Engendering agricultural research', International Food Policy Research Institute (IFPRI), *IFPRI Discussion Paper 00973*, Washington, DC, 76 p.

Merrill-Sands, D., Fletcher, J.K., Starks Acosta, A., Andrews, N. and Harvey, M. (1999) 'Engendering organizational change: a case study of strengthening gender equity and organizational effectiveness in an international agricultural research institute', *CGIAR Gender Program – Working Paper, No 21*, CGIAR Secretariat World Bank, Washington, DC.

Miron, E., Erez, M. and Naveh, E. (2004) 'Do personal characteristics and cultural values that promote innovation, quality, and efficiency complete or complement each other?', *Journal of Organizational Behavior*, 25, pp. 175–199.

Njuki, J., Parkins, J. and Kaler, A. (eds) (2016) *Transforming Gender and Food Security in the Global South*, Taylor and Francis and IDRC, Canada.

Paris, T.R. and Rola-Rubzen, F. (2009) *Impact of Migration and Off-Farm Employment on Roles of Women and Appropriate Technologies in Asian and Australian Mixed Farming Systems*, ACIAR, Canberra, Australia.

Prain, G., Hambly, H., Jones, M., Leppan, W. and Navarro, L. (2000) CGIAR Program on Participatory Research and Gender Analysis, *Internally Commissioned External Review*, CGIAR.

Quisumbing, A.R. (Ed) (2003) *Household Decisions, Gender, and Development a Synthesis of Recent Research*, International Food Policy Research Institute, Washington, DC.

Quisumbing, A. and McClafferty, B. (2006) *Using Gender Research in Development: Food Security in Practice*, International Food Policy Research Institute, Washington, DC.

Sarapura, S. (2012), 'Gender and Investment and Assessment Strategies – the Case of Papa Andina, Peru, Module 1- Innovative Activity Profile 6', *The Agricultural Innovation Systems Sourcebook*, World Bank, Washington, DC, pp. 680.

Sarapura, S. (2013) Gender and Agricultural Innovation in Peasant Production of Native Potatoes in the Central Andes of Peru, PhD Thesis, University of Guelph, Canada, pp. 351.

Sims Feldstein, H. (1995) 'Gender analysis in the CGIAR: achievements, constraints and a framework for future action', CGIAR Gender Program.

Sperling, L., Ashby, J.A., Smith, M.E., Weltzien, E. and McGuire, S. (2001) 'A framework for analysing participatory plant breeding approaches and results', *Euphytica* 122, pp. 439–450.

Tegbaru, A., Fitzsimons, J., Gurung, B. and Hambly Odame, H. (2010) 'Change in gender relations: managerial and transformative approaches of gender mainstreaming in agriculture', *Journal of Food, Agriculture & Environment*, 8(3&4), pp. 1024–1032.

Tegbaru, A., FitzSimons, J., Kirscht, H. and Hillbur, P. (2015) 'Resolving the gender empowerment equation in agricultural research: a systems approach', *Journal of Food, Agriculture and Environment* 13(3&4), pp. 131–139.

Thiele, G. and Devaux, A. (2011) 'Adding Value to Local Knowledge and Biodiversity of Andean Potato Farmers: The Papa Andina Project', in Devaux, A., Ordinola, M. and Horton, D. (eds) *Innovation for Development: The Papa Andina Experience*, International Potato Centre, Lima, Peru, pp. 58–60.

United Nations (2013) *The Millennium Development Goals Report*, New York.

Part IV

Systems and institutional innovation

22 What kinds of 'systems' are we dealing with?

Implications for systems research and scaling

Cees Leeuwis and Seerp Wigboldus

Introduction

In Humidtropics and other development-oriented Consultative Group on International Agricultural Research (CGIAR) programmes, researchers speak a lot about 'systems' and 'systems research', and they distinguish this from other kinds of research such as 'commodity research'. Systems research is typically legitimized with reference to the fact that smallholders do not just grow one crop but rather integrate a range of crops and livestock in their farming system, and that their livelihood also depends on non-farm activities. Moreover, proponents of systems research emphasize that a holistic perspective is required to do justice to enabling and constraining conditions in the broader environment, and to possible trade-offs between production and sustainability objectives. It is argued that understanding the contextual interdependencies in the system is essential for generating development outcomes. As evidenced by the wide range of terminologies available (e.g. cropping systems, farming systems, agricultural systems, innovations systems), systems researchers do not necessarily have a shared understanding of the kinds of systems we are dealing with. To complicate things further, there exist different scientific traditions of 'systems thinking'. Consequently, even when and if researchers talk about systems with similar entities and boundaries, they may subscribe to different views on and conceptualizations of how such systems function, how they change and how they may be influenced through interventions, including research interventions.

This chapter aims to provide some clarity in this discussion. We argue that we may overcome the different systems definitions by recognizing that we are dealing with intertwined configurations in which bio-material relations are mediated by social and symbolic phenomena, and vice versa. Subsequently, we discuss how such configurations may change, and what roles research may play in this when taking into account some special features of intertwined configurations that complicate deliberate attempts to foster change. It is argued that, in order to help realize public values, research needs to become a mechanism which informs adaptation to ever changing conditions, and leverage the emergence of novel bio-material, social and symbolic opportunities. This requires a considerable change in the way research processes are predominantly organized and embedded in society.

Looking at systems as intertwined configurations of the bio-material and social phenomena

In agriculture, systems research has a long tradition (Darnhofer et al., 2012). It has been recognized for an equally long time that the boundaries of systems can and need to be drawn differently depending on the issue at hand. This has led several researchers (e.g. Fresco, 1986) to talk about agriculture as a nested hierarchy of systems (Figure 22.1).

While it was recognized from the outset that people play an important role in these kinds of systems, both the labelling of systems and the kinds of research and modelling conducted under the banner of 'systems research' in the 1980s and 1990s reflect a perspective on systems that emphasizes bio-material phenomena. Since then, new kinds of system approaches have emerged that aimed to make the social dimensions of systems more explicit; examples are concepts like 'livelihood systems' (Van Ginkel et al., 2013; Sinclair, this volume), 'socio-technical systems' (Geels, 2002) and 'social-ecological systems' (McGinnis and Ostrom, 2014). There is also a long tradition of systems thinking under the banner of 'agricultural knowledge and information systems' (Röling and Engel, 1990) and 'innovation systems' (Hall, 2005, 2006) which depicts the interactions within a network of actors that are considered to be critical in bringing about system change (for an overview, see Klerkx et al., 2012). When visiting conferences that present agricultural systems research, one can note that presentations typically centre around one of the categories mentioned above and/ or in Figure 22.1, and also that a large variety of systems terminology is used. Hence, it is clear that individual researchers interpret and define boundaries of systems differently.

Perhaps more importantly, when considering that systems researchers often aim to contribute to change and development, it has been documented (Checkland, 1981) that systems researchers may subscribe to different ontological and epistemological ideas regarding the nature of systems, how we can generate knowledge about them and subsequently how people can work towards change in systems and system change. In other words, there may be a double confusion when researchers not only talk about different systems, but also think differently about systems and how they change. In the context of this paper it goes too far to elaborate the many different strands of systems thinking that have emerged over time. Some key features and sources are provided in Table 22.1 (Leeuwis, 2004: 295–301 for an extended summary).

Clearly, the above indicated diversity can (and does) easily lead to a Babylonic confusion of speech among communities of researchers with different backgrounds, all operating under the banner of systems research. Moreover, it reflects that 'systems' are to a considerable extent a human construct; scientists choose to call something a system in order to make sense of a complex whole, but differ in their purpose and conceptualization of entities, boundaries and key processes (Checkland, 1981). The solution to this is not to ensure that researchers start to think alike and study systems in the same way. This is not only unrealistic but also undesirable, as different points of entry, perspectives

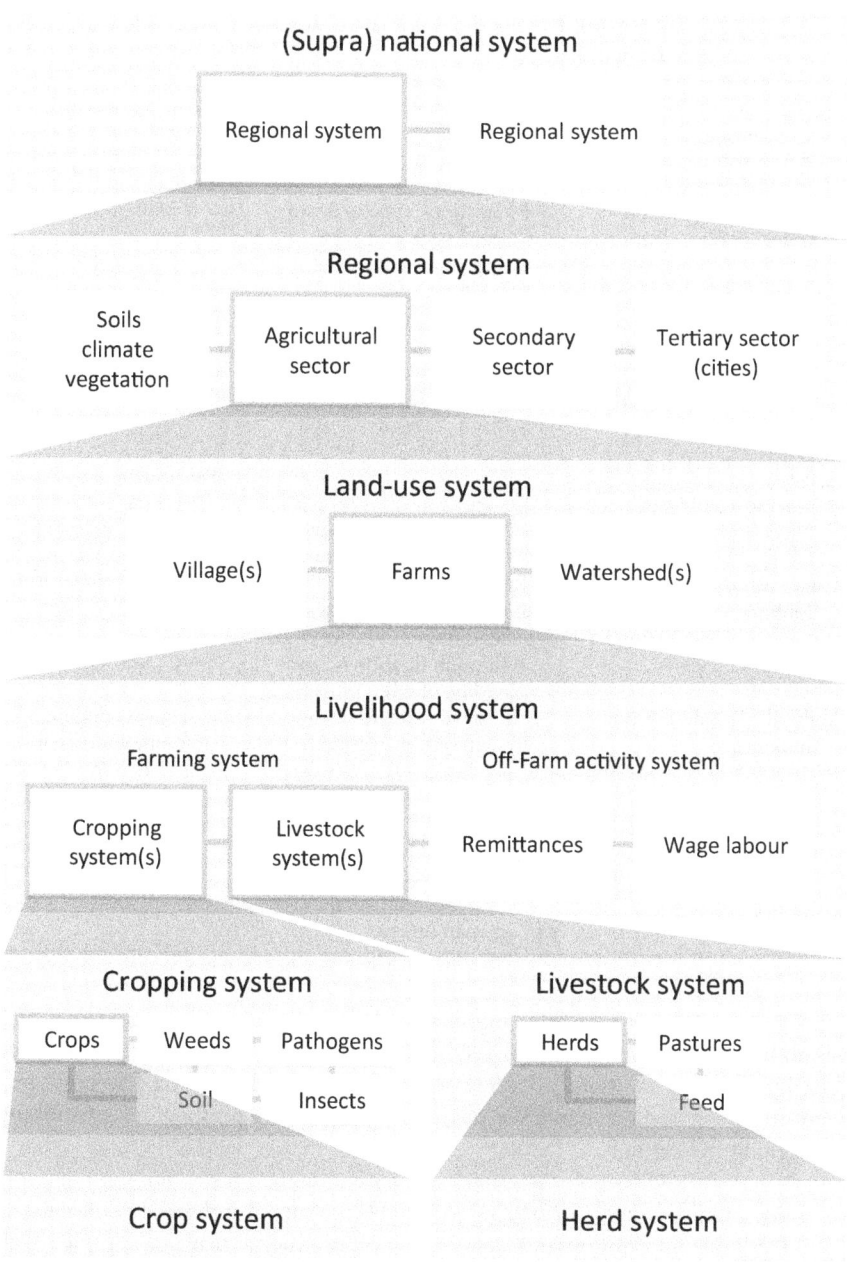

Figure 22.1 Agriculture as a hierarchy of systems

Source: Modified from Fresco, 1986: 47; Hart and Pinchinat, 1982; Odum, 1983.

Table 22.1 Different ways of thinking about natural and social systems, largely independent of system boundary

Type of systems thinking (origin and/or literature sources)	Key metaphor and assumption depicting how systems are seen	Key change strategy implied
Hard system thinking (scientific management, Taylor, 1947)	Machines Interactions in natural and social systems can be known and predicted	Engineer and optimize towards a given goal
Functionalist systems thinking (human relations management, Roethlisberger and Dickson, 1961; structural functionalism, Parsons, 1951)	Organisms Systems are functional wholes, depending on relations between components and environment	Re-balance and adapt in a changing environment
Soft systems thinking (Churchman, 1979; Checkland, 1981)	Meanings Systems consist of people with different worldviews and boundary definitions	Foster dialogue, learning and agreement among actors
Cognitive/autopoietic systems thinking (Luhmann, 1984; Maturana and Varela, 1984)	Psychic prisons Biological and social systems tend to perceive the world through their own logic and be blind to others	Shock therapy by creating a crisis
Political/critical systems thinking (Jackson, 1985; Ulrich, 1988)	Arenas of struggle Systems are characterized by power structures that constrain system change	Coalition building, competition and negotiation
Social/institutional systems thinking (Giddens, 1984; North, 1990)	Rules Formal and informal rules are produced and reproduced in interaction, resulting in certain orders	Change rules and incentive structures

and viewpoints may usefully highlight relevant aspects and dimensions of a complex whole. However, to overcome the different definitions and modes of thinking, it may be useful to agree at a more abstract level on a few basic features of the systems at hand. We propose the following perspective for looking at agricultural systems:

1   Agricultural systems involve interrelated bio-material and social phenomena;
2   Desirable and undesirable system outcomes emerge from (a complex web of) interactions between bio-material and social phenomena at and from different levels;
3   Interactions between bio-material phenomena are often mediated by social phenomena and human practices. For example, whether availability of nutrients from fertilizers leads to higher yield depends on the quality and quantity of human labour invested (Figure 22.2).

4   Similarly, interactions between social phenomena can be mediated by bio-material phenomena. For example: whether or not there is tension between stakeholders upstream and downstream in a water catchment may depend on the technology available for water distribution.

And importantly for scientists:

5   Human understanding and perception (i.e. knowledge) of interactions in the system may provide feedback on the system itself. That is, changes in human thinking and understanding may lead to changes in individual and collective behaviour that affect system dynamics and outcomes.

The last point highlights the importance of how individuals and collectives perceive, think and talk about the system, which we label the 'symbolic' dimension (Leeuwis, 2013). From this perspective, science (including systems research) can be seen as a feedback mechanism that may (or may not) influence the dynamics of the system.

In all, we propose to recognize that we are fundamentally dealing with intertwined configurations in which bio-material relations are mediated by social and symbolic phenomena, and vice versa.

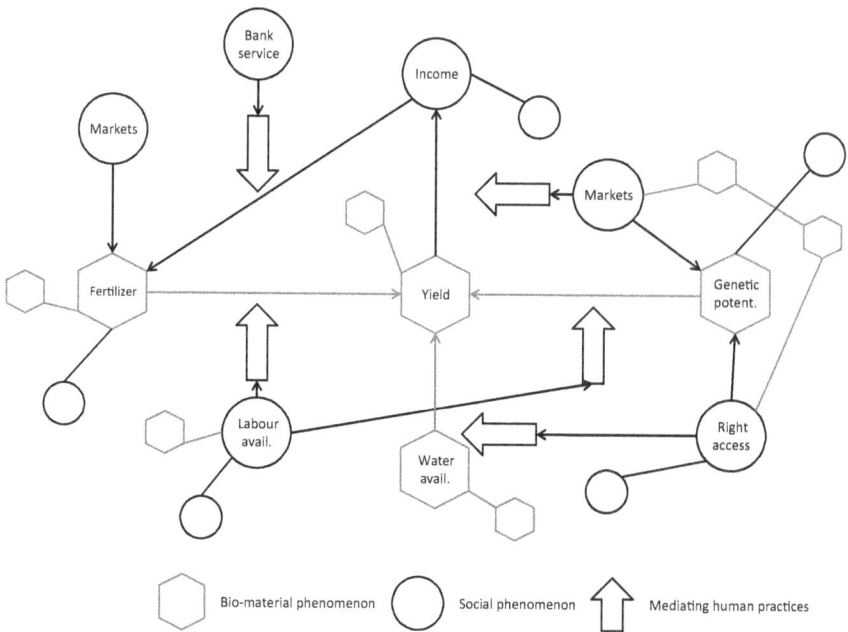

Figure 22.2 Image of the intertwining of bio-material phenomena and social phenomena: how bio-material relations are mediated and shaped by social phenomena and human practices

Change and scaling in intertwined configurations

When we recognize that the systems we deal with are complex configurations, it becomes relevant to ask how such configurations may change, and what roles research may play in this. In relation to these questions it is relevant to be aware of some special features of intertwined configurations that complicate deliberate attempts to foster change.

First of all, complex configurations do not have a central locus from which a system is or can be steered and controlled. Numerous actors and organizations pursue their own goals and projects, and the resulting changes are largely the unintended outcome of numerous intentional actions which interact and interfere with each other in complex ways (Scharpf, 1978; Castells, 2004; Van Woerkum et al., 2011). This is called 'self-organization': the emergence of new patterns and orders without central steering and control (Nicolis, 1989; Leeuwis and Aarts, 2011).

Secondly, it is clear that actors (e.g. farmers) in a configuration often cannot change if others (e.g. stakeholders in a value chain) do not simultaneously change. For example:

- Promising technical strategies for enhancing soil fertility may not be used by farmers unless land-tenure rules and litigation systems are adapted in such a way that it becomes attractive for people to invest in soil fertility (Adjei-Nsiah et al., 2004).
- Improved varieties may not be used unless new input and output market channels are set up, and unless rules of water distribution in irrigation systems or catchments are re-negotiated.

In other words, meaningful change happens in networks of interdependent actors. This is often insufficiently recognized in the agricultural innovation and behaviour change literature, which tends to present 'adoption of innovation' as a largely individual affair (Ajzen and Madden, 1986; Rogers, 1995), and ignores that any change in practice tends to require a range of other changes in practice at the level of a farm and beyond (Van der Ploeg, 1990; Leeuwis, 2004). Innovation thus involves numerous simultaneous changes, and often cannot be usefully reduced to a single practice. In the past, the social and institutional practices and infrastructures that are required to 'make technology work' were seen as 'external conditions', but from a perspective of systems and system innovation it is clear that these should be seen as part and parcel of the innovation challenge (Geels, 2002; Leeuwis, 2004, 2013).

A third complication for bringing about change in systems is that interdependent stakeholders often have diverging interests, and frequently do not agree with each other on the desirable way forward. This makes it difficult to move ahead in a concerted manner. Reaching agreement on a desirable future is further complicated by the fact that stakeholders often face uncertainty about what the immediate and longer term consequences of certain courses of action are, as well as uncertainty about how others on whom they depend will respond.

In connection with the existence of different interests and values, we also see that different stakeholders initiate and promote different technical and socio-institutional solutions. In many ways these options can be seen to 'compete' with each other in a broader and dynamic selection environment, whereby the success of different solutions depends not only on their effectiveness in a technical sense, but also on the relative strength of the coalition that advocates of particular technical and socio-institutional solutions are able to forge (Geels, 2002; Leeuwis, 2013). The aspect of coalition formation is largely overlooked in the classical thinking and practice of extension and technology transfer (Leeuwis, 2004).

What becomes clear from this discussion is that meaningful change in systems happens in configurations, and includes multiple changes in practice (both technical and social) at different levels and in different spheres. As we have seen, technical changes at farm level may well be contingent on the achievement of changes in labour organization, land-tenure policy, market organization and credit provision. This may be obvious, but is still not reflected in dominant ways of thinking in international agricultural research establishments. Agricultural innovation is still often associated primarily with technical innovation, rather than with the institutional innovations (e.g. changes in markets, policy, legal frameworks) that are necessary to enable people across the value chain to make use of new ideas and technical opportunities (Hounkonnou et al., 2012; Leeuwis, 2013). Similarly, a notion like 'scaling' is still often associated with the spreading of a particular technology or practice, rather than with the simultaneous upscaling and downscaling of multiple variables across spheres and levels, with cross-scale effects that are hard to predict and anticipate (Figure 22.3). Yet, it is clear that there exist critical interdependencies and feedbacks

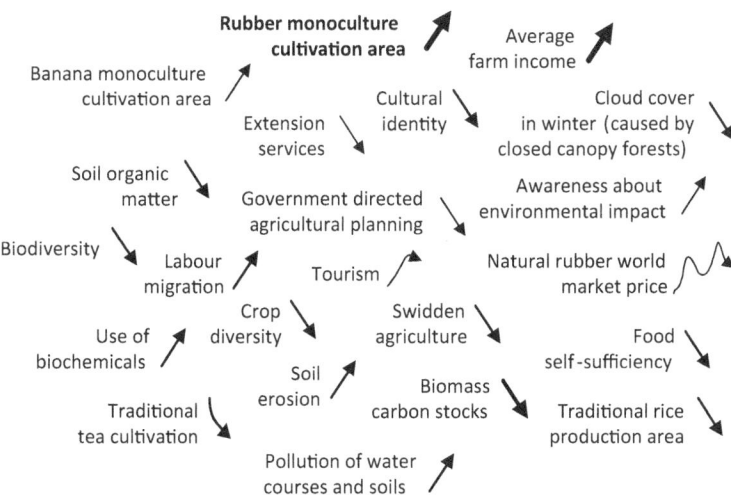

Figure 22.3　Scaling seen as the simultaneous upscaling and downscaling of multiple variables across spheres and levels (example of scaling dynamics related to and relevant for rubber cultivation in Yunnan, China)

between variables and levels, and that these eventually determine whether a particular practice 'goes to scale' and also whether this contributes to realizing development objectives. Arguably, navigating such interdependencies is the key function of multi-stakeholder collaboration (e.g. in the form of innovation platforms) around research (Kilelu et al., 2013; Boogaard et al., 2013).

Towards research as a leverage for change

So what do these insights about system change imply for agricultural system researchers who are under pressure to contribute to societal impact? We have already seen that research can be regarded as having a feedback role in systems. While playing such a role requires at least some degree of interaction and engagement with societal partners in the system, this role can be organized in different ways and through different types of research, each with specific strengths and weaknesses. We describe three basic types that can of course be combined with each other: (a) research aimed at describing and understanding complexity; (b) research aimed at identifying key leverages, opportunities and constraints; and (c) research aimed at enlarging variation in society. In addition, we reflect on how such research may itself be turned into a leverage.

Understanding the complexity: Research can aim to describe complexity and investigate the dynamic interrelations between different variables and phenomena (social and technical) at different levels. Such research can lead to advanced insight in diversity of farming patterns (see e.g. Tittonell et al., 2005, 2009) and may also contribute to insight in trade-offs, in how the possible scaling of one phenomenon is likely to relate to the scaling of other phenomena. A significant amount of systems research is of this type. A problem with this kind of research is that it is notoriously difficult, data intensive and time consuming, and thus prone to the critique that yielding sophisticated descriptions and understandings of systems at high costs does not necessarily induce change in line with development objectives. Moreover, it seems to be based on an implicit assumption that it is possible to describe and model complexity in a way that is externally valid and that such understanding is necessary for the formulation and tailoring of interventions to different contexts and different categories of farmers. Thus, it reflects a relatively strong belief in our capacity to steer and control systems rationally, which is rather questionable in view of the complexities discussed in the previous section.

Identifying key areas of leverage: Research can be directed towards identifying possible or likely leverages for change in systems (Meadows, 2009). This may, for example, take the form of the identification of emerging windows of opportunity through the analysis of simultaneously occurring trends at different levels and in different spheres that may together result in new opportunities (or barriers) for change (see Box 22.1 for an example).

Box 22.1 Identifying windows of opportunity

In the beginning of the 21st century we witness a number of relatively independent developments: (1) rapid population growth in China; (2) increased presence of China in the African continent; (3) improved access to internet in both China and Africa; and (4) the recent signing by China of WTO treaties. In view of these coinciding circumstances, it is not unthinkable that African smallholders might be able to gain access to the Chinese food market. Of course, it would require a network of people to see the opportunity, gain insight in the Chinese food preferences, introduce e.g. soybean production in Africa, find reliable business partners in China, organize transport and permits, and so forth. It may never happen, or it may happen in such a way that smallholders do not benefit. And the wider consequences of it happening are difficult to foresee: will it undermine local food security in Africa, or boast local economies and food production? But the least one can say is that the window of opportunity for accessing the Chinese market is probably enlarged when compared with the previous decades.

Source: Leeuwis, 2013: 23.

The strength of such research is that it connects to and capitalizes on already occurring dynamics of scaling. Alternatively, research could focus on identifying key barriers for change at the level of technology, individuals, organizations or institutions in relation to already identified 'best bet solutions'. Such research can inform the formulation of a range of targeted strategies geared towards leveraging change in a system through building upon opportunities and addressing key constraints. One cannot predict in detail whether and in what way such strategies will be effective, but there are certainly examples where the removal of barriers has led to meaningful change (Hounkonnou et al., 2012).

Piloting multiple technical and institutional options in society: Since one cannot predict in complex configurations which (combinations of) technical and social solutions will eventually work and flourish to address problems, it often makes sense to 'put the eggs in more than one basket'. Experimentation in society with multiple options then serves to create the necessary variation and diversity from which societal stakeholders may choose and select. Research can initiate such experiments and monitor intended and unintended consequences in different spheres, and make trade-offs transparent. Similarly, research can identify naturally occurring societal experiments that are already taking place without being labelled as pilots. At this point in time, the agricultural research establishment tends to have a much greater capacity to engage in technical experiments and pilots when compared to socio-institutional experiments, even though there is recognition of the importance of institutional issues (Van Ginkel et al., 2013; Schut et al., 2015).

Turning research into a leverage of its own: When the purpose of research is to have impact in society, it does not make all that much sense to conduct the kinds of research mentioned above in isolation and/or in an extractive mode. While research carried out in isolation may be effectively communicated about after it is completed, and while its relevance may be enhanced by ensuring that researchers have sufficient understanding of local contexts and knowledge before they start, an important opportunity is likely to be lost when researchers operate largely in their own sphere. This relates to the experience that collaborative research (e.g. research that is embedded in innovation platforms) is a potentially powerful instrument to address the typical challenges that occur at the level of configurations: the absence of central steering and control, interdependence, disagreement and uncertainty. In connection with this it is important to acknowledge that it is not only the eventual scientific results of a research effort that may have impact in society, but also the process of research itself. Joint fact finding is known to be a powerful strategy in conflict management since doing something together may improve human relationships and because joint research may foster common understandings and insights that help stakeholders overcome their differences and address uncertainties they experience. This largely happens during the research process and well before results are finished and published (Schut, 2010; Milgroom et al., 2011). Embedding research in a broader collaborative process can enhance the learning capacity of society, and improve through the symbolic route the chances that systems (i.e. configurations of bio-material and social phenomena) become well adapted to their dynamic environment and that interdependent scaling processes become aligned (Klerkx et al., 2010; Kilelu et al., 2013).

Implications for systems research

Our insights on how complex configurations change and may be informed by different kinds of research has implications for the kinds of systems research pursued in agricultural research for development.

Broadening impact pathways and theories of change: Looking at the role of systems research as a feedback mechanism and process that can leverage change (i.e. have impact) in societal configurations while still 'in-the-making' calls into question the idea that one can usefully make a separation between a 'science phase' and an 'impact phase' in agricultural research. This separation is still often made, and also reflected in the 'impact pathways' and 'theories of change' (Douthwaite et al., 2007; Alvarez et al., 2010) that international agricultural researchers are asked to construct by donors to demonstrate their impact orientation. Typically, such impact pathways or theories of change consist of a narrative and visual diagram that stipulates how 'research inputs' lead to 'research outcomes', and how these 'research outcomes' are taken up by 'next-users' and lead to 'development outcomes' on the side of 'end-users' and eventually to 'impact'. Clearly, this way of framing impact pathways has not only linear connotations, but also suggests a rather firm and unrealistic belief in the possibility to steer and control change in complex configurations of bio-material, social and symbolic phenomena. The equally implicit separation between a 'science' and 'impact' phase conceals the potential value of collaborative research in systems. Moreover, the idea that all research must have impact seems to be at odds with the insight (common in innovation studies) that one needs considerable variation (and hence a certain number of 'failures') to see successful solutions emerge. It is equally unrealistic, however, that such planning logics will disappear any time soon. Hence, it is important to be creative, and incorporate process-oriented operationalizations in the 'impact pathway' language. This may be done by widening the array of 'research inputs', 'research outputs' and 'research outcomes' that are considered relevant (Table 22.2).

As it can easily take much longer than the typical project horizon (3 or 4 years) for meaningful system change to occur, such process categories and indicators can also help to capture progress in change trajectories. In essence, they capture important phenomena that are often overlooked or less visible in change trajectories and impact assessments.

Broadening our view of 'systems research': The various modes of research discussed above also invite the question of what counts as 'systems research', and how it might move ahead. As we have seen, research can play several useful roles in supporting change in complex configurations beyond providing an ex-ante understanding of complex interdependencies in a system. While this kind of 'research on systems' does certainly have value, we have seen that a systems perspective may also inspire other forms of 'research in systems', notably the collaborative exploration of leverages and the piloting of (combinations

Table 22.2 Suggestions to widen relevant categories in impact pathways and theories of change in order to capture dynamics and benefits of collaborative research in systems

Additional 'research inputs'	Processes of demand-articulation, institutional experimentation, visioning, network building, mediation
Additional 'research outputs'	New relationships, improved trust, shared visions, enhanced agreement, strengthened coalitions for change, lessons from failures
Additional 'research outcomes'	Societal pressures building up, weakening of dominant socio-technical regimes, shifting discourses in society, institutional change

of) technical and socio-institutional options. These forms of research are less geared to developing an ex-ante understanding of complexity in systems, and more towards using research as a vehicle for testing and evaluation options in society. Since the options tested may in fact be similar to those tested in more conventional (e.g. 'commodity') programmes, the question should be raised whether and when the label 'systems research' remains valid. In our view this can be the case when: (a) the research is deliberately embedded in a network of interdependent stakeholders; (b) when intended and unintended consequences and effects are assessed from multiple perspectives and angles (stakeholder values, disciplines, system parameters, etc.); and (c) when such effects and consequences are evaluated across various societal and ecological levels and spheres, while considering different possible or likely degrees of uptake. The latter because outcomes may be regarded as positive when uptake is still limited in scale, may have to be assessed differently when uptake is massive. Eventually, such informed and holistic assessments may thus provide feedback that contributes to responsible decision-making and change in complex configurations.

In all, we propose that enhancing the leverage and actionability of 'systems research' through the collaborative piloting and systemic evaluation of combined technical and socio-institutional options should be an important component of a vision on systems research in CGIAR programmes. Such an approach is congruent with credible theories on how change in complex configurations happens, and leaves sufficient space for different conceptualizations of systems boundaries and entities. It requires investment in the development of new methodologies for experimentation with socio-institutional options, since this is currently a poorly developed capacity in the agricultural research establishment.

References

Adjei-Nsiah, S., Leeuwis, C., Giller, K.E., Sakyi-Dawson, O., Cobbina, J., Kuyper, T.W., Abekoe, M. and Van Der Werf, W. (2004) 'Land tenure and differential soil fertility management practices among native and migrant farmers in Wenchi, Ghana: implications for interdisciplinary action research', *NJAS – Wageningen Journal of Life Sciences*, vol 52, no 3–4, pp. 331–348.

Ajzen, I. and Madden, J.T. (1986) 'Prediction of goal-directed behavior: attitudes, intentions and perceived behavioral control', *Journal of Experimental Social Psychology*, vol 22, pp. 453–474.

Alvarez, S., Douthwaite, B., Thiele, G., Mackay, R., Cordoba, D. and Tehelen, K. (2010) 'Participatory impact pathways analysis: a practical method for project planning and evaluation', *Development in Practice*, vol 20, no 8, pp. 946–958.

Boogaard, B.K., Schut, M., Klerkx, L., Leeuwis, C., Duncan, A.J. and Cullen, B. (2013) *Critical issues for reflection when designing and implementing research for development in innovation platforms*, Report for the CGIAR Research Program on Integrated Systems for the Humid Tropics, Knowledge, Technology & Innovation Group (KTI), Wageningen University & Research centre, the Netherlands.

Castells, M. (2004) *The Power of Identity: The Information Age: Economy, Society and Culture Vol. II*, Blackwell, Cambridge.

Checkland, P.B. (1981) *Systems Thinking, Systems Practice*, J. Wiley & Sons, Chichester.

Churchman, C.W. (1979) *The Systems Approach and Its Enemies*, Basic Books, New York.

Darnhofer, I., Gibbon, D. and Dedieu, B. (eds.) (2012) *Farming Systems Research into the 21st Century: The New Dynamic*, Springer Science & Business Media, Dordrecht.

Douthwaite, B., Alvarez, B.S., Cook, S., Davies, R., George, P., Howell, J., Mackay, R. and Rubiano, J. (2007) 'Participatory impact pathways analysis: a practical application of program theory in research-for-development', *Canadian Journal of Program Evaluation*, vol 22, no 2, pp. 127–159.

Fresco, L.O. (1986) 'Cassava in shifting cultivation; a systems approach to agricultural development in Africa', PhD thesis, Wageningen University, Wageningen.

Geels, F. (2002), 'Understanding the dynamics of technological transitions. A co-evolutionary and socio-technical analysis', PhD thesis, Twente University Press, Enschede.

Giddens, A. (1984) *The Constitution of Society: Outline of the Theory of Structuration*, Polity Press, Cambridge.

Hall, A. (2005) 'Capacity development for agricultural biotechnology in developing countries: an innovation systems view of what it is and how to develop it', *Journal of International Development*, vol 17, no 5, pp. 611–630.

Hall, A. (2006) 'Public-private partnerships in an agricultural system of innovation: concepts and challenges', *International Journal of Technology Management and Sustainable Development*, vol 5, no 1, pp. 3–20.

Hart, R.D. and Pinchinat, A.M. (1982) 'Integrative agricultural systems research', in Servant, J. and Pinchinat, A. (eds.), *Caribbean Seminar on Farming Systems Research Methodology*. Point-àPitre, Guadeloupe, FWI, May 4-8, 1980, pp. 555–565.

Hounkonnou, D., Kuyper, T.W., Kossou, D., Leeuwis, C., Nederlof, S., Röling, N., Sakyi-Dawson, O., Traoré, M. and van Huis, A. (2012) 'An innovation systems approach to institutional change: Smallholder development in West Africa', *Agricultural Systems*, vol 108, pp. 74–83.

Jackson, M.C. (1985) 'Social systems theory and practice: the need for a critical approach', *International Journal of General Systems*, vol 10, pp. 135–151.

Kilelu, C.W., Klerkx, L. and Leeuwis, C. (2013) 'Unravelling the role of innovation platforms in supporting co-evolution of innovation: contributions and tensions in a smallholder dairy development programme', *Agricultural Systems*, vol 118, pp. 65–77.

Klerkx, L., Aarts, N. and Leeuwis, C. (2010) 'Adaptive management in agricultural innovation systems: the interactions between innovation networks and their environment', *Agricultural Systems*, vol 103, pp. 390–400.

Klerkx, L., van Mierlo, B. and Leeuwis, C. (2012) 'Evolution of systems approaches to agricultural innovation: concepts, analysis and interventions', in Darnhofer, I., Gibbon, D. and Dedieu, B. (eds.), *Farming Systems Research into the 21st Century: The New Dynamic*, pp. 457–483, Springer Science & Business Media, Dordrecht.

Leeuwis, C. (2013) 'Coupled Performance and Change in the Making', Inaugural lecture, Wageningen University, Wageningen.

Leeuwis, C. (with contributions by A. Van den Ban) (2004) *Communication for Rural Innovation. Rethinking Agricultural Extension*, Blackwell Science, Oxford.

Leeuwis, C. and Aarts, N. (2011) 'Rethinking communication in innovation processes: creating space for change in complex systems', *Journal of Agricultural Education and Extension*, vol 17, no 1, pp. 21–36.

Luhmann, N. (1984) *Soziale Systeme: Grundriss einer allgemeinen Theorie*, Suhrkamp Taschenbuch Wissenschaft 666, Suhrkamp, Frankfurt am Main.

Maturana, H.R. and Varela, F.J. (1984) *The Tree of Knowledge: The Biological Roots of Human Understanding*, Shambala, Boston.

McGinnis, M.D. and Ostrom, E. (2014) 'Social-ecological system framework: initial changes and continuing challenges', *Ecology and Society*, vol 19, no 2, p. 30.

Meadows, D.A. (2009) *Thinking in Systems – A Primer*, Edited by Diana Wright, Earthscan, London.

Milgroom, J.M., Leeuwis, C. and Jiggins, J. (2011) 'Limpopo case: the role of research in conflict over natural resources; informing resettlement negotiations in Limpopo National Park, Mozambique', in van Paassen, A., van den Berg, J., Steingröver, E., Werkman, R. and Pedroli, B. (eds) *Knowledge in Action: The Search for Collaborative Research for Sustainable Landscape Development*, Wageningen Academic Publishers (Mansholt publication series 11), Wageningen, pp. 247–276.

Nicolis, G. (1989) *Self-Organised Criticality: Emergent Complex Behaviour in Physical and Biological Systems*, Cambridge University Press, Cambridge.

North, D.C. (1990) *Institutions, Institutional Change and Economic Performance*, Cambridge University Press, Cambridge.

Odum, H.T. (1983) *Unifying Concepts*, John Wiley & Sons, New York.

Parsons, T. (1951) *The Social System*, Free Press, Glencoe.

Roethlisberger, F.J. and Dickson, W.J. (1961) *Management and the Worker*, Harvard University Press, Cambridge, MA.

Rogers, E.M. (1995), *Diffusion of Innovations*, 4th edition, Free Press, New York.

Röling, N.G. and Engel, P.G.H. (1990) 'IT from a knowledge systems perspective: concepts and issues', *Knowledge in Society: The International Journal of Knowledge Transfer*, vol 3, pp. 6–18.

Scharpf, F.W. (1978) 'Interorganizational policy studies: issues, concepts and perspectives', in Hanf, K. and Scharpf, F.W. (eds) *Interorganizational Policy Making: Limits to Coordination and Central Control*, pp. 345–370, Sage, London.

Schut, M., Klerkx, L., Sartas, M., Lamers, D., Campbell, M., Ogbonna, I., Kaushik, P., Atta-Krah, K., Leeuwis, C. (2015) 'Innovation platforms: experiences with their institutional embedding in agricultural research for development', *Experimental Agriculture*, pp. 1–25.

Schut, M., Slingerland, M. and Locke, A. (2010) 'Biofuel developments in Mozambique. Update and analysis of policy, potential and reality', *Energy Policy*, vol 38, no 9, pp. 5151–5165.

Sinclair, F. (2016) 'Systems science at the scale of impact: reconciling bottom up participation with the production of widely applicable research outputs'.

Taylor, F.W. (1947) *Scientific Management*, Harper & Row, New York.

Tittonell, P., Vanlauwe, B., Leffelaar, P.A., Rowe, E.C. and Giller, K.E. (2005) 'Exploring diversity in soil fertility management of smallholder farms in western Kenya I. Heterogeneity at region and farm scale', *Agriculture, Ecosystems and Environment*, vol 110, pp. 149–165.

Tittonell, P., van Wijk, M.T., Herrero, M., Rufino, M.C., de Ridder, N. and Giller, K.E. (2009) 'Beyond resource constraints: exploring the biophysical feasibility of options for the intensification of smallholder crop-livestock systems in Vihiga district, Kenya', *Agricultural Systems*, vol 101, no 1–2, pp. 1–19.

Ulrich, W. (1988) 'Systems thinking, systems practice, and practical philosophy: a program of research', *Systems Practice*, vol 1, pp. 137–163.

Van der Ploeg, J.D. (1990) *Labor, Markets, and Agricultural Production*, Westview Press, Boulder, CO.

Van Ginkel, M., Sayer, J., Sinclair, F., Aw-Hassan, A., Bossio, D., Craufurd, P., El Mourid, M., Haddad, N., Hoisington, D., Johnson, N., Velarde, C.L., Mares, V., Mude, A., Nefzaoui, A., Noble, A., Rao, K.P.C., Serraj, R., Tarawali, S. and Vodouhe, R. (2013) 'An integrated agro-ecosystem and livelihood systems approach for the poor and vulnerable in dry areas', *Food Security*, vol 5, no 6, pp. 751–767.

Van Woerkum, C., Aarts, N. and Van Herzele, A. (2011) 'Changed planning for planned and unplanned change', *Planning Theory*, vol 10, no 2, pp. 144–160.

23 How can external interventions build on local innovations?

Lessons from an assessment of innovation experiences in African smallholder agriculture

Bernard Triomphe, Anne Floquet, Brigid Letty, Geoffrey Kamau, Conny Almekinders and Ann Waters-Bayer

Introduction

The world, and with it the environment smallholders across developing countries live in and depend on, is fast changing as a result of climate change, changing food security, political turmoil, increasing urbanisation and its impact on consumers' diets and demands, globalisation and environmental concerns, among others. All such changes contribute to re-assessing the values, performance and current practices of economic actors and sectors (Malerba, 2007), including those involved in agriculture and rural development. In such a context, creating and maintaining a dynamic agricultural innovation scene at various scales, from local to national to international, seems critical, even though innovation is neither a panacea nor an end in itself. This will allow smallholders and other rural stakeholders to adapt to and, whenever possible, take advantage of the positive (e.g. new market opportunities) or even negative (e.g. degradation of natural resources) changes affecting their natural and socioeconomic environment, to improve their livelihoods and to achieve a better future for themselves and their children.

Over the past two decades, scholars, development professionals and a wide array of organisations have increasingly paid attention to understand better what innovation is all about and to devise ways of nurturing it and bringing it to scale. In doing so, they have shaped the concepts and approaches related to agricultural innovation systems, and started implementing them in practice (World Bank, 2006; Geels and Shot, 2007; Waters-Bayer et al., 2011; Adekunle et al., 2012; Klerkx et al., 2012; World Bank, 2012; FAO, 2014; Touzard et al., 2014; Schut et al., 2015). These studies have revealed that innovation takes place within heterogeneous networks of researchers, farmers, private entrepreneurs, non-governmental organisations (NGOs), government agents and other stakeholders (Hall and Clark, 2010). In such networks, stakeholders interact in a non-linear, iterative and mostly non-predictable fashion to solve pressing problems, adapt to new conditions or take advantage of new opportunities. The focus and outcome of such interactions usually consist of a mix of technical, organisational and institutional innovations developed and refined 'on the go', often quite different from what the innovations envisioned at the start of the process.

While how innovation operates has become clearer, relatively little has been documented so far about how innovation processes actually unfold in African smallholder agriculture, and about how interventions and Agricultural Research & Development (AR&D) stakeholders can learn from such detailed understanding to implement better strategies aimed at fostering innovation. African smallholders and AR&D actors in Africa face a host of challenging circumstances such as lack of capital, low levels of formal education and poorly functioning markets and institutions (e.g. Hounkounou et al., 2012; Struik et al., 2014). On the positive side, they also face opportunities such as high demographic and economic growth rates, fast development of markets and expanding demand for quality produce, as well as a relative abundance of external funding (Sanginga et al., 2009).

Schematically speaking, two contrasting avenues through which innovation may develop can be recognised. On the one hand, many formal AR&D actors rely extensively on externally-driven and externally-funded interventions to foster innovation among farmers and other local stakeholders, using the well-established linear 'top-down' model of transfer of technology, lately with a participatory twist. For that purpose, they usually introduce ideas, principles, knowledge, technologies and resources, developed and acquired outside of the intervention area as main inputs for developing and disseminating solutions supposedly able to respond to generic objectives such as increasing production, reducing environmental degradation, transforming produce or accessing markets. On the other hand, smallholders and other local stakeholders across the developing world innovate on their own or with limited support (mostly from NGOs) to respond to local challenges and opportunities, as demonstrated by studies spanning several decades (e.g. Richards, 1985; Chambers and Thrupp, 1994; Veldhuizen et al., 1997; Buckles et al., 1998, Sanginga et al., 2009; Waters-Bayer et al., 2015). In some cases, local innovation borrows knowledge, technologies and resources brought through external interventions or also discovered during migrations of farmers outside of their region of origin. However, in many cases, external interventions and projects do not seem aware of, nor do they try to build on, local innovation processes and overall dynamics. Hence, a major concern is how to reconcile and articulate these two forms of innovation, so that their advantages can be combined to ensure greater inclusiveness of local stakeholders in the process, strengthened capacity to innovate as an outcome, and 'sufficient' scaling of results (Leeuwis et al., 2014; Dolinska and d'Aquino, 2016).

Within such a background, the EU-funded project JOLISAA[1] (Joint Learning in and about Innovation Systems in African Agriculture) endeavoured to assess recent and diverse innovation experiences in smallholder farming in Benin, Kenya and South Africa involving multiple stakeholders (Triomphe et al., 2013). Issues JOLISAA addressed included (1) how does innovation unfold over time; (2) what roles different stakeholders play in innovation; (3) what knowledge and other resources each of them contribute; (4) what effects the innovations developed bring; and (5) what conditions favour or impede innovation. JOLISAA also aimed at developing concrete recommendations for policy, research and practice about how best to foster multi-stakeholder innovation in an African context.

This chapter presents key results obtained by the JOLISAA team across its three target countries. After presenting the approach used by JOLISAA for assessing innovation experiences and the diversity of experiences actually assessed, the chapter summarises the key insights into innovation processes gained from a cross-analysis of the case studies JOLISAA selected. It also presents lessons and recommendations about how best to assess and support innovation and make it more inclusive, with an emphasis on how to document and support local innovation.

Approach to the assessment of innovation experiences

JOLISAA undertook its assessment in five major, partly overlapping phases: (1) development of an analytical framework; (2) inventory of innovation cases; (3) collaborative case assessment; (4) cross-analysis of cases; and (5) development of policy recommendations.

Analytical framework

To facilitate subsequent cross-analysis among cases and countries, JOLISAA started by developing a common analytical framework for describing and assessing the various experiences. The framework was divided into two successive sets of guidelines and instruments: one for the inventory (Triomphe et al., 2013) and one for the collaborative assessment (Triomphe et al., 2012). It draws on the innovation system concept and perspective (Hall et al., 2003; World Bank, 2006; World Bank, 2012; Touzard et al., 2014), by focusing on stakeholders, their roles and their interactions, and on the 'enabling' environment in which innovation unfolds. We also refer to innovation processes (how a specific innovation unfolds), dynamics (different types of innovations involving different stakeholders unfolding with no or loose links between them) and landscape (the various scales at which actors may act and interact around different types of innovation) when not specifically referring to formal innovation systems. Our framework focuses on innovation type, nature and domain; stakeholders, their roles and interactions; innovation triggers and drivers; innovation history; and results and outcomes obtained (Table 23.1).

Inventory of innovation experiences

The main criteria for considering cases for inclusion in the three national inventories of agricultural innovation experiences were: (1) smallholder and other resource-poor rural stakeholders were actively involved; (2) at least three different types of stakeholders were involved; and (3) the innovation process was, at least, three years old and had gone beyond the initial stages. Cases were sought through

Table 23.1 Main categories and variables used in the assessment framework

Theme/dimension/variable	What JOLISAA[1] wanted to know about it
Local innovation context	What were the key agro-environmental and socioeconomic features that may have shaped or influenced the innovation process?
Innovation: type, nature, domain	What was the diversity of innovations tackled?
Stakeholders' roles and interactions	Who have been leading or active stakeholders?
	What type of coordination took place among stakeholders?
Role of local knowledge	What role has local knowledge played in the innovation process?
Innovation triggers and drivers	What have been the key triggers and drivers of the innovation process?
Innovation dynamics	What have been the key phases that the innovation process went through?
Scale at which innovation is taking place	Did the innovation process take place mainly at local, regional or national scale, or at several scales?
Results and 'impact' obtained	What have been the effects so far, positive or negative, intended or not, in the different dimensions?

[1] JOLISAA: Joint Learning in and about Innovation Systems in African Agriculture: EU-funded FP7 project
Source: Adapted from Triomphe et al., 2013.

literature search, interactions with resource persons in universities, research institutes and networks within the national agricultural innovation landscape, drawing on JOLISAA national team members' prior knowledge of specific innovation cases, and/or seeking innovation within a given region, area or farming system in each country (Triomphe et al., 2013). Field visits were also made to supplement the available documentation. Numerous cases were identified but not documented, because of the lack of (access to) documentation or unwillingness of key informants to share information and because getting the information would have been too costly. The outputs of the inventory are two-fold: short qualitative semi-structured narratives describing the 57 cases and a spreadsheet in which each case is characterised through a series of semi-quantitative descriptors. Both outputs cover the main categories identified for the analytical framework.

Collaborative case assessment

Out of the 57 cases inventoried in the three countries, the JOLISAA team selected 13 for collaborative case assessment (CCA), in which representatives of local stakeholders were involved alongside JOLISAA researchers and MSc. students. The 13 cases selected (Table 23.2) represented the seemingly richest and

Table 23.2 Diversity of cases selected for collaborative assessment within the JOLISAA[1] project framework

Country	Domain: Natural resources management	Domain: New value chains
South Africa	Developing rainwater harvesting techniques for growing field and vegetable crops (~ 10 years) Developing a participatory extension approach (~ 15 years)	Bulk buying of inputs through credit and saving schemes (4 years)
Benin	Integrated soil fertility management for new high value products (~ 15 years) Intensification in indigenous aquatic agricultural ('hwedos') systems (several decades)[a]	Emergence of parboiled rice value chains (~ 10 years) Development of multiple soybean value chains (~ 40 years)[b]
Kenya	Using by-products for soil rehabilitation and securing access to lime (10 years) Management of an invasive tree (*Prosopis sp.*) for charcoal and fodder in semi-arid areas (~ 30 years)[c]	Activation of a natural resource (aloe) in semi-arid Baringo (~ 30 years)[d] Introducing a new crop (Gadam sorghum) as in input into beer production and other processed foods in a semi-arid area (~ 8 years) Developing a mango value chain (~ 20 years) Adding value though solar cooling of milk (~ 5 years)

Notes: years indicate time frame considered for assessing the innovation process
[1] JOLISAA: Joint Learning in and about Innovation Systems in African Agriculture: EU-funded FP7 project
Selected innovation main story lines in a nutshell:
[a] *Hwedos Benin:* The hwedos systems were initially an endogenous management system for taking advantage of a flooding plain environment to catch fish, but they have recently evolved into mixed horti-aquatic systems as well as undergone intensification of their aquatic component through introduction of fingerlings, for example
[b] *Soybean Benin:* Soybean was first introduced with women by NGOs as a home-grown and processed baby food to fight against infant mortality. Small to medium scale businesses producing baby food for the market were then developed. Gradually small-scale women enterprises developed new processing methods to introduce soybean in local foods such as tofu and 'mustards' directed at low-income consumers, soybean representing a cheap substitute for traditional ingredients. Soybean has also become a cash crop with several alternative markets, which also substitute cotton seed for the oil industry or find its way in new value chains
[c] *Prosopis Kenya:* The *Prosopis* tree was introduced to counter rapid deforestation and degradation of a semi-arid area of Kenya: it soon became invasive, and options for its management and transformation into commercial grade charcoal, including changes in rules for charcoal production in sensitive environments, were gradually developed in response
[d] *Aloe Kenya:* As wild aloe became harvested for export markets, it became necessary to develop more sustainable harvesting methods and even introduce cultivation techniques to counter the risk of over-exploitation. There was also a formal attempt at developing a certified value chain, which met with mixed success
Source: Adapted from Triomphe, Floquet et al., 2014.

most complementary sets of experiences, as well as the ones that had been the most dynamic over recent years and for which key stakeholders were interested in joint learning about their experiences. These cases were assessed with respect to the actual roles and contributions of the different actors, the nature of linkages between them, the history and dynamics of the innovation process over time in relation to the wider political, economic and institutional environment in which they were embedded (in order to identify key triggers and drivers of innovation), and the role of local knowledge and creativity. The assessment was also forward-looking: it identified recommendations for moving the innovation process forward.

Collaborative case assessment methods used included, among others, a mix of collective and individual semi-structured interviews, focus group discussions with key stakeholders, multi-stakeholder assessment and validation workshops, direct observations and bibliographic review of grey literature related to the cases (Triomphe et al., 2012).

Cross-analysis and policy formulation

These two steps were mostly overlapping, as policy formulation was based on the results of cross-analysis. Despite the differences in how the case assessments were conducted in the three countries, a meaningful cross-analysis was still possible, as common major themes were reported systematically in each case study. After the case assessments were completed, the cross-analysis was conducted at country level and across Benin, Kenya and South Africa. At both levels, it focused mainly on the role of different stakeholders in the innovation process and the nature of linkages developed among them, the innovation triggers and drivers. How the innovation developed, often beyond the initial intention and support of external stakeholders, was also scrutinised. At global level, the JOL-ISAA research team synthesised the results from each country, re-assessed the different cases according to key themes and distilled evidence-based generic lessons and recommendations. The lessons and recommendations were submitted to critical feedback through electronic and face-to-face discussions with a large group of researchers and practitioners from different horizons before being eventually validated and published as country specific or global policy briefs (Waters-Bayer et al., 2013).

Key results and discussion

Between them, the 57 cases inventoried and the sub-set of 13 CCA cases covered a wide diversity of experiences in terms of type (technical, organisational, institutional), domain (cropping, livestock-keeping, fishery, processing and marketing), scale (local, regional, national) and duration of the innovation process (a few years to several decades). Several key features are discussed below: the diversity of stakeholders involved in innovation; the diversity of innovation triggers; the relevant time frames for making sense of innovation processes; the multiple dimensions of innovation; the relationship between innovation processes and externally-funded projects and, last but not least, local innovation. Illustrations of the results are taken mostly from the four cases presented in a nutshell at the bottom of Table 23.2.

Who were the stakeholders?

The stakeholders involved in a given innovation case typically included a mix of individual farmer-innovators, one or more community-based or farmer organisations, research, extension services, NGOs, private entrepreneurs, government and externally-funded projects. Depending on the specific case and phase of innovation, leading and active stakeholders varied. For instance, researchers, an NGO or a project might be very active in the initial stages (on-farm experimentation, building capacity, facilitating interactions, etc.), while farmers and their organisations or a business stakeholder tended to become more active in later phases. In many cases, one of the stakeholders (typically an externally-funded project) played the role of intermediary (Klerkx and Leeuwis, 2008) to facilitate interaction among the stakeholders. Formal research did not usually initiate or play a leading role in many innovation cases; rather, ideas and initiatives came from different sources, including farmers. Policymakers and private sector actors were seldom among the active stakeholders. This may reflect that conventional R&D actors (research, extension) still dominate initiatives focusing on smallholder agriculture, as well as the relative scarcity of specific pro-innovation public policies in the three countries. It could also reflect a sample bias, due to the limited connections of national JOLISAA teams with 'non-conventional' R&D partners such as NGOs or entrepreneurs. In addition, JOLISAA found few truly farmer-led innovation processes, probably because such cases were less visible and less likely to be documented (see below).

Innovation triggers

Most cases had a mix of different triggers for innovation. Degradation of natural resources (e.g. declining soil fertility, dwindling supply of water, disappearing forest) was a common trigger. Others included seizing a local or global market opportunity, creating or improving a value chain, introducing an improved technology or practice (e.g. new livestock breed, new way of processing rice). Changes in policy were rarely mentioned as triggers, however, yet they played a significant role as drivers (positive but also negative) of the overall process.

Relevant time frame for assessing innovation

In most cases, the relevant time frame for understanding the innovation process easily spanned at least one, and often several, decades. Over time, the innovation processes often seemed to go through successive phases (Figure 23.2) at an uneven pace – sometimes very rapid, sometimes almost dormant – under the influence of external and internal factors (e.g. resource availability, changing drivers in the overall environment, entrance and exit of key individuals and stakeholders, etc.). Consequently, innovation stories tend to be rather complex, with different stakeholders having different perceptions of what has happened and why. The soybean case in Benin (Floquet et al., 2014) illustrates the intricate intertwining of innovation types and phases over time, as well as the evolving nature of innovations developed by different stakeholders groups (Figure 23.1; Table 23.2).

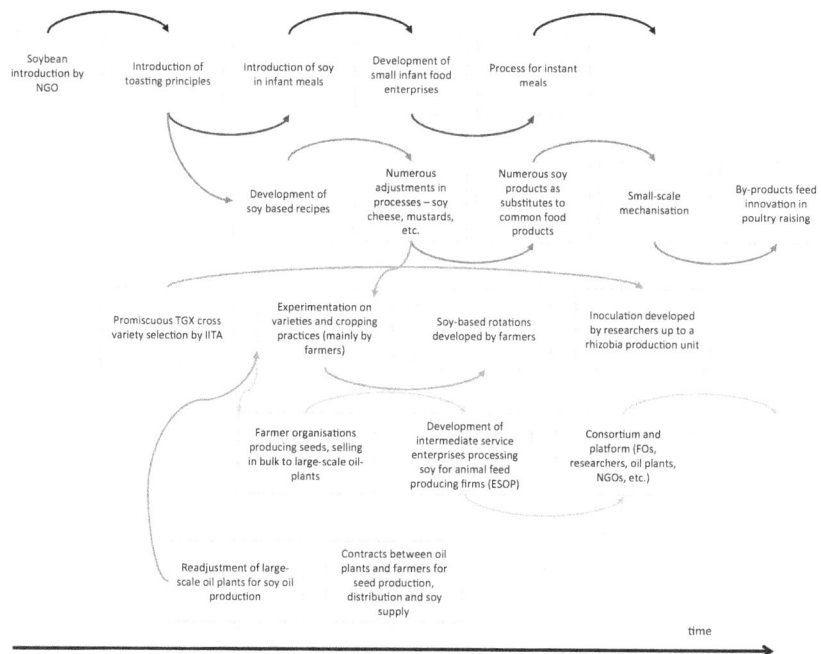

Figure 23.1 Technical, organisational and institutional innovations inducing each other within the soy innovation process in Benin

Source: Adapted from Floquet et al., 2014.

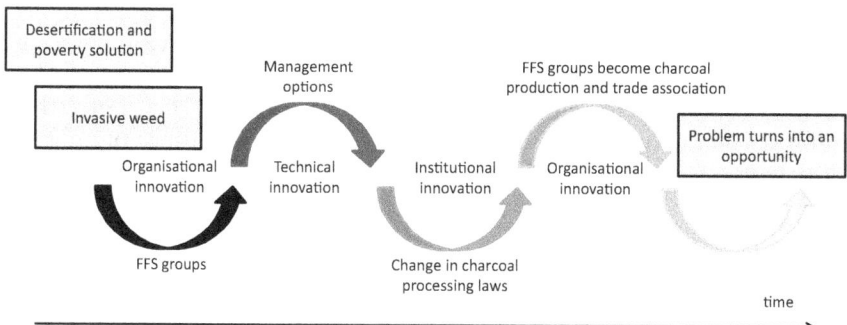

Figure 23.2 Sequencing of technical, organisational and institutional innovation in the *Prosopis sp.* case in Kenya

Source: Adapted from Chengole, Welimo et al., 2014.

Dimensions of innovation

Outcomes in terms of innovations resulting from a given innovation process typically exhibited several interwoven and interdependent dimensions: technical (e.g. a new variety), organisational (e.g. farmers acting collectively to acquire inputs or sell their produce) and institutional (e.g. new coordination mechanism), as the 'simple' *Prosopis* case illustrates (Figure 23.2; Table 23.2) (Chengole, Welimo et al., 2014). These various dimensions emerged organically over time as the innovation process unfolded from a specific entry point (often a new technology). New dimensions usually resulted from new stakeholders coming on board, or from stakeholders starting to change their practices and, in so doing, needing to make other transformations or wanting to take advantage of the evolving environment in which they operated.

Innovation and externally-funded projects

Many innovation cases that were well documented and well known had a strong link with externally-funded projects (Triomphe, Floquet et al., 2014). The chaotic abundance, and succession of 'projects', aiming to stimulate innovation is typical of developing countries. As public funding for innovation is scarce across Africa, public institutions and NGOs depend heavily on external support to carry out innovation-related activities, while smallholders are usually too poor to pursue innovation at a significant scale on their own. Projects can be important for creating innovation dynamics embedded in a temporary favourable environment (akin perhaps to an 'innovation niche' as defined by Geels and Shot, 2007), shielding the process from the usual inhibiting or disabling factors and drivers. In doing so, they may thus allow a minimum critical mass to be reached or initial bottlenecks to be overcome. However, projects often artificially promote short-term use of technologies that may not be sustainable, trigger opportunistic behaviour from some stakeholders, lead to an aid mentality and overlook more endogenous, low-cost and potentially more sustainable innovation pathways and outcomes. Projects may also have difficulties in formulating objectives and designing activities that are truly in line with the demands and needs of local stakeholders. Finally, most projects typically seem to underestimate what it takes to implement an exit strategy able to prevent the collapse of the emerging, yet fragile, innovation process the project has nurtured. Another unexpected consequence of this overabundance of projects is that researchers and other formal AR&D actors tend to be relatively blind to innovations that have happened outside formal projects and arrangements.

Taking into account local innovation?

While projects had a predominant influence in many documented cases, a major issue coming out of the JOLISAA assessment is how much publicly-supported and funded innovation processes take into account the local innovation landscape and dynamics. Yet, such innovations might be essential for ensuring the eventual success of an innovation process and for sustaining its momentum. Floquet et al. (2014) showed in the case of the 'hwedo' aquatic systems that innovation may take place with hardly any external support and that innovation may accelerate once an intervention was over (case of the emergence of multiple soybean value chains), often as a result of changing triggers and drivers in the external environment (Table 23.2). In the aloe case in Kenya (Table 23.2), the effort to build a certified aloe value chain, driven by external R&D actors, interacted only little with the locally-driven aloe innovation process for a number of reasons, despite the notable achievements of the latter (Figure 23.3) (Chengole, Belmin et al., 2014). This shows a recurrent feature, i.e. that the persons in charge of external intervention may forget and fail to build on local innovation, either because it remains unknown to them (for lack of documentation), or at times because it is explicitly disregarded because of its supposed weaknesses. In the aloe case, the local innovation around wild aloe sap harvesting and marketing was deemed undesirable in the name of its supposed potential negative impact on natural resource degradation, yet it was never properly assessed.

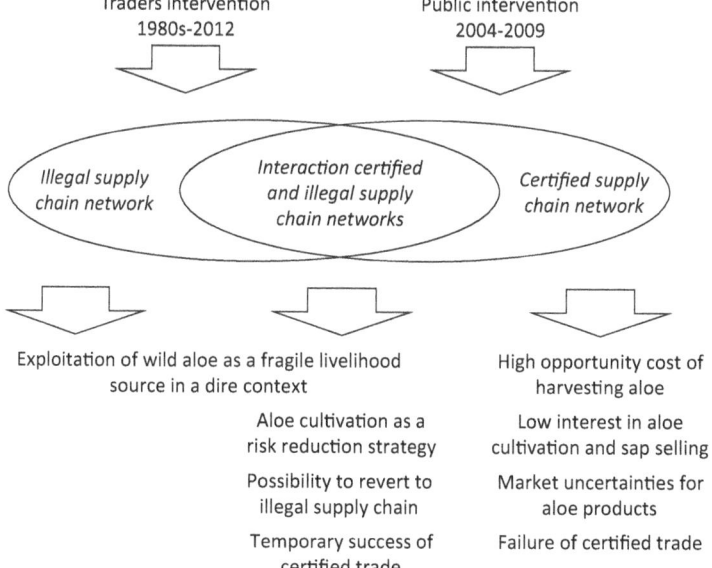

Figure 23.3 Partially interacting illegal and public value chains for processing aloe sap in semi-arid Kenya

Source: Adapted from Chengole, Belmin et al., 2014.

Key implications and recommendations

The assessments conducted under the JOLISAA framework provide useful lessons for policymakers, researchers and development practitioners about what innovation dynamics and processes are all about, how to assess them and how to support them in ways that build upon the knowledge, creativity and existing linkages of smallholders. In doing so, the aim is to render smallholders more resilient to rapid and even sudden changes.

Major lessons and recommendations drawn by JOLISAA were summarised in a policy brief (Waters-Bayer et al., 2013). They include the following:

Build on innovation 'in the social wild'

JOLISAA case studies confirm that many smallholders actively innovate individually and collectively in many directions to solve problems, improve their farming and income, and grasp opportunities, with little or no support from public R&D institutions or after externally-driven, usually short-term public R&D interventions terminate. Yet many such initiatives take place 'under the radar' of most AR&D actors, or more globally 'in the social wild' (outside of the formal AR&D sector) (Sherwood et al., 2012). Hence they are usually ignored or even shunned by state, non-*state*, private sector and even farmer organisations trying to develop and diffuse agricultural technologies. They are also difficult to identify from outside and hence most take place undocumented.

JOLISAA recommendation is that policymakers, researchers and practitioners should do their utmost to recognise, build on and strengthen existing local innovation processes and local energies, rather than ignoring them altogether or trying to substitute them, as a starting point for fostering sustainable, locally-led and locally-supported innovation processes. The rationale for doing that is that chances of success and scaling up would then increase, as interventions would benefit from an already socially well-rooted process among local stakeholders rather than trying to induce it from scratch, an uncertain and slow enterprise. Adequate documentation of existing practice and innovation is a necessary starting point before intervention. Public R&D organisations might want to invest in long-term monitoring of a range of strategic situations and environments to draw a comprehensive and up-to-date picture of selected ongoing innovation processes. A change of attitude may also be needed, so that any external documenters acquire a more positive view of local innovation and local knowledge; this may require changes in the entire agricultural educational, research and extension system.

Support unpredictable innovation processes

Several JOLISAA cases (soybean in Benin, *Prosopis* in Kenya) illustrate that innovation trajectories extend in unpredictable ways over the long-term. What may have seemed the 'right' way to go at one moment in time may reveal itself to be premature, a dead end or not possible within a given environment, or it may create unexpected new problems (e.g. Chengole, Welimo et al., 2014). New stakeholders emerge or intervene. Possibilities and conditions that were never dreamed of materialise and spur a new cycle of innovation or allow fast uptake of a, until then, dormant innovation (Floquet et al., 2014). At the same time, intervention projects frequently lack the proper instruments or flexibility for dealing effectively with the unpredictability of social change.

The JOLISAA recommendation is that interventions should adopt an open-ended approach to planning their activities, rather than a rigid log frame-type approach in which an intervention team designs activities covering several years from the onset. This would require the donors' commitment and adapted modalities for funding. Long-term involvement in and support of innovation should be negotiated or at least anticipated from the start, rather than adhering uncritically to the logic of short-term funding cycles. This may mean that innovation teams strive systematically to identify a portfolio of donors rather than depending on one single donor (to bridge several project cycles and opportunities) and also plan interventions and support (including brokering mechanisms or organisations) that do not rely excessively on external funding for their implementation but rather on commitments and institutional mechanisms put in place by national governments. This may also imply relying much more on the initiatives and resources of local stakeholders.

Address the multiple dimensions of innovation

A major lesson derived from cross-analysing the JOLISAA case studies, highlighted also by other authors (Hall et al., 2010, Sulaiman et al., 2011; Kilelu et al., 2013), is that the various dimensions of innovation (technological, institutional and organisational) co-evolve, i.e. interplay and complement each other along an innovation trajectory, without one dimension being intrinsically more important than the other. Eventual success and scaling up of innovation seems to result from a proper attention to and sequencing of these different types of innovations over time. This goes counter to commonly held views and practices of biophysical researchers and other professionals, who tend to give priority to technological innovation, or business-oriented professionals who put a premium on managerial and organisational innovation.

To take this lesson on board, JOLISAA recommends that a broad conception of mixed innovation bundles (rather than a focus on individual innovations) should be ensured right from the beginning of an intervention, including its diagnostic phase. Time should be taken for considering carefully and flexibly the proper sequencing of innovations of various types as the innovation process unfolds. To improve the process, periodically assessing it with the relevant stakeholders to identify missing or needed types of innovation, or to decide collectively on the proper timing for addressing technical, organisational or institutional changes, should be an integral part of sound approaches to monitoring and supporting innovation.

Overall, such lessons and recommendations have been shared in different forms and fora since JOLISAA was terminated in 2013, and add to those already found by other programmes such as the Convergence of Science – Strengthening of Innovation Systems (also known as COS-SIS) (cf. Struik et al., 2014), or the Research into Use programme funded by the UK Department for International Development (Suleiman et al., 2011). Time will tell if, and how much, some of these lessons and recommendations are indeed taken up somehow and contribute to changing the approaches to fostering innovation of donors, governments and public or not-for-profit R&D institutions for the better, and to eventual impact.

Conclusions and implications

By assessing a series of case studies involving African smallholder farmers, JOLISAA was able to understand better the nature of innovation in a diversity of contexts. JOLISAA results confirm that numerous and diverse multi-stakeholder innovation initiatives have taken place in recent years or are still ongoing across the three African countries of study, something others have observed elsewhere in Africa (e.g. Sanginga et al., 2009; Adekunle et al., 2012; Hounkounou et al., 2012; Sanyang et al., 2014; Struik et al., 2014). While external interventions through public AR&D actors remain a dominant feature of the African innovation landscape, a sizable, barely visible nor properly documented proportion of innovation processes and dynamics rely on local initiatives taken by farmers and other rural stakeholders, with little support from formal R&D institutions, as was documented on a global level by Waters-Bayer et al. (2015). Identifying, assessing and supporting better such initiatives and helping local stakeholders to strengthen their capacity to innovate (Leeuwis et al., 2014) are among the critical lessons and recommendations JOLISAA has produced.

Despite the associated challenges and at times a cruel lack of adapted incentive structures, many of the AR&D actors seem increasingly aware of the need for, and benefit from, tighter and better collaboration with smallholder farmers and their organisations, as well as with each other. Strengthened, more balanced and more inclusive collaboration among stakeholders should allow dealing better with complex problems and challenges that cannot be effectively handled otherwise, such as sustainable production and processing, secure access to new markets, climate change, food security and poverty reduction.

Finally, JOLISAA results also show the value, as well as the significant challenges, of using an innovation system perspective in uncovering key factors related to the nature and features of 'real' innovation processes, even though implementing such approaches on a large scale may prove challenging. Hopefully, more and more researchers and practitioners will be willing and better able to prepare themselves to meet and overcome such challenges in the future, something which will require significant improvements of initial and continuing education programmes, as well as changes in the incentives and rewards systems and in the rules of operation of many AR&D and donor organisations.

Acquiring and applying such capacity is key to increasing our detailed knowledge of the dynamics of contemporary African agriculture and with it our collective ability to improve the pace, relevance and reach of many innovation initiatives.

Acknowledgements

This work was carried out as part of the JOLISAA project (www.jolisaa.net) under Framework Programme 7 of the European Commission. The opinions expressed herein are the sole responsibility of the authors. The authors thank other JOLISAA consortium members (Jolanda van den Berg, Bernard Bridier, Todd Crane, Rock Mgonbo, Teresiah N'gan'ga, Nicoliene Oudwater, Gerrit Rootman, Nour Selemna, Joe Stevens, Rigobert Tossou and Davo S. Vodouhe) and their partners in Benin, Kenya and South Africa for providing data and useful comments on the many intermediary reports and documents which were used for developing the content for this paper.

Note

1 See www.jolisaa.net for a comprehensive overview of the project's approach and results.

References

Adekunle, A.A., Ellis-Jones, J., Ajibefun, I., Nyikal, R.A., Bangali, S., Fatunbi, O. and Ange, A. (2012) *Agricultural Innovation in Sub-Saharan Africa: Experiences from Multiple-Stakeholder Approaches*. Forum for Agricultural Research in Africa (FARA), Accra.

Buckles, D., Triomphe, B. and Sain, G. (1998) *Cover Crops in Hillside Agriculture: Farmer Innovation with Mucuna*, IDRC and CIMMYT, Ottawa, 218 p.

Chambers, R. and Thrupp, L.A. (Eds.) (1994) *Farmer First: Farmer Innovation and Agricultural Research*, KARTHALA Editions, Paris, France.

Chengole, J.M., Belmin, R., Welimo, M., Kamau, G. and Triomphe, B. (2014) Understanding the Innovation System in Aloe Domestication and Exploitation in Baringo County, Kenya. p. 72–75. In: Triomphe, B., Waters-Bayer, A., Klerkx, L., Schut, M., Cullen, B., Kamau, G. and Leborgne, E. (Eds.), Innovation in Smallholder Farming in Africa: Recent Advances and Recommendations. *Proceedings of the International Workshop on Agricultural Innovation Systems in Africa* (AISA), 29–31 May 2013, Nairobi, Kenya. Cirad, Montpellier.

Chengole, J.M., Welimo, M., Belmin, R., Ng'ang'a, T. and Kamau, G. (2014). From a Desired to a Despised and Later Desired Tree: The Case of Prosopis Introduction and Management in Marigat, Baringo County, Kenya. p. 76–79. In: Triomphe, B., Waters-Bayer, A., Klerkx, L., Schut, M., Cullen, B., Kamau, G. and Leborgne, E. (Eds.), Innovation in Smallholder Farming in Africa: Recent Advances and Recommendations. *Proceedings of the International Workshop on Agricultural Innovation Systems in Africa* (AISA), 29–31 May 2013, Nairobi, Kenya. Cirad, Montpellier.

Dolinska, A. and d'Aquino, P. (2016) Farmers as agents in innovation systems. Empowering farmers for innovation through communities of practice, *Agricultural Systems.* 142: 122–130.

FAO (2014) *The State of Food and Agriculture: Innovation in Family Farming*. FAO, Rome, 161 p.

Floquet, A., Vodouhê, G., Michaud, A., Bridier, B., and Vodouhê, S.D. (2014). How Innovation Processes Unfold Along Unexpected Trajectories – The Case of Soy in Benin. p. 105–110. In: Triomphe, B., Waters-Bayer, A., Klerkx, L., Schut, M., Cullen, B., Kamau, G. and Leborgne, E. (Eds.), Innovation in Smallholder Farming in Africa: Recent Advances and Recommendations. Proceedings of the International Workshop on Agricultural Innovation Systems in Africa (AISA), 29–31 May 2013, Nairobi, Kenya. Cirad, Montpellier.

Geels, F.W. and Shot, J. (2007). Typology of socio-technical transition pathways. *Research Policy*. 36: 399–417.

Hall, A. and Clark, N. (2010) What do complex adaptive systems look like and what are the implications for innovation policy? *Journal of International Development*. 22(3): 308–324.

Hall, A., Sulaiman, V.R., Clark, N. and Yogoband, B. (2003) From measuring impact to learning institutional lessons: an innovation systems perspective on improving the management of international agricultural research. *Agricultural Systems*. 78: 213–241.

Hounkonnou, D., Kossou, D., Kuyper, T.W., Leeuwis, C., Nederlof, S., Röling, N., Sakyi-Dawson, O., Traoré, M. and van Huis, A. (2012) An innovation systems approach to institutional change: smallholder development in West Africa. *Agricultural Systems*. 108: 74–83.

Kilelu, C.W., Klerkx, L. and Leeuwis, C. (2013) Unravelling the role of innovation platforms in supporting co-evolution of innovation: contributions and tensions in a smallholder dairy development programme. *Agricultural Systems*. 118: 65–77. http://dx.doi.org/10.1016/j.agsy.2013.03.003.

Klerkx, L. and Leeuwis, C. (2008) Matching demand and supply in the agricultural knowledge infrastructure: experiences with innovation intermediaries. *Food Policy*. 33: 260–276.

Klerkx, L., van Mierlo, B. and Leeuwis, C. (2012) Evolution of Systems Approaches to Agricultural Innovation: Concepts, Analysis and Interventions. pp. 457–483. In: Darnhofer, I., Gibbon, D. and Dedieu, B. (Eds) *Farming Systems Research into the 21st Century: The New Dynamic*, Springer, Dordrecht, Netherlands.

Leeuwis, C., Schut, M., Waters-Bayer, A., Mur, R., Atta-Krah, K. and Douthwaite, B. (2014). Capacity to innovate from a system CGIAR research program perspective. Penang, Malaysia: CGIAR Research Program on Aquatic Agricultural Systems. Penang: *Program Brief*: AAS-2014–29.

Malerba, F. (2007) Sectoral systems of innovation: a framework for linking innovation to the knowledge base, structure and dynamics of sectors. *Economics of Innovation and New Technology*. 14(1–2): 63–82.

Richards, P. (1985). *Indigenous Agricultural Revolution: Ecology and Food Production in West Africa*, Westview Press, Boulder, CO, 192 p.

Sanginga, P., Waters-Bayer, A., Kaaria, S., Njuki, J. and Wettasinha, C. (Eds) (2009) *Innovation Africa: Enriching Farmers' Livelihoods*, Earthscan, London.

Sanyang, S., Pyburn, R., Mur, R. and Audet-Bélanger, G. (2014) *Against the Grain and to the Roots: Maize and Cassava Innovation Platforms in West and Central Africa*, LM Publishers, Arnhem, Netherlands.

Schut, M., Klerkx, L., Rodenburg, J., Kayeke, J., Hinnou, L.C., Raboanarielina, C.M. and Bastiaans, L. (2015) RAAIS: rapid appraisal of agricultural innovation systems (part I). A diagnostic tool for integrated analysis of complex problems and innovation capacity. *Agricultural Systems*. 132: 1–11.

Sherwood, S.C., Schut, M. and Leeuwis, C. (2012) 'Learning in the Social Wild: Encounters Between Farmer Field Schools and Agricultural Science and Development in Ecuador', p. 102–137. In: Ojha, H.R., Hall, A. and Sulaiman, R. (Eds), *Adaptive Collaborative Approaches in Natural Resources Governance: Rethinking Participation, Learning and Innovation*, Routledge, London, UK.

Struik, P.C., Klerkx, L. and Hounkonnou, D. (2014) Unravelling institutional determinants affecting change in agriculture in West Africa. *International Journal of Agricultural Sustainability*. 12(3): 370–382.

Sulaiman, R.V., Hall, A. and Reddy, V.T.S. (2011) *Missing the target: lessons from enabling innovation in South Asia*, Wallingford, UK (www.dfid.gov.uk/r4d/PDF/Outputs/ResearchIntoUse/riu11discuss25asia.pdf).

Touzard, J.M., Temple, L., Faure, G. and Triomphe, B. (2014) Systèmes d'innovation et communautés de connaissances dans le secteur agricole et agroalimentaire. *Innovations* 1(43): 13–38.

Triomphe, B., Floquet, A., Kamau, G., Letty, B., Vodouhé, D.S., N'gan'ga, T., Stevens, J., van den Berg, J., Selemna, N., Bridier, B., Crane, T., Almekinders, C., Waters-Bayer, A., Oudwater, N. and Hocdé, H. (2013) What does an inventory of recent innovation experiences tell us about agricultural innovation in Africa? *Journal of Agricultural Education and Extension*. 19(3): 311–324.

Triomphe, B., Floquet, A., Kamau, G., Letty, B., Vodouhé, D.S., Stevens, J., van den Berg, J., Crane, T., Almekinders, C., Selemna, N., Bridier, B., Oudwater, N. and Waters-Bayer, A. (2014) 'Multistakeholder Innovation Processes in African Smallholder Farming: Key Lessons and Policy Recommendations from Benin, Kenya and South Africa', p. 45–56, in Triomphe, B., Waters-Bayer, A., Klerkx, L., Schut, M., Cullen, B., Kamau, G. and Leborgne, E. (Eds), Innovation in Small Holder Farming in Africa: Recent Advances and Recommendations. *Proceedings of the International workshop on Agricultural Innovation Systems in Africa* (AISA), 29–31 May 2013, Nairobi, Kenya. Cirad, Montpellier.

Triomphe, B., van den Berg, J., Kamau, G., Floquet, A., Letty, B. and Bridier, B. (2012) *JOLISAA Approach and Guidelines to Collaborative Case Assessment*, JOLISAA project, Cirad, Montpellier.

Triomphe, B., Waters-Bayer, A., Klerkx, L., Schut, M., Cullen, B., Kamau, G. and Leborgne, E. (Eds.) (2014) Innovation in small holder farming in Africa: recent advances and recommendations. *Proceedings of the International workshop on Agricultural Innovation Systems in Africa* (AISA), 29–31 May 2013, Nairobi, Kenya. Cirad, Montpellier, 229 p.

Veldhuizen, L.V., Chaparro, F., Torres, R., Barquero, I., Cox Balmaceda, M., Bisang, R. and Kaimowitz, D. (1997) Farmer's research in practice, lessons from the field (No. E14 88), Ministerio de Comercio Exterior, San José, Costa Rica.

Waters-Bayer, A., Kristjanson, P., Wettasinha, C., van Veldhuizen, L., Quiroga, G., Swaans, K. and Douthwaite, B. (2015) Exploring the impact of farmer-led research supported by civil society organisations. *Agriculture and Food Security*. 4(1): 4.

Waters-Bayer, A., Triomphe, B. and Oudwater, N. (Eds) (2013) 'Building on local dynamics: Five policy recommendations for enhancing innovation by African smallholder farmers', *JOLISAA global policy brief*. JOLISAA project, Montpellier (France), 6 p.

Waters-Bayer, A., van Veldhuizen, L., Wongtschowski, M., Wettasinha, C., Triomphe, B., Mekonnen, F., and Fenta, T. (2011) Farmer-Managed Innovation Funds Drive Multi-Stakeholder Learning Processes. p. 15–18. In: *Proceedings, International Conference on Innovations in Extension and Advisory Services*, Nairobi, Kenya, 15–18 November 2011, CTA and GFRAS, Wageningen.

World Bank (2006) *Enhancing Agricultural Innovation: How to Go Beyond the Strengthening of Research Systems*, World Bank, Washington DC.

World Bank (2012) *Agricultural Innovation Systems: An Investment Sourcebook*, World Bank, Washington, DC.

24 Constraints and opportunities in using multi-stakeholder processes to implement integrated agricultural systems research

The Humidtropics case

Lisa Hiwasaki, Latifou Idrissou, Chris Okafor and Rein van der Hoek

Introduction

The world currently faces complex agricultural problems that cannot be addressed effectively by a single institution or a simple technology transfer process. The collective intelligence and efforts of multiple stakeholders such as farmers, the private sector (traders, processors, financial organizations, and so forth), researchers, non-governmental organizations (NGOs), and policy makers are needed to develop and scale up innovative solutions (Geels, 2002; CGIAR-ISPC, 2015). The theory of change of the CGIAR research program for integrated systems of the humid tropics (Humidtropics) is based on the hypothesis that a region's inherent potential is best realized through an integrated systems approach, involving collective action across stakeholder groups (Humidtropics, 2012).

Since activities began in 2013, Humidtropics core research partners, together with local partners, established research for development (R4D) Platforms, Innovation Platforms (IPs), and learning alliances in four Action Areas – East and Central Africa, West Africa, Central Mekong, and Central America and the Caribbean. Each action area consists of a number of action sites (often being a part of a country). These multi-stakeholder platforms (MSPs) were set up to enable multi-stakeholder engagement and facilitate the process of integrated systems at different levels, using various approaches, and dealing with a diverse range of issues within systems, such as productivity, natural resources management, inputs, markets, finance, nutrition, policy, and youth and gender. Key stakeholders are brought together for communication, problem analysis, joint decision-making, action (implementation, evaluation, and learning) (Hemmati, 2002). In agricultural systems R4D, MSPs are critical to interventions, as several components of the system and different stakeholders are involved. MSPs provide space for these actors to present different perspectives, debate issues, evaluate options, and implement collective actions (Schut, Klerkx, Sartas et al., 2015).

This chapter shares our analysis of using MSPs in Humidtropics and the lessons learned from our experiences of facilitating multi-stakeholder processes to generate innovative and sustainable solutions to complex agricultural challenges.

Background: Multi-stakeholder platforms for agricultural research for development

Multi-stakeholder processes have been promoted by the CGIAR since the early 2000s as a way to foster cross-sectoral collaborative action that ensures that agricultural R4D has impacts that improve the livelihoods of the rural poor (Ashby, 2003; Lundy et al., 2005; Thiele et al., 2011; CGIAR-ISPC, 2015). Multiple stakeholders – research organizations, donors, international and development organizations, government and policy makers, the private sector, and farmers – with diverse interests are brought together to tackle common agricultural problems and to promote collective learning.

In Humidtropics, two levels of MSPs were established: R4D Platforms and IPs. R4D Platforms were set up at the action site[1] level to set up priorities, guide implementation, and scale up proven innovations. Action sites used various tools such as rapid appraisal of agricultural system (RAAIS),[2] EX-ante Tool for RAnking POLicy AlTErnatives (EXTRAPOLATE),[3] field visits complemented with results from Situation Analysis,[4] and participatory workshops to further analyze constraints and prioritize entry points. These entry points form the basis for the R4D Platform to develop integrated systems research projects. Innovation Platforms were set up at the field site level to identify, develop, and test 'best-bet'[5] options. Innovation Platforms are task-oriented in that they evaluate and validate specific system interventions adapted to specific constraints applicable to the field site. It should be stressed that different approaches and tools were adapted depending on the action area, reflecting the different circumstances and contexts in which we work.

An important initiative to kick-start R4D activities was a small-scale funding mechanism called Platform Research Projects (also known as 'Cluster 4'). These projects were seen as an instrument to generate broad partnership engagement in integrated systems research, and were led by local partners such as NGOs, universities, national research institutions, farmer cooperatives, and municipalities.

Methodology

Data for this chapter were collected using various methods, including: participant observation, focus group discussions, interviews, and reports. The primary source of information was the Action Area Coordinators who are the authors of this chapter. Action Area Coordinators are at the heart of the multi-stakeholder processes by being involved in the design, implementation, and evaluation of the MSPs. Data were also collected by Humidtropics' Monitoring and Evaluation Officers, facilitators of action sites, field sites and MSPs, and other MSP members such as researchers, farmers, private sector partners, universities, and government representatives. Their reflections and thoughts on the topic were captured through formal and informal discussions and interviews.

Data were also collected from technical, monitoring, and meeting reports provided by the R4D and IP facilitators and reports on activities is superfluous, such as results of tools used in MSPs, and other research implemented before and after MSPs' launch. Documents related to multi-stakeholder processes such as students' theses, reports of visiting scientists, and scientific papers were also important sources of information.

Analytical framework

Numerous case studies indicate that for MSPs to be successful, they should: (1) have clearly defined common objectives; (2) be composed of the right stakeholders with appropriate representatives; (3) engage stakeholders with strong incentives and internal motivation; (4) be based on long-term partnerships and trust; (5) have secure resource(s); and (6) have a capable facilitator who understands the complexities of multi-stakeholder processes and appropriately manages power dynamics and conflict among the stakeholders (Lundy et al., 2005; Thiele et al., 2011; Boogaard et al., 2013; Brouwer et al., 2013; Schut, Klerkx, Sartas et al., 2015). Additionally, reflexive monitoring and evaluation can support the continuous learning process, and enhance MSPs' adaptive capacity in light of changing contexts (Lundy et al., 2005; Schut et al., 2014). Sustainability of MSPs beyond project funding is also an important concern (Lundy et al., 2005; Thiele et al., 2011; Boogaard et al., 2013).

In the following sections we describe various multi-stakeholder processes that are being used for systems research within Humidtropics, followed by analysis according to the above six factors.

Experiences of multi-stakeholder platforms in Humidtropics action areas

East and Central Africa

The East and Central Africa (ECA) Action Area covers a range of humid and sub-humid tropical highlands, with six action sites: Burundi, Democratic

Republic of Congo (DRC), Rwanda, Western Kenya, Western Ethiopia, and Lake Victoria Basin of Uganda. Following the Action Area inception workshop in May 2013, each action site held its own inception workshop, where field sites and R4D Platforms were identified and established after stakeholder mapping. R4D Platforms are facilitated by a staff of a national facilitating organization, who is hired part-time as action site facilitators. Action site facilitators support the multi-stakeholder processes and activities at both action site and field site levels.

The six R4D Platforms and 12 IPs in the ECA (see Table 24.1) are operational and carry out integrated R4D with varying degrees of success, as reflected in some platform monitoring and project reports. In general, there is appreciation of the value of multi-stakeholder processes to identify and prioritize issues of mutual interest and to find solutions to site-specific agricultural challenges. With a better understanding of their fundamental interdependencies, some value chain actors now show more interest and commitment to platform activities, which strengthen the cause for collective action. For example, financial institutions who give credit to farmers appreciate the fact that farmers need the support of produce buyers, as farmer groups cannot repay back loans if they do not sell their produce at good prices. Similarly, farmers need credit to purchase inputs from input dealers to be able to produce appreciable quantities. Farmers' knowledge is a critical input factor in R4D, while the expected output is to help farmers be more productive with their limited resources. Some MSP experiences and lessons learned, coming from two action sites – Rwanda and DRC – are illustrated below.

In Rwanda, the R4D Platform has a fair representation of some key stakeholders such as research organizations, NGOs, farmers, and government. The private sector has not been consistently represented at Platform meetings. Some stakeholders are represented in both the R4D Platform and IP, which enables better interface between the two Platforms, shared understanding of issues on the ground, improved linkages, resource sharing, and mobilization of policy support for interventions. For example, the Kadahenda IP has received recognition and support of the government for its work on the potato system. By participating in research actions, farmers now see themselves as co-generators of knowledge and not as mere recipients of knowledge, and table their demands directly to research organizations. Another important lesson is the need to focus attention on 'quick wins', which meet farmers' pressing needs. By addressing planting material needs of farmers first, it was relatively easy to win Kadahenda farmers' commitment for broader integrated systems intervention. To be sustainable, MSPs require a significant amount of time to build relationships and trust amongst actors and require self-financing mechanisms. Although Humidtropics was introduced to partners as a 15-year program, the decision to phase it out by the end of 2016 demotivated national partners, which affected the nascent trust – a critical factor to establish a solid basis for self-financing

Table 24.1 Multi-stakeholder platforms set up in Humidtropics

East and Central Africa Highlands

Name and level of MSP	Name of platform	Facilitating organization	Stakeholders involved	Set up month/year	Frequency of meetings
R4D Platforms National (action site level)	Burundi R4D Platform	Institut des Sciences Agronomiques du Burundi (ISABU)	Representatives of international research institutes, national research institutes including universities, private and public sectors, government agencies, local, national, and international NGOs, and farmer organizations	June 2014	Formally meet 4 times a year or when the need arises
	DRC R4D Platform	Plate-forme DIOBASS au Kivu (DIOBASS)		April 2014	
	Ethiopia R4D Platform	Oromia Agricultural Research Institute (OARI)		April 2015	
	Kenya R4D Platform	Western Region Agriculture Technology Evaluation (WeRATE)		October 2014	
	Rwanda R4D Platform	Rwanda Agriculture Board (RAB)		April 2014	
	Uganda R4D Platform	Makerere University (MAK)		March 2015	
Innovation Platforms Local (field site level)	Murayi–Carire IP (in Gitega Field Site, Burundi)	Institut des Sciences Agronomiques du Burundi (ISABU)	Representatives of international research institutes, national research institutes including universities, private and public sectors, government agencies, local, national, and international NGOs, and farmer organizations	June 2014	Every month if necessary, or when the need arises
	Chokola IP (in Mushinga Field Site, DRC)	Plate-forme DIOBASS au Kivu (DIOBASS)		April 2014	
	Diga and Jeldu IPs (Ethiopia)	Oromia Agricultural Research Institute (OARI)		April 2015	
	Busia, Vihiga, Kakamega, and Kisumu IPs (Kenya)	Western Region Agriculture Technology Evaluation (WeRATE)		June 2014	
	Kadahenda and Musanze IPs (Rwanda)	Rwanda Agriculture Board (RAB)		May 2014	
	Kiboga-Kyankwanzi and Mukono–Wakiso IPs (Uganda)	Makerere University (MAK)		February 2014; January 2014	

West Africa Humid Lowlands

Name and level of MSP	Name of platform	Facilitating organization	Stakeholders involved	Set up month/year	Frequency of meetings
R4D Platforms Regional (action site level)	Cameroon R4D Platform	Institut de la Recherche Agricole pour le Développement (IRAD)	Farmer organizations, NGOs and civil society, research centres and universities, private sector (agricultural products processors, micro-finance institutions, traders, transporters, etc.), and government	February 2014	Usually quarterly, but depends on the importance of issues requiring meeting of the platform members
	Nigeria R4D Platform	Obafemi Awolowo University (OAU)		January 2014	
	Ghana R4D Platform	Crop Research Institute-Council for Scientific and Industrial Research (CRI-CSIR)		March 2015	
	Ivory Coast R4D Platform	Ecole Supérieure d'Agronomie-Institut National Polytechnique Houphouët-Boigny (ESA-INPHB)		February 2015	
Innovation Platforms Local (field site level)	Central Region Field Site (Cameroon); Littoral–South–Western Region Field Site (Cameroon); Western–Northwest Region Field Site (Cameroon)	Key Farmers; North West Farmer Organization (NOWEFOR); Centre for Assistance in Sustainable Development (CASD-NGO)	Farmer organizations, NGOs and civil society, research centres and universities, private sector (agricultural products processors, micro-finance institutions, traders, transporters, etc.), and government	September 2014	Usually monthly, but depends on the importance of issues at stake requiring meeting of the platform members
	Iwo-Ayedire Field Site; Atakumosa–Ife East Field Site (Nigeria); Ibarapa East–Ido Field Site (Nigeria); Ogo-Oluwa and Ori-Ire Field Site (Nigeria)	OFFER Centre; Agricultural Development Programmes		May 2014	
	Offinso Field Site (Ghana)	Ministry of Agriculture		May 2015	
	Soubré Field Site (Ivory Coast)	Agence Nationale d'Appui au Développement Rural (ANADER)		June 2015	

(Continued)

Table 24.1 (Continued)

Central Mekong

Name and level of MSP	Name of platform	Facilitating organization	Stakeholders involved	Set up month/year	Frequency of meetings
R4D Platforms Regional (province(s) of a country within an action site)	Northwest Vietnam R4D Platform	Forest Science Center for the Northwest, supported by ICRAF Vietnam	National and local research institutes, CGIAR and non-CGIAR international research organizations, provincial and district-level government representatives and extension workers, local mass organizations (women's and youth groups)/CSOs, farmer groups	August 2013	Twice a year, or depending on need
	Central Highlands R4D Platform (Vietnam)	Western Highlands Agriculture and Forestry Science Institute (WASI), supported by CIAT Laos		September 2014	
	Nan R4D Platform (Thailand)	Chiang Mai University (CMU) supported by AVRDC Taiwan/Thailand		May 2014	
	Xishuangbanna R4D Platform (China)	Yunnan Institute of Tropical Crops, supported by ICRAF China		September 2014	
Innovation platform Local (field site level)	Innovation Platform for commercial vegetables in the Mai Sơn district, Son La province (Vietnam)	Mai Sơn Division of Agriculture and Rural Development, supported by AVRDC Vietnam	National research organizations, provincial and district government and extension workers, local mass organizations (women's and youth groups), farmers	December 2014	Platform to meet twice a year, core members to meet 4 times a year

Central America and the Caribbean

Name and level of MSP	Name of platform	Facilitating organization	Stakeholders involved	Set up month/year	Frequency of meetings
R4D Platforms National	Northern Nicaragua R4D platform	CIAT-Bioversity	Farmer organisations, NGOs and civil society, research centres and universities, private sector (agricultural products processors, micro-finance institutions, traders, transporters, etc.), and government	April 2013	Twice a year.
	Haiti-Dominican Republic R4D Platform	LWR (Lutheran World Relief) (Haiti), IDIAF (Instituto Dominicaco de Investigaciones Agropecuarias y Forestales) (Dominican Republic)		April 2015	
Innovation platforms Local	Estelí-Condega mixed systems Learning Alliance; Jinotega-El Cuá coffee-based systems Learning Alliance; Rancho-Grande-Waslala cocoa-based systems Learning Alliance (Nicaragua)	ASDENIC (Asociación de Desarrollo Social de Nicaragua), FUMAT (Fundación Madre Tierra), Cuculmeca, Soppexcca, FEM (Fundación Entre Mujeres), Aldea Global	Farmer organisations, NGOs and civil society, research centres and universities, private sector (agricultural products processors, micro-finance institutions), and government	January 2014	Monthly and depending on the importance of issues at stake requiring meeting of the alliance.
	Dondon and Grande Rivière du Nord Field Site (Haiti)	LWR (Lutheran World Relief)		April 2015	
	Dajabón, Santiago Rodríguez, Santiago, La Vega (Dominican Republic)	IDIAF (Instituto Dominicano de Investigaciones Agropecuarias y Forestales)		April 2015	

mechanisms. Another lesson learned is the need for continued sensitization of actors in order to sustain their interest and commitment to cooperate. This is particularly important as different individuals often represent stakeholders at meetings, thereby disrupting information flow between the MSP and the actors represented.

Facilitating MSPs, which include actors with diverse interests, knowledge, and experience, is a challenging task. A knowledgeable and motivated facilitator is needed to guide the MSP in the right direction. Rwanda R4D Platform benefits from the support of the Rwanda Agricultural Board through one of its senior scientists who serves as the action site facilitator. The institutional interest of the Board and the Ministry of Agriculture in the activities of the Platforms, as well as the technical knowledge and motivation of the action site facilitator, are key success factors.

The R4D Platform in DRC has insignificant government and private sector presence. The Platform Research Project on cassava-grain legume production system was started under the leadership of the Institut National d'Etude et Recherche Agronomiques (INERA). Weak institutional framework affects the functioning of the Platform, including the uptake of research outputs. The Platform falls short of the main ingredients described above that gave the Rwanda MSPs impetus to excel. In the absence of strong government support, the agricultural sector has relied largely on donor-funded projects, and as a result, many government agencies rely on external support. Similarly, the number of local NGOs that depend on donor-funded projects has increased over the years. Consequently, Platform meetings are dominated by actors that are self-motivated. Also, the high rate of turnover of Platform members has made it difficult for them to fully grasp the essence of the MSP and integrated systems research.

West Africa

The West Africa (WA) Action Area is comprised of four countries: Cameroon, Nigeria, Ghana, and Ivory Coast. Within each country, action sites covering an area representing the agro-ecological diversity of the humid tropics region were delineated. Establishment of the four R4D and nine IPs (see Table 24.1) started with a mapping exercise to identify the key stakeholders of the MSP, after which members were identified based on their potential contributions and gains from their participation.

Agricultural integrated systems improvement in the WA Action Area started with the Situation Analysis followed by the identification of entry points using RAAIS with the R4D Platform members, and the results were validated with the IPs. For example, in Ivory Coast, sustainable intensification of food crop production was identified as the main entry point. In this farming system, almost all the land is occupied by tree-crops such as cocoa, coffee, rubber, and oil palm with cocoa as the dominant tree-crop, which are generally owned by men. Only marginal lands with low soil fertility are left to women for food crop production. Thus, the R4D and IP members identified the main food crops grown in their region and agreed to test options for their sustainable intensification. The 'mother-baby approach'[6] was used to test improved varieties of cassava, maize, vegetables, and legumes in different intensification options (e.g., soil with no fertilizer application, fertilization with manure or chemicals), which enabled the farmers and other MSP members to learn together through the mother plots. The farmers then tested selected sub-sets of technologies from the mother plots themselves in the 'baby' plots established in their own farms. The same approach is being used in Cameroon where MSP members are improving the cocoa-based farming system by testing best-bet options in the integration of vegetables, cassava, plantain, and fruit trees in cocoa farms with farmers.

However, some challenges were encountered while promoting MSPs. In Nigeria, the members did not fully understand the functioning mechanisms of MSPs and continued to request logistic support (transport and accommodation where applicable) as a precondition for their participation in Platform meetings and activities. Instead of learning and partnership building driving participation in MSPs, members are accustomed to projects providing funding for all activities they participate in. However, this is not the case in other countries where members even made in-kind contributions or provided their own funds to carry out Platform activities. In Cameroon, the main challenge concerns lack of synergy among research and development partners to design common research and interventions. R4D thus appears to be an aggregation of activities carried out by these partners, instead of integrated actions to improve the farming system. In both cases, good facilitation of the processes among MSP members is required to sensitize them on how MSPs operate and what members should expect from their participation. It is clear that the ability of facilitators to convince the members to have realistic expectations from their participation, and coordinate actions of stakeholders towards the achievement of the common goal of the Platform, is very important.

Central Mekong

At the official launch of the Central Mekong (CM) Action Area in May 2013, three transboundary action sites which share common agro-ecological and social challenges were identified.[7] In order to strengthen the capacity of facilitators and others who support the five MSPs in the CM Action Area (see Table 24.1), the Action Area Coordinator organized two 3-day capacity development workshops. A document intended to guide establishment and improving the functioning of MSPs was also developed as a guide to Platform facilitators and core partners supporting them. Despite such efforts, engaging and managing multi-stakeholder processes remain difficult. Ensuring that MSPs function well is challenging given the cultural and institutional contexts of the region, as described below.

The R4D Platform for the Northwest Vietnam in the Green Triangle Action Site is an MSP that struggled initially, but currently plays an important role in R4D taking place there. After its official launch the Platform members participated in Situation Analysis, which identified entry themes. These were further jointly narrowed to entry points, which in turn formed the basis for a Platform Research Project that focus on agricultural diversification through intercropping coffee-fruit trees-grass strips, and fruit trees-vegetables, in a predominantly maize mono-cropping system. Since the re-launch of the Platform in early 2015, and engagement of a paid, part-time facilitator in mid-2015, the R4D Platform has been successful in bringing together stakeholders to discuss and guide interventions to improve the agricultural system of the region. Furthermore, an IP on Commercial Vegetables Production was launched, and the two Platforms share many members and interact closely. In 2016, the Platform Research Project in Northwest Vietnam consolidates activities of both the R4D Platform and the IP.

What contributed to the success of the Northwest Vietnam R4D Platform is the active support of the lead core partner (ICRAF) for the CM Action Area, who pushed to get the multi-stakeholder process going and funds MSP meetings. Engagement of an appropriate and knowledgeable Platform facilitator well-connected with the relevant government and non-government entities is another factor. The Platform Research Project works as a good mechanism to bring together the numerous research institutes working in the region to implement research and share its results.

Unfortunately, although three other MSPs were launched in 2014, they did not continue in 2015 due to various factors: lack of funding to implement R4D activities; lack of understanding by core partners responsible for supporting the MSPs of the need to allocate financial and human resources for MSPs; and limited understanding of the objectives of MSPs by their facilitators and members. The example of the CM Action Area demonstrates that considering the cultural and institutional contexts of the region, minimum requirements for MSPs to function are financial support, capacity-building, and engagement of and continued backstopping provided to Platform facilitators. Even when these are met, other challenges remain, such as the top-down manner in which the multi-stakeholder processes function, driven by government and national research institutions. No MSP in CM Action Area managed to fully engage the private sector, thus a key stakeholder is missing in MSPs' activities. Further, it is difficult to address, let alone challenge, existing power dynamics through MSPs, thus important issues such as inclusion of women and ethnic minorities are compromised. In order to incite collective action across broader stakeholder groups as was envisioned by Humidtropics, a different approach that fully takes into account the local realities of the Central Mekong may have been more successful.

Central America and the Caribbean

The Central America and the Caribbean (CAC) Action Area consists of three action sites: Northern Nicaragua (Nicanorte), Haiti-Dominican Republic border region, and the border region between Honduras, El Salvador, and

Guatemala. The discussion here mainly draws from the experience in Nican-orte as this is the most developed action site.

The process of establishing MSPs in Nicanorte began with a Situation Anal-ysis at the national level with the participation of a range of organizations. Key to this process was the analysis of these organizations, their relationships, and their experiences with innovation. The organizations formed the basis of a national R4D Platform (see Table 24.1), which selected three field sites, each focusing on a different land-use system central to agricultural production in the Nicaraguan humid tropics.

After the Situation Analysis at the national level, a similar analysis was con-ducted at each field site[8] followed by a series of five participatory workshops focusing on different dimensions of local agricultural systems. During this pro-cess, entry themes were identified and IPs were established. The entire process was finalized with a convergence workshop between the three IPs, at which more specific entry points were identified in a participatory exercise.

Since their establishment, the IPs have mainly focused on implementing (inter-)Platform Research Projects with the support of Humidtropics core part-ners, and developing their own theories of change. They are facilitated by local consultants familiar with their respective territories and scientists from the CAC Humidtropics team. The R4D Platform has also continued to develop with the participation of a range of national organizations, and functions mainly as an advi-sory body and a mechanism for scaling up innovations emerging from the IPs.

The overall experience of establishing the MSPs in Nicaragua has been favourable. It has strengthened relationships between organizations both within and across the territories. The participatory workshops proved to be very effec-tive in generating positive working relationships between organizations both within and among the three territories, many of which previously did not work together, and this has carried over to the work on Platform Research Projects. Second, the organizations are becoming more efficient learning organizations, seeking to improve their capacity to innovate in local systems, and at the same time they are becoming increasingly relevant collective learning spaces for rural communities. Through their participation in Platform Research Projects, IP members learn new methods and skills, which reflect strategies for making better management decisions and addressing challenges at the farm and com-munity level using participatory approaches.

Despite these important advances, one important challenge remains secur-ing the participation of the public and private sectors in the MSPs. Due to the highly centralized nature of the Nicaraguan state, involvement of government organizations such as the national agricultural research institute, ministries of agriculture and environment at local and regional levels, and municipal govern-ments is poor.

Discussions: Synthesis of challenges and lessons learned

In the above section, we illustrate the diverse ways MSPs function in the dif-ferent Action Areas. Although various tools are used in the process of setting up and managing the MSPs, the guiding principles are the same: identify key stakeholders to build the MSP and engage effective facilitators, analyze the system to select entry points, and identify and test best-bet options to get best-fit options to improve the system. Significant outcomes resulted from these multi-stakeholder processes: organizations that had previously not collaborated brought together to tackle systems issues; constraints and opportunities of the systems identified and analyzed by multiple stakeholders; partnerships developed in value chains; innovations tested for adaptation and adoption; and increased emphasis on cross-cutting issues such as gender. Concrete outputs include knowledge based on Situation and Territorial Analysis, theories of change, integrated systems research projects ('Cluster 4'), and other project initiatives. In Table 24.2 below, we analyze the factors that contribute to our mostly positive experiences with MSPs that resulted in such achievements, along with the challenges that were experienced in their implementation. We then examine five lessons that have emerged out of our experiences.

First, although R4D Platforms were hypothesized to be an ideal mechanism to tackle larger and complex systems issues, it has been difficult to communicate that MSPs are a tool to foster agricultural innovations, which inevitably entails a long-term process. It was a challenge for the members of MSPs at the higher (national) level to understand the benefits of participating in Platforms which do not show quick results, especially when many of the same stakeholders were used to seeing each other in contexts where material or financial incentives are provided. On the other hand, IPs, especially those that were successful in involving the private sector and which focus on concrete issues, resulted in quick and tangible benefits. It is thus important to reinforce the link between the IPs and R4D Platforms, for instance through joint meetings, at which IPs present their results to stakeholders at the R4D and national level.

Second, the significant role researchers played in the Humidtropics MSPs impacted the ability of non-technology-oriented innovations to be born out of the process. Because researchers play multiple roles in the multi-stakeholder process facilitated under Humidtropics – scientists/experts, facilitator, project manager – the activities and meeting agendas of MSPs were often set and led by researchers. It was thus easy for the non-researcher members of MSPs to be confused about what researchers represent and what role they should play in MSPs (Boogaard et al., 2013). The role of CGIAR to launch and support MSPs needs to be carefully considered and measures put in place to ensure the innovations that could emerge out of them are not biased. This can be addressed by ensuring CGIAR centres take advisory and facilitating roles, rather than a leading role, in the process of facilitating innovation. Securing sources of funding other than the CGIAR or other R4D institutes would also make it possible for all MSP members to work together freely and independently.

Third, funds, even modest amounts, for use by MSPs to implement R4D are critical to energize the MSPs. As the Platform Research Projects were led by local organizations, they played a key role in generating broad partnership engagement in R4D activities. Moreover, these Platform Research Projects

Table 24.2 Analysis of multi-stakeholder platforms in the Flagships against success factors

Success factors	ECA	WA	CM	CAC
Have clearly defined common objectives	Yes, assisted by RAAIS	Yes	Yes, but the objectives were set by researchers	Yes
Composed of the right stakeholders, with appropriate representatives	Yes. However, private sector representation fluctuates at meetings. In addition, no consistent representation by specific officers	Yes, but over-representation of government and research institutes, few representatives of the private sector	Partly, but over-representation of government and research institutes, no representation of the private sector	Partly, but under-representation of private sector and government
Stakeholders engaged with strong incentives and internal motivation	In some action sites like Rwanda and Uganda, there was a strong internal motivation to stay engaged, while in sites like DRC, the quest for 'what is in it now for me' was a strong motivating factor initially	Not at all sites. In some sites financial incentive is key to stakeholders' participation whereas in others stakeholders even contribute in-kind to realize activities of the platform	Partly, but stakeholders expect financial support (for transportation and accommodation) for participation, and material inputs	Yes, members have been motivated to participate in the MSPs because of the history of popular education in the region and participatory approaches to learning
Based on long-term partnerships and trust	Informal relationships existed amongst some of the members and some have been working together in other development projects for quite some time. These pre-existing relationships helped the platforms to consolidate quickly in most of the action sites	Some stakeholders were involved in projects in the past, some have met for the first time	Some members were brought together for the first time while others had long-term relationships already	Yes, already partly present at the onset of activities
Have secure resource(s)	MSPs are in the process of establishing mechanisms for self-financing, including developing joint projects and specific commodity value chain initiatives	Not yet. Some MSPs are developing some value chain activities that will contribute to self-financing	Core partners that continued to allocate funds for meetings of MSPs led to their success. Thus sustainability of these MSPs is an issue	Mechanisms established, and proposal development started for self-financing
Have a capable facilitator	Yes, in 5 out of 6 countries	No in most of the cases. Facilitation is still an issue and most of the facilitators don't have the skills	Yes, in the 2 MSPs that are still functioning	Yes, in all MSPs

were effective in filling gaps in existing R4D activities (Schut, Klerkx, Sartas et al., 2015). It should be noted, however, that these projects have the potential to undermine the willingness of Platform members to invest their own resources in R4D activities; thus, allocation of such funds need to be carefully considered.

Fourth is the challenge of operationalizing a common multi-stakeholder process in a global research program. In April 2014, Humidtropics organized a week-long capacity development workshop, involving Action Area Coordinators, other Humidtropics staff, and some R4D Platform facilitators.[9] Through the workshop, participants obtained a common understanding of the terminologies and methods in Humidtropics to launch, support, and facilitate MSPs. This played a key role in building the capacity of Platform facilitators, which we recognize as key to successfully using MSPs to improve the integrated agricultural system. As important as this was, it was not sufficient. An important lesson learned from the Humidtropics experience is the need for some sort of guidelines that go beyond setting common principles, something that actually helps operationalize MSPs across all Action Areas. Although one of the Humidtropics strategic research themes is on assessing the contribution of MSPs to enhance institutional innovation and improving stakeholders' capacity to innovate, insufficient support has been provided to operationalize MSPs.

Finally, closely related to the above point, when operationalizing MSPs, it is necessary to be flexible and adopt divergent approaches and processes that take into account different local realities. Action Area Coordinators establish and manage MSPs in ways that reflect the great diversity across Action Areas: the social, political, and institutional contexts, the amount of funding available, the size of area and population covered, and the number of partners engaged. This resulted in varied experiences and results of using MSPs across Action Areas and sites, which had the potential to greatly enrich the learning processes and sharing of experiences across Action Areas. Unfortunately, time and resource constraints have not made it possible for this to take place.

Conclusions

In this chapter, we shared experiences from our unified efforts to transform theory and concepts of multi-stakeholder processes into effective actions on the ground through establishing and supporting MSPs in Humidtropics. We also described and analyzed success factors and challenges of using MSPs to implement integrated agricultural R4D and to foster agricultural innovations. Despite some challenges, we note that significant achievements, both research and development, have resulted from these processes. Development outcomes are mainly related to establishment of multi-stakeholder alliances to develop and scale-up innovations. Research (for development) outputs include identification and testing of best-fit options, which will continue beyond Humidtropics. We believe these outcomes will contribute to our collective effort towards generating innovative and sustainable transformations of the livelihoods of the rural poor. It is clear that fostering innovation in the agricultural system is a complicated process that requires long-term commitments and partnerships, which unfortunately will not be realized under Humidtropics, due to the CGIAR decision to close the program by the end of 2016.

Notes

1 Most Action Areas (also known as placed-based Flagships) designated action sites at the national level, and field sites within action sites at county level, based on well-defined protocols. For the Central Mekong Action Area, action sites were transboundary.

2 RAAIS (Rapid Appraisal of Agricultural Innovation System) is a diagnostic tool that can guide the ex-ante analysis of complex agricultural problems, and identification of entry points that enhance the innovation capacity of the agricultural system in which the complex agricultural problem is embedded. For more information, see Schut, Klerkx, Rodenburg et al. (2015).

3 EXTRAPOLATE (EX-ante Tool for RAnking POLicy AlTErnatives) is a decision support tool that assesses the impact of different policy measures. The tool facilitates discussion of the relevant issues and helps decision makers identify policy measures that could be applied in a specific situation to achieve outcomes that further specific policy objectives. For more information: http://www.fao.org/ag/againfo/programmes/en/pplpi/dsextra. html.

4 Situation Analysis was one of the first activities undertaken in an action site. Its objectives are to broadly characterize important system aspects in an action site, to develop a shared understanding of the issues that need to be addressed among partners, and to initiate and facilitate stakeholder engagement in the R4D Platform development. For more information, see Cadilhon et al. (2015).

5 Humidtropics defines 'best-bet' option as a prioritized intervention influenced by relevant typologies within the field site to a constraint that will be evaluated at the field site. They are innovations that address constraints and challenges related to an entry theme, which, after being tested through a participatory evaluation processes, can become 'best-fit' options, for scaling up.

6 'Mother' trials represent researcher-managed trials, testing various technology options under farm conditions and with full engagement and involvement of farmers. 'Baby' trials are based on farmers' selections of sub-sets of the technologies (from the 'mother' trial), of particular relevance and interest to them, for testing and adaptation under their own (farmer) management, in their own individual farms. Researchers collect different types of information and data from the two types of trials, in relation to best-bets and best-fits.

7 The 'Green Triangle' Action Site is composed of Northwest Vietnam, Northern Lao PDR, and Honghe Prefecture, Yunnan, China; the 'Golden Triangle' is composed of Northwest Lao PDR, Northern Thailand, Eastern Myanmar, and Xishuangbanna Prefecture, Yunnan, China; and the 'Development Triangle' is composed of Southern Lao PDR, Northeast Cambodia, and Central Highlands, Vietnam.

8 For more information on Situation Analysis, as well as IP projects and news, see http://alianza-cac.net.

9 For more information on the workshop, see http://clippings.ilri.org/2014/05/05/cap-dev-humidtropics/ and Schut et al. (2014).

References

Ashby, J.A. (2003) 'Introduction: Uniting Science and Participation in the Process of Innovation-Research for Development', in: Pound, B., Snapp, S., McDougall, C. and Braun, A. (eds.), *Managing Natural Resources for Sustainable Livelihoods: Uniting Science and Participation*, Earthscan, London; IDRC, Ottawa, pp. 1–19.

Boogaard, B., Klerkx, L., Schut, M., Leeuwis, C., Duncan, A. and Cullen, B. (2013) 'Critical issues for reflection when designing and implementing Research for Development in

Innovation platforms', *Report for the CGIAR Research Program on Integrated Systems for the Humidtropics*, Knowledge, Technology & Innovation Group (KTI), Wageningen University and Research Centre, Wageningen, the Netherlands, p. 42.

Brouwer, H., Hiemstra, W., van der Vugt, S. and Walters, H. (2013) 'Analysing stakeholder power dynamics in multi-stakeholder processes: insights of practice from Africa and Asia', *Knowledge Management for Development Journal* 9(3): pp. 11–31.

Cadilhon, J., Child, K., Raneri, J., Robinson, T., Staal, S. and Teufel, N. (2015) 'Guidelines for conducting a Situational Analysis (SA) for the Humidtropics Research Program', Unpublished document. Downloaded from Alfresco, 4 February 2016.

CGIAR-ISPC (2015) *Good practice in AR4D partnership: Guidance paper, September 2015 (draft)*, CGIAR Independent Science & Partnership Council, Rome, Italy, p. 86.

Geels, F.W. (2002) *Understanding the Dynamics of Technological Transitions: A Co-Evolutionary and Socio-Technical Analysis*, Twente University Press, Enschede, The Netherlands, p. 318 p.

Hemmati, M. (2002) *Multi-Stakeholder Processes for Governance and Sustainability: Beyond Deadlock and Conflict*, Earthscan, Earthscan Publications Ltd., London-Sterling, VA, p. 327.

Humidtropics (2012) 'Proposal for CRP1.2: The CGIAR Research Program on Integrated Systems for the Humid Tropics.'

Lundy, M., Gottret, M. and Ashby, J. (2005) 'Learning Alliances: An Approach to Building Multi-Stakeholder Innovation Systems', ILAC Brief.

Schut, M., Dror, I., Arkesteijn, M. and Ekong, J. 2014. 'Understanding, facilitating and monitoring agricultural innovation processes', *Report of the First Humidtropics Capacity Development Workshop*, Nairobi, Kenya, 29 April-2 May 2014, Nairobi, Kenya: ILRI. Available at https://cgspace.cgiar.org/handle/10568/35683.

Schut, M., Klerkx, L., Rodenburg, J., Kayeke, J., Raboanarielina, C., Hinnou, L.C., Adegbola, P.Y., van Ast, A. and Bastiaans, L. (2015) 'RAAIS: rapid appraisal of agricultural innovation systems (part I). A diagnostic tool for integrated analysis of complex problems and innovation capacity', *Agricultural Systems* 132: pp. 1–11. http://dx.doi.org/10.1016/j.agsy.2014.08.009.

Schut, M., Klerkx, L., Sartas, M., Lamers, D., McCampbell, M., Ogbonna, I., Kaushik, P., Atta-Krah, K. and Leeuwis, C. (2015) 'Innovation platforms: experiences with their institutional embedding in agricultural research for development', *Experimental Agriculture*, pp. 1–18. http://dx.doi.org/10.1017/S001447971500023X.

Thiele, G., Devaux, A., Reinoso, I., Pico, H., Montesdeoca, F., Pumisacho, M., Andrade-Piedra, J., Velasco, C., Flores, P., Esprella, R., Thomann, A., Manrique, K. and Horton, D. (2011) 'Multi-stakeholder platforms for linking small farmers to value chains: evidence from the Andes', *International Journal of Agricultural Sustainability* 9(3): pp. 423–433.

25 Learning System for Agricultural Research for Development (LESARD)

Documenting, reporting, and analysis of performance factors in multi-stakeholder processes

Murat Sartas, Marc Schut, and Cees Leeuwis

Introduction

Rationale

Multi-stakeholder processes, where a set of interdependent stakeholders interact and organize activities to achieve a set of goals and targets collectively, were implemented in agricultural research for development (AR4D) for a few decades. Some recent examples of MSPs in AR4D are innovation platforms, learning alliances, and participatory value chain development processes. Moreover, the utilization of systems approaches using multi-stakeholder processes (MSPr) as a mode of intervention has been increasingly experimented with (Klerkx, Van Mierlo, and Leeuwis, 2012; Schut et al., 2015). Recent research findings from so-called developing countries indicate that MSPs increase the impact of AR4D interventions. However, these studies either apply a black box approach, i.e. not inform about how intervention worked (Bloom and Skloot, 2010), or their results are hardly generalizable (Hall et al., 2001).

Quantitative assessments focusing on the performance of MSPs in AR4D interventions are mainly based on comparing pre- and post-intervention data (Duflo et al., 2014; Pamuk et al., 2014). Although these identify whether interventions have made a significant impact, they fail to indicate sufficient evidence on the specific performance factors of the interventions and their impacts on the process. Qualitative assessments do offer information on specific elements of the intervention. These provide insight into the key factors that influence the contribution of the process to AR4D interventions such as clear demonstration of the utility of the process policy and leadership commitment (Anandajayasekeram, 2011). However, qualitative assessments alone do not sufficiently report on contextual factors such as proximity to markets and speed of population growth, and their results are considered insufficient in providing generalizable evidence for decision-making in other contexts (Spielman et al., 2008). In brief, research approaches that provide quantitative evidence about performance factors of MSPs in AR4D interventions are scarce.

Most of the AR4D system interventions, such as Humidtropics (the CGIAR Research Program on Integrated Systems for the Humid Tropics), use two interwoven principles: a systems approach and the MSPs (CGIAR, 2012). In basic terms, one of the principles of systems approaches to AR4D is to optimize multiple systems outcomes. For instance, in addition to the conventional objectives of improving yields, incomes, and environmental services, Humidtropics targets nutrition, capacity to innovate, and gender empowerment. In Humidtropics, MSPs are implemented through innovation platforms (operating at local, community levels) and research for development platforms (operating at higher – often regional or national – systems levels). The MSPs in Humidtropics form a central role in the identification of, experimenting with, learning from, and scaling of the Humidtropics innovations.

Effective learning in systems research requires not only comprehension and operationalization of the systems approach but also of the performance of MSPr; this requires a defined approach and a set of tools and indicators. For instance, systemness of the intervention, i.e. whether the interventions consider the synergies and tradeoffs of different components within the system, is an important indicator that is not considered in AR4D interventions without a systems approach. Moreover, since Humidtropics aims at development impact, all activities are expected to contribute to process, leading to outcomes and outputs that lead to impact. In brief, effective learning from AR4D system interventions with MSPs could significantly benefit from a data gathering protocol for both quantitative and qualitative data, as well as data management that contributes to the ongoing processes aiming to achieve development outcomes and impact.

It is this need to integrate quantitative and qualitative data, provide evidence about performance factors for MSPs in AR4D interventions, and to contribute to the performance of these interventions, that has led to the development of a new approach, Learning System for Agricultural Research for Development (LESARD), which is the subject of this chapter. The approach was designed and tested in Humidtropics project.

LESARD and beneficiaries

LESARD is an action-oriented data management and decision-making support system for AR4D operating through MSPs. It has two major objectives. First, it attempts to measure the performance of the MSP and provide generalizable evidence of the performance factors for the MSP in achieving development outcomes. It follows integrated data collection and analysis methods and attempts to discover what works in AR4D. Second, it aims to improve the effectiveness and functioning of AR4D interventions with MSPs for achieving development outcomes. It provides information about the MSP performance and contribution towards achieving AR4D objectives. For instance, it informs AR4D intervention teams on the convergence of perspectives of different stakeholders on a regular basis and periodically reports on the contributions of specific AR4D activities such as agronomic trials to the overall anticipated development outcomes. By combining *research* and *development* aspects of AR4D, LESARD aims to contribute to the evidence base on design, implementation, and evaluation of AR4D interventions with MSPs, especially AR4D systems interventions aiming for transformative changes in poor farmers' livelihood systems.

The objective of this chapter is to position LESARD in the existing evidence base on AR4D interventions, to reflect on the development and testing of its methods, and to briefly describe the underlying principles required for effective and efficient learning from AR4D interventions with MSPs. Although we do embed LESARD in different literature tracks, i.e. monitoring and evaluation, information management, and innovation studies, we limit conceptual and theoretical discussions. This is to provide information to a broader audience of donors, designers, managers, practitioners, and MSPr participants of AR4D interventions, as well as to researchers from different disciplines.

Methods

This section describes the development and testing of LESARD in Humidtropics action sites in Uganda, Rwanda, Burundi, and eastern Democratic Republic of Congo (DRC).

Development of LESARD

The development of LESARD started with a rapid literature review and workshops during the inception period of Humidtropics in late 2013 and early 2014. This review identified four key references that informed LESARD: Njuki et al. (2010), Van Mierlo et al. (2010), Lundy et al. (2013), and Pali and Swaans (2013). These references constitute the fundamentals of LESARD. The fundamentals were updated and elaborated during three workshop events: 'Planning workshop for the institutional innovation and scaling component of Humidtropics' in Wageningen, the Netherlands; 'Expert meeting on participatory agricultural research: approaches, design and evaluation' in Oxford, the UK; and 'Workshop on conceptualizing and metrics of capacity to innovate' in Amsterdam, the Netherlands. Following the review and consultations, a test version of LESARD was developed.

Testing of LESARD

LESARD was piloted and participatory tested in two stages. Initial piloting was done in the Uganda Humidtropics action site by utilizing two innovation platforms located in Kiboga-Kyankwanzi and Mukono-Wakiso. Initial piloting was implemented between June 2014 and September 2014. Provisional theory of change (ToC), results framework, and tools identified in the development phase were tested for their usefulness to document, report, and analyze different platform events, i.e. platform meetings, field monitoring visits, and trainings. Feedback obtained from monitors[1] who implemented LESARD and other members of Humidtropics in Uganda was used in updating the LESARD approach, tools, and procedures. In October 2014, a new version of the LESARD was further tested in action sites in Uganda, and LESARD testing commenced in Rwanda, Burundi, and DRC, also using the revised tools. This second testing period was finalized in December 2014, and LESARD started to be implemented fully in the four Humidtropics action sites in Uganda, Rwanda, Burundi, and DRC.

Results

Components of LESARD

Following the testing periods, LESARD components were updated. LESARD in its current form covers four components, each of which will be briefly described in the following sections.

Theory of change and results framework

The first component of LESARD is the results framework (RF) including a generic ToC for AR4D systems interventions with MSPs. The ToC represents the concepts and their relations regarding the contribution of the MSP to development outcomes. RF ensures the coherence of the different components of LESARD (Figure 25.1). In the ToC, a subset of the actors in AR4D landscape (I) makes a decision to participate in the MSP (II). They engage in the MSP (III), identify collectively priorities, and experiment and pursue the innovations that can achieve their individual and/or collective objectives. If the MSP is effective, it produces AR4D outputs (IV) such as disease resistant varieties that contribute to achieving development outcomes (V). This part of the TOC is an ongoing process in the agricultural innovation systems.

AR4D interventions target three major steps in the ToC and attempt to influence the process capturing steps I to V. First, they encourage the subset of actors' decision to participate in the MSP (A). They try to create interest in the intervention as well as in the MSP and provide incentives to participate such as providing e.g. a cash allowance. Second, interventions support the MSP (B), for instance through logistical support for the events and research support for more effective identification and experimentation of innovation potentials. Third, interventions provide development outputs that can contribute to the outputs of the MSP (C). Provision of agricultural inputs such as high-quality seed kits and training in better use of inputs supporting the innovations developed by MSPs, are two examples.

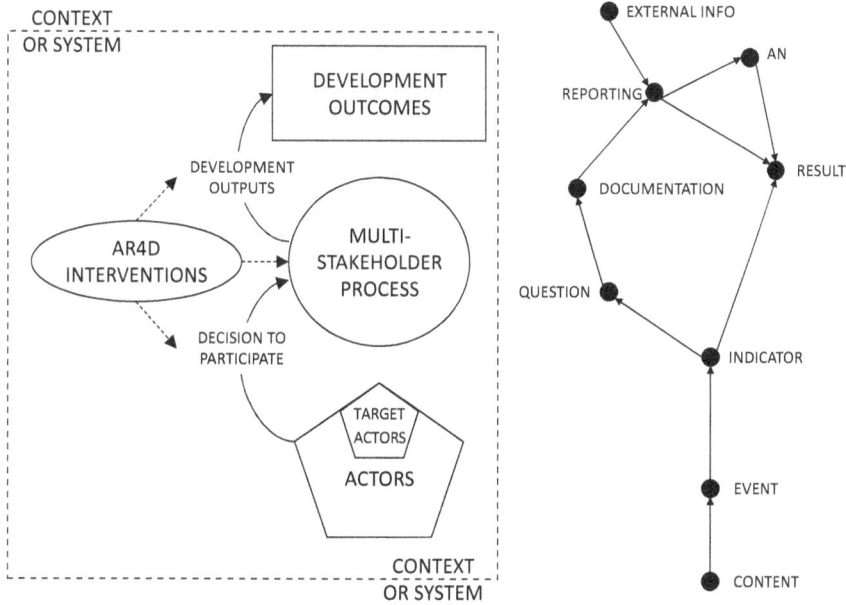

Figure 25.1 Theory of change for agriculture for development (systems) interventions with multi-stakeholder processes (left) and result framework for LESARD (right)

The ToC provides the framework and the content of the result framework (RF). RF was formulated using a logic model. Initially different events, where LESARD ToC is relevant, are identified. Depending on the events, indicators are determined. Some indicators, such as the planting of seedlings, could be captured by pictures and can inform the RF directly. For all other indicators, a question informing the indicator is formulated. Afterwards, the most suitable documentation tool and best reporting tool for the learning question are identified. Some reports can provide the result directly. For instance, the number of trainees attending an entrepreneurship training can be reported using training reports. Some other reporting tools are used for further analyses that present the results. For example, data collected by surveys are reported by online databases. In some of the cases, information generated outside of the LESARD is utilized, and is fed into the reports. For instance, a publication about other project activities in the area can provide information. Finally, the results are provided to different audiences such as the academic world, policy makers, and extension officers, which ultimately contributes to the targeting of outputs related to the interventions such as publications, policy decisions, or better extension practices (Box 25.1).

Box 25.1   An example from LESARD, action-orientedness of Humidtropics multi-stakeholder platforms in Uganda

LESARD reports on 72 different indicators and provides information about performance factors for achieving different development outcomes in Burundi, DRC, Rwanda, and Uganda. An important indicator, action-orientedness of Humidtropics multi-stakeholder platforms in Uganda in 2014, is used as an example to illustrate how LESARD operates (Figure 25.1) and how it contributes to *development* and *research* objectives of AR4D interventions with MSPs.

How LESARD operates?

Action-orientedness refers to the willingness of the MSP actors such as farmers and local government officials to engage in experimentation and scaling of potential innovations that contribute to development outcomes. Since MSPs aim to identify, experiment, learn about, and scale innovations, action-orientedness is a performance factor relevant to MSPr *theory of change*. Action-orientedness is relevant for all events except the action events, which by definition have the highest action-orientedness such as planting and field monitoring. It is investigated in the majority of the *events* in the MSPr. The *indicator* used to measure action-orientedness is engagement level of the event, i.e. the average of individual engagement levels. Individual engagement level is a number between 1 and 6 where 6 indicates high action-orientedness and 1 presents low action-orientedness. To determine the engagement level, each event participant *is asked* 'What is your objective attending this event?'[2] The question is *documented* by using the dynamic learning agenda, *reported* by Google forms, and *analyzed* using a combination of text and statistical analyses. *The result* of the analysis contributes to reflecting and improving action-orientedness of the MSPr and targets all the stakeholders in the process. The indicator ultimately contributes to all *development outcomes* targeted by the AR4D interventions.

How does LESARD contribute to research and development objectives of AR4D interventions with MSPs?

Figure 25.2 presents a graph generated in the process of validation and reflection by the researcher and stakeholders collectively. Initially the researcher made an analysis and prepared the figure without specification of the drivers causing kink points. The graph was presented in a

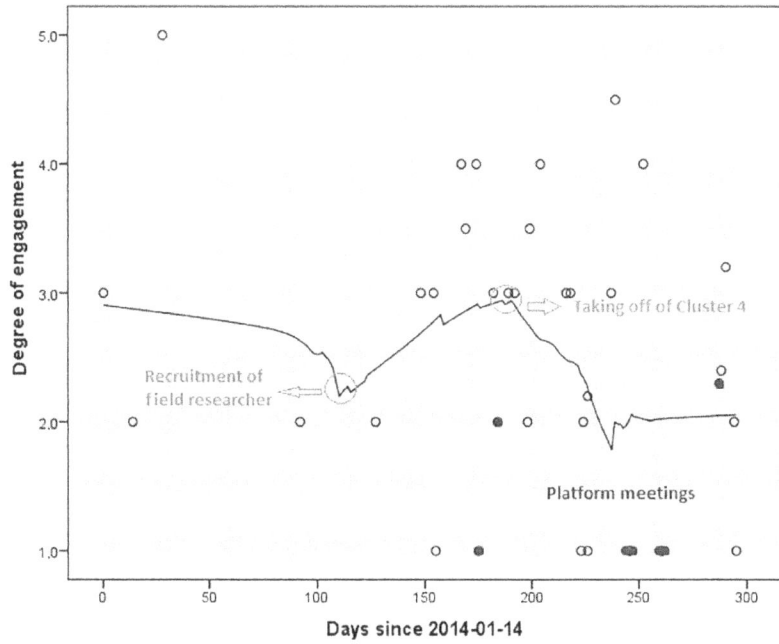

Figure 25.2 Degree of engagement is a scale classifying the involvement of stakeholders. It ranges from observing (1) to leading (6). The degree was calculated by a reference system where different words of action such as 'work
together' and 'conduct' are mapped to specific levels as 'owning' (5). For
each event, represented by circles, dynamic learning agendas filled by
individual stakeholders as well as the meeting minutes were used to calculate average degree of engagement of the event, represented by a single
circle in the graph. Filled circles represent the platform meeting

reflection meeting where different stakeholders participated. A facilitated discussion about what could the kink points be led the participants
to identify those points as recruitment of field researchers and introduction of platform initiated research as indicated in the Figure 25.2 by
cluster 4. In the reflection meeting participants first identified drivers,
therefore contributing to the validation of the analysis and thus creating a *research* output. Second, they engaged in a discussion of how they
could improve the action–orientedness given the results in the figure. In
other words, the graph triggered a learning process contributing to the
improvement of the performance of the MSPr or generated a *development* output.

Documentation toolkit

The documentation toolkit includes the tools necessary to capture data. LESARD has a multi-layer documentation approach to investigate indicators of interest in the RF.

- The first layer concerns the structured surveys, results of which can be reported and analyzed relatively quickly. These include (1) an event log, a simple form aiming to capture factual information on events such as date and location; (2) a dynamic learning agenda, where individual perspectives of participants, i.e. their objectives, priorities, are captured before and after each event; (3) participant lists; and (4) participant profiles aiming to profile the individuals and organizations in the process.
- The second layer consists of text-based tools. They include meeting minutes, reports of periodic reflection meetings, and other reports of the non-meeting type of events such as field monitoring activities.
- The third layer covers photos, audio, and video capturing tools of major events.
- The fourth and last layer consists of e-mails and interview records.

For development and science stakeholders, the first layer is the main source of data about the indicators. If the data provided by the first layer is not satisfactory, data generated by the second layer should be utilized. The same logic is applied to the following layers if the data are still unsatisfactory. For accountability purposes, minutes and reports, i.e. the second layer, is more useful. For the provision of basic dissemination of information, photos, audio, and video data, i.e. the third layer, are more useful and relevant (e.g. for policy and other scaling actors).

Reporting toolkit

The reporting toolkit for LESARD has two major tool categories or toolkits. The primary reporting toolkit consists of (1) a simple folder structure, i.e. naming rules and folder hierarchies, identified by major users among the stakeholders; (2) Google Calendar; and (3) online repositories. Folder structure guides how the information is stored. Google Calendar is used to compile and notify the stakeholders about all MSP events. It provides open access to all stakeholders and is administered by monitors, facilitators, and organization staff. Online repositories have a storage and reporting function. They not only store data but also implement basic analyses such as descriptive statistics and basic text analysis. LESARD utilizes data-event type for naming the folders and all relevant materials are stored in these folders. Google Forms and Google Drive are used as an integrated online repository. Dropbox was also used due to its effective synchronization capabilities. When internet access is problematic, external hard disks are utilized for storage. The secondary reporting toolkit covers e-mails and web-based platforms providing different organization and communication services (e.g. WhatsApp). When the stakeholders are not familiar with the primary toolkit and they have not been trained, e-mails are used to report data.

Analysis toolkit

The analysis toolkit has two layers. The primary layer contains tools for descriptive statistics, statistical analysis, content analysis, and social network analysis. Depending on the focus indicators and documentation opportunities, one tool or a few of them are selected. Descriptive statistics (e.g. on stakeholder group participation, the percentage of male/females) are used by monitors, facilitators, and organizers of the MSP. Other tools require specific knowledge and are used mainly by the researchers. Some such knowledge intensive tools, such as content analysis, require continuously involved researchers during the implementation.

Descriptive statistics tools are applied for informing a broader set of stakeholders and used to provide quick feedback. Automated Google Forms reports and Microsoft Excel are the main software tools. Statistical analyses, specifically IBM SPSS software, is utilized for understanding trends in specific performance factors and to detect potential causes of major changes in those performance factors. Text analysis is the most commonly used content analysis method. QSR Nvivo and Atlas.ti software are used for comparing and contrasting perspectives of different stakeholders. Social network analysis tools are the last tools in the toolkit. Gephi was used to map existing stakeholder networks and the changes in them following the MSPr. Different tools can be combined when required, e.g. when targeting scientists who are generally interested in more details. For instance, text analysis and statistical analysis were combined to understand the action-orientedness of the MSP. The secondary analysis toolkit covers mind maps, timelines, spatial modelling, and econometric modelling. Whenever one of the tools in the primary set is not fit for the indicators or the stakeholder indicates a strong preference for them, these secondary tools are utilized. The software programs Xmind, Microsoft Excel, and Arc GIS are the basic tools in the secondary analysis toolkit.

LESARD currently covers 72 indicators that inform about performance factors of AR4D interventions with MSPs. These 72 indicators belong to 8 sub-categories. These 8 categories represent components of the ToC as well as some important aspects that define the characteristics of systems interventions, i.e. system mind set and scale of the activities and outcomes. LESARD indicators cover a range of data types among qualitative and quantitative spectrums generated from different sources such as simple lists, conversations, budget tables, text analysis, and social network analysis (Table 25.1).

LESARD performance principles

This section covers the performance principles that influence the effectiveness and efficiency of learning from AR4D interventions with MSPs. The principles

Table 25.1 Sample from LESARD indicators

Subcategory	Coverage	Example indicators
Context	Contextual factors	Place in urban-rural gradient Population size trend
Actor typology	Participating actors and other stakeholders	Value chain position Centrality in collaboration network
Process success factors	Factors of success related to organization and conducting of the MSP	Facilitation quality Action-orientedness
Event typology	Events organized in the MSP	Time of event Location of event
Development outcomes	Outputs and outcomes targeted and produced by the MSP and interventions	Related Sustainable Development Goals (SDG) Related SDG target
Scales	Scale of the activities and outcomes	Administrative scale Multi-scale index
Intervention modalities	Specific activities and targets of intervention	Targeted ToC aspect Sources of finance
System mind set	Perspectives of managers and implementer and other stakeholders about activities	Mentioned interactions among SDG Recognized limitations and infeasible options

were hypothesized and tested using the Humidtropics experience. The current LESARD version effectively follows these.

Platform meetings are just one mechanism of MSPs

The drivers and immediate causes of change in complex social systems are very difficult to detect. Using only the main event, the multi-stakeholder platform meeting, provides limited information about these drivers and immediate causes. For instance, in Humidtropics, on several occasions, a subgroup of MSP stakeholders developed a partnership outside of the formal platform meetings and planned changes with their own objectives. These spillover impacts were hardly reported in the platform meetings. Thus, it is critical to document other events such as field activities, researcher meetings, and sub-group meetings for learning in addition to the platform meetings. Although targeting different events does not guarantee capturing all drivers and immediate causes of change, it does improve the likelihood of capturing them substantially.

Both factual and perception data need to be gathered

Systems research and development interventions target not only physical but also behavioural change in complex livelihood systems. Anticipating change and the immediate drivers of change a priori is quite difficult to achieve. For instance,

MSPs might follow spontaneous opportunities and this might lead to different impact pathways from those envisioned. In these circumstances, it is very difficult to understand performance factors based on only tangible and physical, i.e. factual, outcomes. However, stakeholders who are participating in most of the MSP events such as facilitators, organizers, and monitors develop their opinions, and their thinking evolves as the process evolves. These opinions and changes in thinking provide important insights into what works, what does not and why. The inclusion of these perceptions on performance through semi-structured periodic reflections can complement the factual data and make an important contribution to the learning about MSPs. For instance, initial high engagement of farmers, local government, and local staff of national research systems in Humidtropics in Uganda was considered first to be a response to the availability of funding by some actors involved in the intervention on regional and global levels. Nonetheless, the real fund allocation did not confirm that Uganda had more resources as compared to other action areas in Burundi, DRC, and Rwanda. However, reflection with the Humidtropics Uganda team revealed that it was not the availability of funds, but the flexible approach to how funds could be used that was the real driver of the engagement of farmers, local government, and national system researchers. Without the reflection with the Humidtropics team, it would have been very difficult to capture this perceptive evidence.

Learning approach needs to provide short-term feedback

In some of the AR4D interventions, stakeholders are exposed to different data collection activities. Different research teams target the same households for a better understanding of system interactions, i.e. tradeoffs and synergies. Moreover, continuous monitoring is necessary for stronger association of the performance factors of MSPs with development outcomes. These two factors led to a research fatigue among the stakeholders of Humidtropics, which was also commonly observed in other interventions (Clark, 2008), with respondents becoming reluctant to complete the surveys and answering questions they do not perceive to be of value. However, our testing showed that once respondents are informed about the results of the data collection and how that can benefit collective decision-making on *current* MSP issues, their willingness to answer the questions and provide data increased dramatically. Among the MSPs in the four countries where LESARD was tested, the participants of those that received short-term feedback in reflection meetings showed a higher willingness to contribute actively to LESARD. In other words, providing short-term feedback to stakeholders can substantially increase the ease of data collection for learning in MSPs.

Toolkits need to be easy to use and cheap

Capacity development is an integral part of MSPs in AR4D interventions. Moreover, the sustainability of the outcomes of the interventions requires the AR4D landscape as a whole to improve in identifying, experimenting, learning, and scaling of innovations, i.e. to improve capacity to innovate (Leeuwis et al., 2014). Learning approaches and tools used in MSPs make a contribution to the capacity to innovate of the stakeholders through providing incentives, knowledge, and access to information tools that can increase the efficiency of collective decision-making and action. Therefore, access of the different MSP stakeholders to learning approaches, publicly available toolkits, and open access data analysis software is important for achieving improvement in the capacity to innovate. For instance, although its reporting and analysis capabilities were more limited, Microsoft Excel was much more utilized by the Humidtropics teams than the more comprehensive and more advanced (and expensive) statistical analysis tool such as SPSS, STATA, and MATLAB packages. In brief, easy to use, openly accessible, and cheap tools have a higher chance to contribute to the capacity to innovate of the stakeholders since their chances of adoption and use are much higher.

Documentation, reporting, and analysis of data needs to be conducted in a coherent manner

A general observation is that in most of the research concerning the monitoring and evaluation of learning processes, only a small share of the data collected is presented as results. Another issue is that most of the stakeholders have very limited access to the generated data and that data gathering is often incoherent, which complicates its analysis and reliability. These issues combined decrease the effectiveness and efficiency of the learning processes substantially. For instance, one of the major reasons for the long delay in reporting and analysis of baseline and situation analysis in Humidtropics was the underutilization of automated reporting, analysis tools, and lack of consideration of the extensive human and financial resourced needed to report and analyze the data. LESARD utilizes a coherent approach to the documentation, reporting, and analysis of learning materials to improve effectiveness and efficiency. To reach better effectiveness and efficiency LESARD maps each documentation, reporting, and analysis tool to an indicator and a learning question. If the tools do not result in a research question and an indicator, or require resources beyond boundaries of the interventions, they are updated or removed. This approach helps minimize redundant data collection activities.

Discussion and way forward

This chapter introduced LESARD: an action-oriented data management and decision-making support system for MSPs in AR4D. Guided by its tested performance principles and through coherent combination of a ToC, RF, documentation, reporting, and analysis toolkits, LESARD has the potential to contribute to not only the effectiveness and functioning of MSPs in achieving development outcomes but also to provide evidence on generalizable performance factors of MSPs in AR4D.

Development and testing improved the performance of LESARD. However, there are still a few challenges limiting LESARDs. First, its implementation requires access to a diverse set of research and ICT skills by research partners and project managers. Second, especially in its introduction period, the commitment of several different people in AR4D interventions is essential. In the absence of a researcher champion, who will provide legitimacy, and a monitor, who will participate in the different events, required continuity and coherence in documentation of LESARD have been hard to achieve. Continuity and coherence in documentation of LESARD are necessary and have been hard to achieve within the Humidtropics MSP due to the absence of researchers and monitors.

LESARD aims to contribute to effective systems research in AR4D by directly targeting research and development objectives of AR4D interventions at the same time. By considering these two objectives and by offering rigorous evidence and action outputs simultaneously, LESARD is an innovative approach. By further addressing the mentioned LESARD challenges, developing user materials for practitioners, providing proof of concept in AR4D science literature, and producing leaner LESARD modules for specific learning purposes, LESARD aims to become an important product in agricultural systems research.

Notes

1 Monitors refers to staff that are responsible for the documenting and reporting of the platform events. In Humidtropics, they were different from M&E officers who were responsible for program M&E. Monitors were recruited, trained, and backstopped by the process research team including authors and reported back to stakeholders of the process, while M&E officers reported to Humidtropics management through separate channels.
2 Responses to the question contained an action word such as present or lead. These words were classified into the six groups by the engagement these words imply. For instance, 'present' is a one directed information provision so signals a lower engagement than the word 'lead', which implies understanding, ownership, and willingness to contribute more.

References

Anandajayasekeram, P. (2011) *The Role of Agricultural R&D within the Agricultural Innovation Systems Framework*. Retrieved from http://citeseerx.ist.psu.edu/viewdoc/download?doi=10.1.1.261.846&rep=rep1&type=pdf.

Bloom, P.N., and Skloot, E. (2010) *Scaling social impact: new thinking*, Palgrave Macmillan, New York.

CGIAR (2012) *CRP 1.2: Proposal for Integrated Systems for Humid Tropics Research Program*. Retrieved from http://library.cgiar.org/bitstream/handle/10947/2554/crp_1.2_humid_tropics_proposal_jan24_2012.pdf?sequence=1.

Clark, T. (2008) 'We're over-researched here! 'exploring accounts of research fatigue within qualitative research engagements', *Sociology*, 42(5), pp. 953–970.

Duflo, E., Keniston, D., and Suri, T. (2014) Diffusion of Technologies within Social Networks: Evidence from a Coffee Training Program in Rwanda.

Hall, A., Bockett, G., Taylor, S., Sivamohan, M.V.K., and Clark, N. (2001) 'Why research partnerships really matter: innovation theory, institutional arrangements and implications for developing new technology for the poor', *World Development*, 29(5), pp. 783–797, doi:Doi 10.1016/S0305–750x(01)00004–3.

Klerkx, L., Van Mierlo, B., and Leeuwis, C. (2012) 'Evolution of systems approaches to agricultural innovation: concepts, analysis and interventions', in Darnhofer, I., Gibbon, D., and Dedieu, B. (eds) *Farming systems research into the 21st century: the new dynamic*, Springer, Dordrecht, Netherlands, pp. 457–483.

Leeuwis, C., Schut, M., Waters-Bayer, A., Mur, R., Atta-Krah, K., and Douthwaite, B. (2014) Capacity to innovate from a system CGIAR research Program perspective. *CGIAR Research Program on Aquatic Agricultural Systems. Program Brief: AAS-2014–29*, Penang, Malaysia.

Lundy, M., Le Borgne, E., Birachi, E., Cullen, B., Boogaard, B., Adekunle, A., and Victor, M. (2013). *Monitoring innovation platforms*, ILRI a member of CGIAR Consortium, Addis Ababa, Ethiopia.

Njuki, J., Pali, P., Nyikahadzoi, K., Olaride, P., and Adekunle, A. (2010). *Monitoring and evaluation strategy for the sub-Saharan Africa challenge program*, Accra, Ghana.

Pali, P., and Swaans, K. (2013) Guidelines for innovation platforms: Facilitation, monitoring and evaluation, ILRI Manual 8, ILRI, Nairobi, Kenya.

Pamuk, H., Bulte, E., and Adekunle, A.A. (2014) 'Do decentralized innovation systems promote agricultural technology adoptionα Experimental evidence from Africa', *Food Policy*, 44, pp. 227–236, doi:10.1016/j.foodpol.2013.09.015.

Schut, M., Klerkx, L., Sartas, M., Lamers, D., Campbell, M.M., Ogbonna, I., Kaushik, P., and Leeuwis, C. (2015), 'Innovation Platforms: experiences with their institutional embedding in agricultural research for development', *Experimental Agriculture*, available on CJO2015, doi:10.1017/S001447971500023X.

Spielman, D.J., Ekboir, J., Davis, K., and Sanginga, P. (2008) 'Developing the art and science of innovation systems enquiry: alternative tools and methods, and applications to sub-Saharan African agriculture', in Sanginga, P., Waters-Bay, A., Kaaria, S., Njuki, J., and Wettasinha, C. (eds) *Innovation Africa: enriching farmers' livelihoods*, Earthscan, London, UK, pp. 72–85.

Van Mierlo, B., Regeer, B., Van Amstel, M., Arkesteijn, M., Beekman, V., Bunders, J., de Cock Buning, T., Elzen, B., Hoes, A.C., and Leeuwis, C. (2010) *Reflexive monitoring in action. A guide for monitoring system innovation projects*, Communication and Innovation Studies, WUR, Wageningen/ Amsterdam; Athena Institute, VU, Netherlands, 9085855993.

Index

Note: **bold** is used to indicate a figure; *italic* is used to indicate a table; n refers to a note.